Advance Praise for

Evolution's Edge

"A convincing analysis of our predicament and an inspiring synthesis of new ideas..."

— Satish Kumar, founder of Schumacher College and editor of *Resurgence Magazine*

"Thank goodness this is not another scary, hopeless book about the future. Graeme Taylor doesn't mince words in laying out the threats to our survival... But at the same time, he lays out the course we must steer to avoid disaster, and how each of us can jump aboard before it is too late. Read it and pass it on to your kids."

— Mary-Wynne Ashford, speaker, author *Enough Blood Shed: 101 Solutions to War, Violence and Terror,* and Past President of the (Nobel Peace Prize-winning) International Physicians for the Prevention of Nuclear War

"In placing the sun at the centre, Copernicus shifted humanity's frame of thinking in a way that made the modern industrial world possible. In *Evolution's Edge* Graeme Taylor cogently explains why humanity's survival now demands a shift to a systems frame of thinking and design. Essential reading for leaders, decision makers and those concerned with the survival of humanity."

— Richard Sanders, ecological economist

"There are many books being written about the negative impacts of climate change and impending global catastrophe. Not all of them couple the diagnostics with concrete proposals for transformation. This one does, so it succeeds in being both a cautionary tale and in providing grounds for optimism. I commend it to analysts and activists everywhere."

— Kevin Clements, Director of the Australian Centre for Peace and Conflict Studies and Secretary General of the International Peace Research Association

"*Evolution's Edge* provides a timely and readable overview of humanity's predicament. Unlike other works, however, it summarises many existing 'seeds of transformative change' and thus provides grounds for informed hope and action. This is an inspiring book that should be read by all those who not only seek a better world but also want to be active in co-creating it."

— Richard Slaughter, educator, author,
Past President of the World Futures Studies Federation,
and Director of Foresight International

"For this...era no piece of writing on any topical developmental issue can surpass this masterly intellectual and practical work. This book...[is] an intellectual compass. Read it and you find your...direction."

— Owen Daka, international human rights advocate
and governance consultant

"*Evolution's Edge* paints a poignant picture — Earth is teetering on the brink of the sixth mass extinction of species.... Taylor writes as an optimist in the face of this scenario. He calls for a clear and urgent response from the people's of Earth to heed what scientists from many disciplines are saying and to act now. This book provides leadership to those engaged in the transformation process to a sustainable society ...."

— Graham Willett, geoscientist and Director, Corporation Technologies
and Living Business Systems

"*Evolution's Edge* makes it clear that a transformation in our societies isn't a choice — our lives depend on it. The practical, moral and historical imperative outlined here compels us to put 'limits' at the heart of humanity's values, and to recognise that our societies and economies are subsystems of the environment."

— Helen Kersley, financial expert,
Advocacy International Research Fellow

Evolution's Edge

The Coming Collapse and Transformation of Our World

GRAEME TAYLOR

Illustrated by Fereshteh M. Sadeghi

New Society Publishers

Cataloging in Publication Data:
A catalog record for this publication is available from the National Library of Canada.

Cover design by Diane McIntosh. Image: iStock

Printed in Canada. First printing June 2008.

Paperback ISBN: 978-0-86571-608-7

To order directly from the publishers, please call toll-free (North America) 1-800-567-6772, or order online at: www.newsociety.com

Any other inquiries can be directed by mail to:

New Society Publishers
P.O. Box 189, Gabriola Island, BC V0R 1X0, Canada
(250) 247-9737

New Society Publishers' mission is to publish books that contribute in fundamental ways to building an ecologically sustainable and just society, and to do so with the least possible impact on the environment, in a manner that models this vision. We are committed to doing this not just through education, but through action. This book is one step toward ending global deforestation and climate change. It is printed on Forest Stewardship Council-certified acid-free paper that is **100% post-consumer recycled** (100% old growth forest-free), processed chlorine free, and printed with vegetable-based, low-VOC inks, with covers produced using FSC-certified stock. Additionally, New Society purchases carbon offsets based on an annual audit, operating with a carbon-neutral footprint. For further information, or to browse our full list of books and purchase securely, visit our website at: www.newsociety.com

NEW SOCIETY PUBLISHERS
www.newsociety.com

Dedication

We are made of stardust. Many of the atoms in our body were formed in the explosions of ancient suns, the products of the mystery and magic of the Creation and its Creator. We would not be here if not for the web of life that surrounds and supports us, an intricately interconnected wonder that started to evolve before there were humans, before there was the planet Earth, in the beginnings of space and time.

We are joined to the past by our parents and to the future by our children. They carry the hopes and dreams of humanity, and they will inherit the Earth. This book has been written for them. May it contribute to the preservation of the beauty and diversity of life and to the transformation of our world into a place of peace, abundance and joy.

Contents

Acknowledgments xi
INTRODUCTION: THE EVOLUTIONARY CHALLENGE 1
At the cutting edge 1
From tipping points to transformation 3
Changing the world 4
The design of this book 8
Part 1: Collapse — the dominant trend
CHAPTER 1: OVERSHOOT 13
The debate over limits to growth 13
Growing populations 15
Growing consumption 18
Available global resources 20
Ecological footprints 23
Overshoot 27
CHAPTER 2: GROWING GLOBAL CRISES 29
Vital signs 29
Growing energy shortages 30
Increasing climate change 35
Growing water shortages 41
Growing food shortages 44

Accelerating rates of extinction 51
Growing crises 56
Rich world, poor vision 59
Spaceship Earth 61
Chapter 3: The Unsustainable Global Culture 63
The fundamental problem 63
Greeds not needs 64
Socially structured behaviors 65
Fear and addiction 66
Real and artificial scarcity 69
Structural inequality and structural violence 71
Selfishness and exploitation 74
Hierarchy and discrimination 75
Cultures of violence 77
War and domination 78
Cultural destruction 83
How much is enough? 84
Chapter 4: The Need for a New Model 87
The historical expansion of the industrial system 87
The evolution of the industrial system 94
The origins of the industrial worldview 95
The development of industrial social structures 95
The development of industrial technologies 97
Designed for constant growth 98
An obsolete model 99
Chapter 5: Cascading Crises and System Failure 103
Why earlier civilizations have collapsed 103
The causes of collapse 104
The loss of resilience 106
Parameters and perfect storms 107
The consequences of system failure 110
Why we can't see the dangers 112
Part 2: Transformation — the emerging trend
Chapter 6:
Sustaining Development or Developing Sustainability? 117

What is sustainable development? 117
What are biophysical needs? 120
What are human needs? 121
Genuine Progress Indicators 122
The requirements of a sustainable societal system 126
CHAPTER 7: TRANSFORMATIVE MATERIAL TECHNOLOGIES 129
New scientific paradigms 129
The emerging worldview 131
The promise of new technologies 132
Renewable energies 134
Conservation 136
Nanotechnologies, biotechnologies and biomimicry 137
Decentralized, distributed production 141
Computers and the Internet 142
Whole-systems design 142
Ecodesign 145
The limits of market forces 146
Planning for sustainable development 148
Flipping the paradigm 150
The evolutionary race 152
CHAPTER 8: TRANSFORMATIVE IDEAS AND SOCIAL MOVEMENTS 155
History is now 155
The process of social transformation 156
Growing awareness and rising expectations 161
Changing global values 162
The power of social movements 167
The emerging integral worldview 170
The spread of deep democracy 173
CHAPTER 9: CONSTRUCTIVE AND DESTRUCTIVE RESPONSES TO CRISES 175
Forecasting the future 175
Probable responses to growing energy shortages 178
Probable responses to growing water and food shortages 179
Probable responses to growing economic crises 183
Probable responses to conflicts over scarce resources 185
Probable responses to climate change and the loss of biodiversity 186

Chapter 10: Future Scenarios 189
The two main trends — collapse and transformation 189
The three possible future scenarios 191
The strengths and weaknesses of the global system 192
Constructive and destructive interventions 200
Scenario 1: Business as usual 203
Scenario 2: Adjusting the existing system 204
Scenario 3: Transformational change 206
The dynamics of societal change 206
Possible time frames for future scenarios 208
The risks and costs of action versus inaction 210
Constructive change is possible 211
Chapter 11: The Design of a Flourishing Earth Community 213
Health is wholeness 213
Conscious self-organization 214
Governance and peace 216
Individual and societal health 219
The culture pattern of a sustainable civilization 223
Chapter 12. Tools for Transformation 225
The power of love 225
The moral imperative 226
Helping people become engaged 227
The Earth Charter 228
Building global agreement 229
Supporting constructive change 232
Include and transcend 233
Truth, trust and transformation 235
Courage and commitment 237
Supporting evolution: the magical formula 238
Love and faith: the gift of our ancestors 241

Endnotes 243

Index 291

About the Author and Illustrator 307

Acknowledgments

WITHOUT THE SUPPORT AND TEACHINGS OF MY PARENTS, Mary and Alastair Taylor, this book would never have been written. *Evolution's Edge* builds on my father's theories on the evolution of societal systems. The book would also not exist without the graphic skills and constant encouragement of my wife, Fereshteh Sadeghi. I am likewise grateful to my brother Duncan, who contributed invaluable ideas and editorial advice, and to my brother Angus, who generously volunteered to take on the initial editing.

You would also not be reading this if it were not for Chris and Judith Plant, Ingrid Witvoet, Betsy Nuse, Mary Jane Jessen, Sue Custance, Greg Green, Elizabeth Hurst and all the other wonderful people at New Society Publishers. We are in a race between education and extinction, and these heroes are walking their talk and working hard to give people the vision and tools to build a better world.

In addition I need to express my gratitude to two saints, Mother Meera and Mata Amritanandamayi, who transformed my life and gave me faith. And of course I want to thank my children Lars and Zara Taylor and all those who enrich my life with their presence and love.

And my thanks to you for taking the time to read this book. May you enjoy it!

Introduction:
The Evolutionary Challenge

At the cutting edge

ALBERT EINSTEIN SAID THAT PROBLEMS CANNOT BE SOLVED at the same level of awareness that created them.[1] Because the global system creates problems like war, poverty and environmental destruction, it cannot solve them. But they could be solved at a different level — by a new type of planetary civilization with different views, values and social institutions.

Evolution's Edge uses societal evolution — the process by which societies reorganize themselves in more complex forms with new capabilities — to explain why the next level of civilization has already begun to emerge. It explains how we can support this evolutionary process — the transformation of our unsustainable Industrial Age into a sustainable Information Age.[2]

At the cutting edge of evolution, changing conditions and competition leave few options: species and societies either evolve or die off. Human societies have been evolving for hundreds of thousands of years. Evolutionary change results when either random mutation (in plants and animals) or conscious invention (in human societies) produce new structures with new capabilities. The need for environmental relevance means that useful changes are preserved, while useless changes disappear. At each new biological and social stage new and more complex forms and functions emerge.

We are the products of many successful evolutionary transformations: inorganic evolution from subatomic particles to complex molecules; biological evolution

Figure 1: BEST Futures [3]

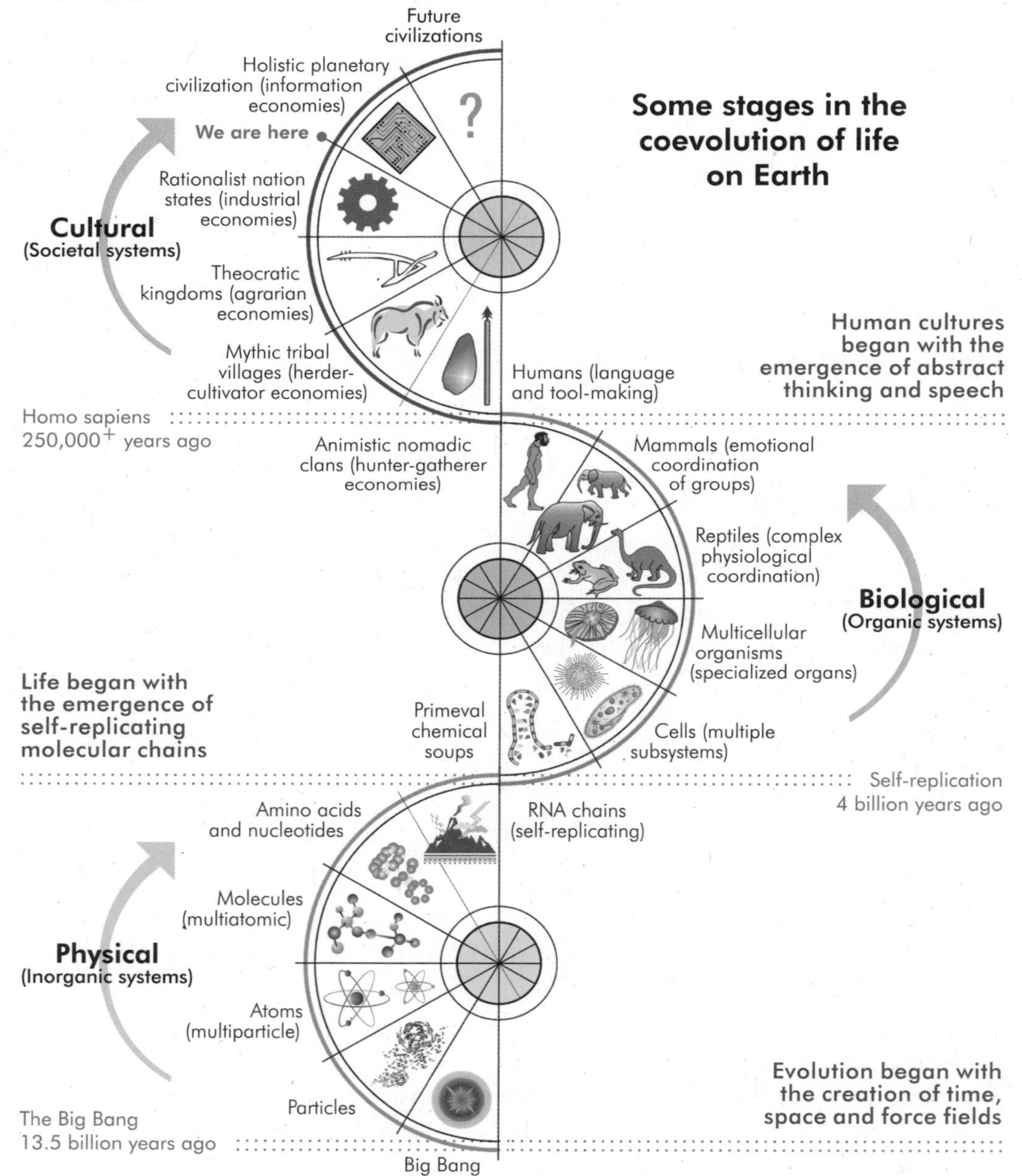

from single-cell organisms to humans; social evolution from hunter-gatherer societies to industrial civilizations. Now we are in the middle of another evolutionary leap. However, our long history does not guarantee future success: most of the species and most of the civilizations that have ever existed on earth are extinct. Because industrial civilization is rapidly degrading the global environment, we have reached a critical point where the survival of humanity is threatened.

The problem is that limitless expansion is not possible on a finite planet. The danger is that our growth-based global system will collapse as critical resources become scarce and major ecosystems fail. The hope is that new ideas, values and technologies will enable us to avoid disaster and create a better world. Humanity has no choice: if global civilization is to survive, it must evolve into a completely new type of societal system. A consumer society cannot be transformed into a conserver society without structural change.

From tipping points to transformation

In front of us are both an immense challenge and a wonderful opportunity. The challenge is to avoid the catastrophic collapse of our natural and social worlds. The opportunity is to finally end humanity's ancient addiction to war and greed and to create a peaceful and healthy civilization. This is possible because the same forces that are driving us to self-destruction are creating the conditions for constructive change.

Human societies have been evolving for more than 200,000 years. Nomadic families of hunter-gatherers armed with stone spears have developed into industrialized nations armed with nuclear missiles. In the process, occasional contacts between isolated bands have developed into constant exchanges among international networks. Globalization marks the beginning of a tremendous shift past tribal and national boundaries towards a planetary civilization. But it also marks the end of unexplored frontiers and the end of major resource discoveries. With the shrinking of time and space, our species has begun to realize that it lives on a finite planet with limited resources.[4]

Globalization is triggering a profound shift in human consciousness. On one hand we are being forced to realize that we cannot do anything we want — the price of continuing to exploit nature and each other will be our own destruction. On the other hand, we are learning that our differences are less important than our commonalities — because we are all humans, if our species succeeds, our children and grandchildren will lead happy lives; if it fails, they will inhabit a dying world.

Although our future is threatened, this is a hopeful book. This is a time when we can — and must — make a *great turning*.[5] We believe that the coming global crisis is a critical but inevitable part of the social evolution of our species. Our species has not failed — rather we risk being the victims of our own success. The Industrial Age has not been an evolutionary error, but a necessary stage in human development. It has encouraged the growth of science and technology; it has given most people better and longer lives. However, these benefits have come with enormous environmental and social costs, and the industrial system has now outlived its usefulness.

The continuing development of both destructive and constructive capabilities creates two trends:

- the dominant trend towards collapse — unsustainable consumption and environmental destruction.
- the emerging trend towards transformation — sustainable ideas, values and technologies.

These two trends are the major forces shaping the world today.[6]

There is no guarantee that all the necessary elements of a sustainable system will develop quickly enough to prevent irreversible environmental and social damage. Major evolutionary transformations only occur after a critical number of useful paradigm changing developments (functional mutations) have taken place within a biological or social system. If these new system components are compatible, their interactions can begin to change the form and function of the entire system.

All of the key social and technological components of a sustainable system will have to be present before it will be possible for our consumer society to transform itself into a conserver society. For this reason we need to actively support their development. In order to do this we need to understand not only the evolutionary process but also the requirements of a sustainable system. The purpose of *Evolution's Edge* is to help us determine how we can best support the constructive transformation of our world.

Changing the world

The challenge is not just to change our values and social institutions, but to change them quickly enough to avoid environmental and social disaster. But how can a world system based on power, violence and inequality become peaceful and just?

Global problems often appear to be too large and complex to understand, let alone manage. This is because human societies, like weather systems, are open systems with chaotic and complex dynamics. However, since all open systems operate within definable parameters and follow predictable patterns, appropriate theories can be used to explain and predict the dynamics of both weather systems and societal systems. The key to analyzing and managing global change is to recognize that our industrial civilization is not only a dynamic system (with all the characteristics of dynamic systems) but also a living and evolving societal system. Evolutionary systems theory provides us with powerful tools from both the natural and social sciences for analyzing complex global problems.[8]

My father Alastair M. Taylor, a historian and political geographer, was the first to use evolutionary systems theory to explain the historical evolution of societal systems and worldviews.

While previous societal systems (*historical ages*) took thousands of years to develop, we have only a few years left in which to transform our civilization. Fortunately, we do not have to start from square one. Because the shift to a holistic society began over a hundred years ago, many of the key components of a sustainable societal system are already present.[9] Moreover, our species is constantly learning new skills and becoming increasingly adaptable.

At the same time as our civilization has become unsustainable, our species has acquired the ability to redesign living systems. We now understand biological and social processes well enough to make scientific interventions such as genetic modification and cultural interventions such as marketing. Scientists have now identified the basic components and codes of biological systems and are racing to create artificial life.[10]

Understanding how living systems work is both powerful and dangerous knowledge. While it can be used in irresponsible and destructive ways, it can also be used constructively to help us design a sustainable societal system. Because evolution is about innovation (the emergence of new forms and functions), it is possible for humans to accelerate evolutionary processes. We can support the emergence of a sustainable civilization through consciously inventing and constructing critical technical and cultural components.

Of course there are profound differences between physical and living systems. Physical systems are externally created while living systems are self-organizing. Societal systems maintain themselves, reproduce themselves and change themselves.

> You never change things by fighting the existing reality. To change something, build a new model that makes the existing model obsolete.
>
> — Buckminster Fuller, architect and visionary (1895-1983)[7]

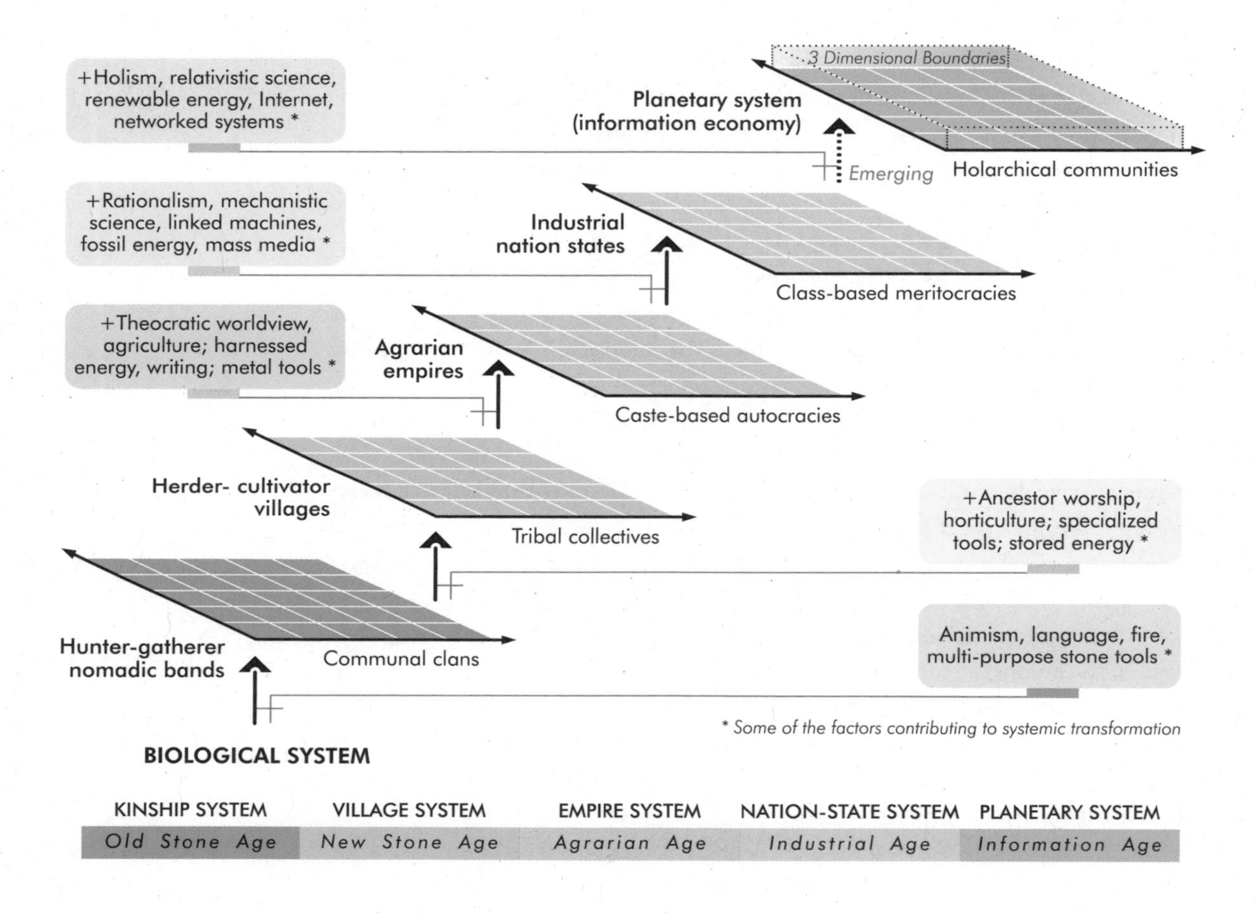

Figure 2: A Taylor[11]

This means that in order to be successful, societal interventions must build on and support existing processes. If the interventions result in useful innovations (functional mutations), they are likely to be adopted and spread throughout the system.[12]

The purpose of *Evolution's Edge* is to contribute to the design and self-organization of a logical, workable solution to our planet's major problems.

- First we will need a clear statement of the general problem and the general solution — a declaration of the mission and vision.
- Next we will need to examine what works and what doesn't work in our current system in order to distinguish bad practices (what causes problems) from best practices (what supports solutions).
- Then we will need to link the best practices together in synergistic ways that support social networking (social self-organizing) and the emergence of functional new structures.

Sound easy? Well, it won't be. The devil is in the details — how the network connects together. The vision has to be right and the components have to have a design that enables them to self-organize into a sustainable system. In fact, in order to design a better social network we need not only better theories but an entirely new paradigm. But we are already more than half way there — better theories and the new paradigm have already been developed.

When President John Kennedy announced on May 25, 1961 that the US would put humans on the moon within a decade, many doubted whether his ambitious goals could be achieved. Although the technology did not yet exist for a trip to the moon, the project was launched because it was theoretically possible (and in the view of the US administration, a strategic necessity).[13] The project achieved its goals ahead of schedule: the first astronaut landed on the moon on July 20, 1969.

The challenge that humanity faces now is to rapidly transform our unsustainable global system into a sustainable system. The survival of our species is a more urgent and important task than the space race, although it is in many ways a similar project. Like going to the moon, we only have a general idea of how we will do it. And although many difficult problems remain to be solved, we already have the basic theoretical skills that we will need to solve them.

We can expect to encounter enormous resistance. Technological innovations — like railways and cars — have always had to overcome initial derision and opposition, and social innovations — like democracy and public education — have been strongly opposed. Vested interests have always argued that progressive changes will cause economic ruin and social chaos. The same arguments are now being raised against efforts to protect the environment and to introduce renewable technologies. As always, these arguments are self-serving and irrational. Because the global economy is no longer sustainable, the complete transformation of the existing system is not an option, but a requirement. Creating a more efficient and equitable economy will not cause a global depression — it is the only possible way to avoid economic collapse and sustain economic growth.

The design of this book

Although much has been written on why our unsustainable system should change, little has been written on how it can change. We know only too well that there is always enough money to finance wars and buy luxuries, but never enough to feed the hungry or preserve rare species. What we need to know now is how we can change the institutions that support exploitation and competition into structures that support conservation and cooperation.

The first part of *Evolution's Edge* describes in detail the coming collapse of our natural and social worlds. It shows why our current global system is environmentally and culturally unsustainable, why the design of the industrial system is now obsolete, why policy adjustments and new technologies will not be able to prevent its inevitable collapse. We look at why earlier civilizations have failed and how cascading crises can provoke system failure.

The second part of *Evolution's Edge* describes how a new and very different type of societal system has begun to evolve and how we can support this process. We explain the paradigm changing factors that cause societies to evolve into more complex and conscious systems; because these factors are now developing and interacting with each other, the potential exists for rapid social evolution. We examine emerging technologies, ideas, values and social organizations; we analyze their role in supporting the evolution of a sustainable system. We then analyze the major developing global trends and probable future scenarios.

We also outline the design requirements of a sustainable planetary system and provide practical tools for supporting the development of its essential social and technological components. We explain how we can change destructive behaviors and empower ourselves and our social networks. Since societies are organized through culture, we also describe why an integral worldview and constructive values are the key to the emergence and integration of a functional holistic system. Finally we outline a common, cooperative vision for uniting a movement for constructive global change.

Evolution's Edge provides a rough map of the terrain that humanity will have to cross on the way to a sustainable future. Since mapping the future is an inexact science, this map will inevitably contain many errors. So read this book carefully and critically. We do not claim to have any final answers. The world's complex problems cannot be solved by a single solution or a single group. It will require the collective wisdom and combined efforts of billions of people to heal our dying planet. The alternative to our cancerous consumer society is not a new dogma and another monoculture, but a critically aware and culturally diverse holistic civilization.

This book builds on the work of countless dedicated activists and scientists from every part of the world. Its purpose is to explain why the collapse of the current global system is inevitable, why constructive change is possible, and how we can support the evolution of a better world.

Part

1

Collapse
— the dominant trend

We stand at a critical moment in Earth's history, a time when humanity must choose its future. As the world becomes increasingly interdependent and fragile, the future at once holds great peril and great promise. To move forward we must recognize that in the midst of a magnificent diversity of cultures and life forms we are one human family and one Earth community with a common destiny. We must join together to bring forth a sustainable global society founded on respect for nature, universal human rights, economic justice, and a culture of peace. Towards this end, it is imperative that we, the peoples of Earth, declare our responsibility to one another, to the greater community of life, and to future generations.

— Preamble to The Earth Charter (2000)[1]

1 Overshoot

> There is something fundamentally wrong with treating the Earth as if it were a business in liquidation.
>
> — Herman E Daly, former World Bank Senior Economist[1]

The debate over limits to growth

For many economists, arguments that the world is about to run out of resources are old, tired and thoroughly discredited. After all, in 1980 the biologist Paul Ehrlich made a famous bet with the economist Julian Simon regarding limits to growth.[2] Ehrlich predicted that growing resource shortages would drive up the price of five minerals over a decade, while Simon predicted that market forces (investments in new mines, etc.) would increase supplies and lower prices. Ehrlich lost the bet — by 1990 the prices had fallen for all five commodities.

Mainstream economic theory believes that economic growth is beneficial. It argues against the idea of biophysical limits to growth on the grounds that resource shortages raise prices, and rising prices provide an incentive for discovering and developing new supplies. Rising prices also encourage research into more efficient ways to produce and utilize resources and stimulate the development of cheaper substitutes. Because market forces and human ingenuity will always be able to meet the growing demand for resources through exploration, increasing efficiencies and substitution, there are no limits to growth. The proof of this logic is easy to see — the world economy is booming like never before.

> Anyone who believes that exponential growth can go on forever in a finite world is either a madman or an economist.
>
> — Kenneth Boulding (1910-1993), former President of the American Economics Association[4]

Case closed. Or is it? Is it possible to discover more fresh water, more wild fish, more topsoil? Is there a substitute for a livable climate, for an oxygen-producing rainforest, for a unique species, for pristine wilderness? Can we continue to produce ever-increasing amounts of food, wood or oil? Can our planet absorb ever-increasing amounts of pollution? Do enough resources exist to double world consumption, and then double that, and then double that — endlessly? Or are there real limits to growth?[3]

The debates between the pro-conservation and pro-growth camps started many years ago. They were popularized by the publication in 1962 of Rachel Carson's book *Silent Spring*. Its descriptions of the destructive effects of pesticides on the environment led to bans on DDT and other dangerous chemicals. Ehrlich's book on the implications of endless population growth also had a powerful impact and supported the introduction of contraceptives. In 1972 the debates heated up with the publication of *Limits to Growth*, a study that used computer simulations to model alternative future scenarios.[5] The study concluded that the planet's physical limits mean that the endless growth of production and consumption is dangerously unsustainable.

Although many scientists and social activists have promoted these views, they have been largely ignored by politicians, business leaders and the media. The global economy has been expanding for sixty years, and most people cannot see any reason why it won't continue to grow and prosper. However, environmental problems have increasingly attracted public concern. A tipping point has been climate change, which has pushed environmental issues onto political and business agendas around the world.

But the fact that there is a growing international consensus on the need to prevent destructive global warming does not mean that there is agreement on the need to limit consumption. The debate over the health of the world continues with the two sides taking diametrically opposed positions. One group of specialists (orthodox economists), says that the patient has a few minor problems but is generally healthy and getting better all the time. The other specialists (holistic ecologists) say that the patient has terminal cancer and will die without immediate surgery and a radical change in lifestyle.[6] Unfortunately, the doctors aren't arguing over the health of some stranger. They are talking about our planet and our lives.

This chapter examines the major issues involved in the debate over limits to growth. While I have attempted to use the most accurate data and the latest science available at the time of writing, I don't pretend to be impartial — I believe that all the evidence points to the need to limit resource consumption. So don't accept my analysis without reading opposing opinions and checking the facts for yourself. We need to know the truth because our futures and our children's futures are at stake.

Warning: you may find the next few chapters slow reading because they are full of data. I have included a lot of facts and references because many of the topics I am discussing (e.g. climate change and future energy supplies) are hotly debated. Please bear with me — because these are important issues, we need to fully understand them.

Growing populations

Let's start by examining Paul Erhlich's concern: population growth. Figure 1 shows how the growth of human populations has been accelerating. It took almost all of human history for the world's population to reach five million people around 10,000 B.C.E. (Before the Common Era). It then took almost 12,000 years (to 1800 C.E.) to increase to around one billion people. In the next 100 years (to 1900 C.E.) another 650 million people were added to the total population. Then in only one more century (to 2000 C.E.) human populations grew by another 4.5 billion people. And in the next 50 years (by 2050 C.E.) another three billion people are expected to be added to the total size of the world's population.[7]

Why did it take 200,000 years for the global population to reach five million, but now will only take 50 years to increase our numbers by another three billion? The reason is a combination of declining mortality rates (due to better technology, improving nutrition and life expectancy) and compound interest: the more people there are, the more babies they have.

Fortunately, the population of the world is not expected to increase indefinitely. People usually have large families in situations where there are high rates of child mortality and few social services, because having many children ensures that some will survive and grow up to help their parents. As health, education and standards of living increase, the economic benefits of having many children decline. In fact the best way to reduce fertility rates is to provide poor women with better education.[8]

With rising standards of living, the average size of families is dropping in most of the world. Many rich countries now need immigration to keep their populations

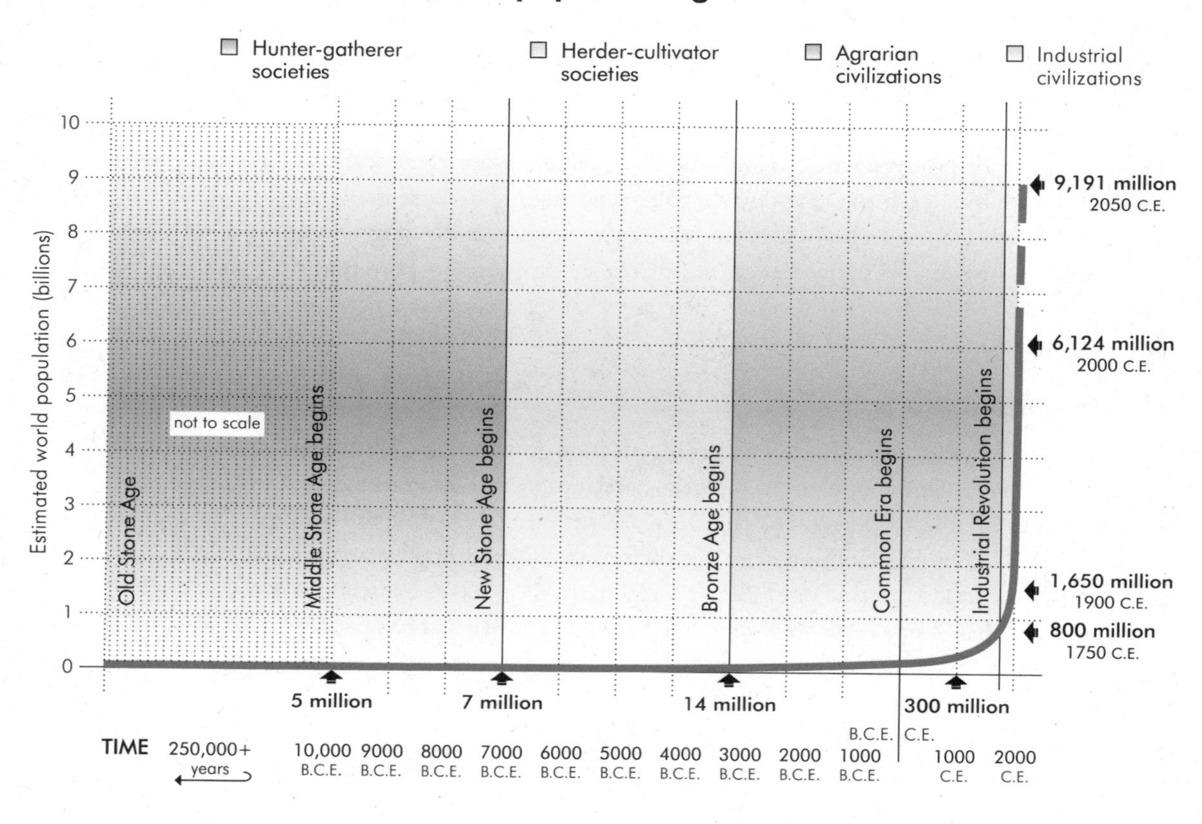

Figure 1:
BEST Futures [9]

from declining. However, populations are still growing in many poorer countries, and the world is expected to have around 9.2 billion people in 2050. Depending on the technological and policy choices that are made in the coming decades, population growth may level off and begin to decrease as early as 2040, or continue to increase to a high of 12 billion in 2100. The median forecast for 2100 is around 10 billion.[10]

The good news is that the population bomb is not going to explode. Rather than increasing indefinitely, the number of people on Earth will probably begin to decline later this century. The bad news is that human populations will continue to increase for many years to come.

A particular problem is that almost all of the increase — another 2.6 billion people by 2050 — will occur in poor countries. This means that the proportion of the world's people living in less developed countries will increase from 80% in 2000 to 88% in 2050.[11]

Rapid population growth puts tremendous demands on the resources and infrastructure of poor countries. For example, a lack of work and educational opportunities encourages expanding rural populations to migrate to the cities. World-wide, urban populations are increasing by 180,000 people *every day*, with most of that growth occurring in the developing world.[12] If current trends continue, the one billion people that are now subsisting in urban slums will have increased to three billion by 2050. And in order to survive, each one of them will need food, energy, water, housing, sewage, education, health care, transport, employment and other goods and services.

Does the world have the resources to provide for not only the needs of the world's growing population, but also their wants? After all, people want more than

Figure 2: BEST Futures [13]

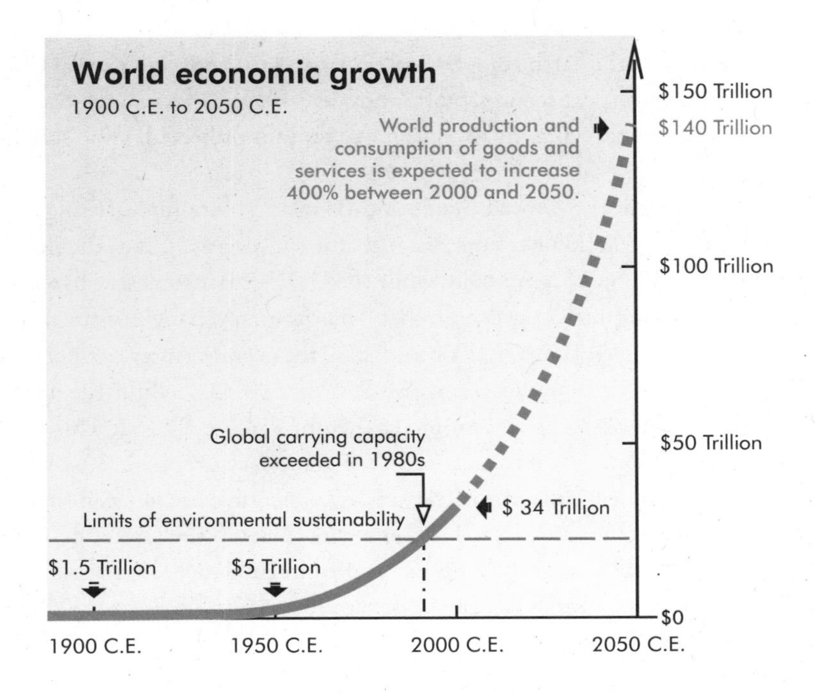

bare subsistence — in every country people want to live the lifestyle that people enjoy in the developed world. These demands are creating enormous environmental and social pressures.

Growing consumption

While global population is projected to increase by 50% from 2000 to 2050, the world economy is projected to increase by approximately 400%.[14] This means that if current rates of growth are maintained, the global economy will double every 25 years.

> The most powerful force in the universe is compound interest.
>
> — Albert Einstein, theoretical physicist (1879-1955)[15]

Once more we run into the problem of compound interest. For the global economy to continue to expand as forecast, world production and consumption of goods and services will have to increase 2 times between 2000 and 2025, 4 times by 2050, 8 times by 2075, 16 times by 2100 and so on.[16] Is limitless material growth possible on our finite planet, or will it inevitably run out of resources?

Many leading ecologists believe that the global economy has already run out of resources — that it passed the point of sustainable growth in the 1980s.[17] This does not mean that there are no resources left, but that the global economy is consuming more resources each year than the planet can replace.

We will examine the earth's biophysical limits and the consequences of unsustainable growth in the next section. But for now let's assume that enough resources exist to allow the global economy to continue to grow as projected. Who stands to benefit the most from this growth?

The economies of the United States and Bangladesh are forecast to grow at similar rates to the year 2050: estimates are that the US Gross Domestic Product (GDP) will increase by 2.8% annually, while the GDP of Bangladesh will increase by 3.0% annually. But this doesn't mean that Americans and Bangladeshis will end up with similar incomes. Between 2005 and 2050 the average American per capita GDP is projected to increase from $36,854 to $93,323, while the average Bangladeshi's income is projected to increase from $388 to $1,163. This means that the average American will consume $56,469 more each year by 2050, while the average Bangladeshi will consume an extra $775. Of course not all economies are projected to grow at the same rate. The per capita GDP in the United Kingdom is expected to increase at a rate of 2.1% per year from 2005 to 2050 (from $26,831 to $57,970) while incomes in Mozambique are forecast to decrease at 0.8% per year (from $339 to $77).

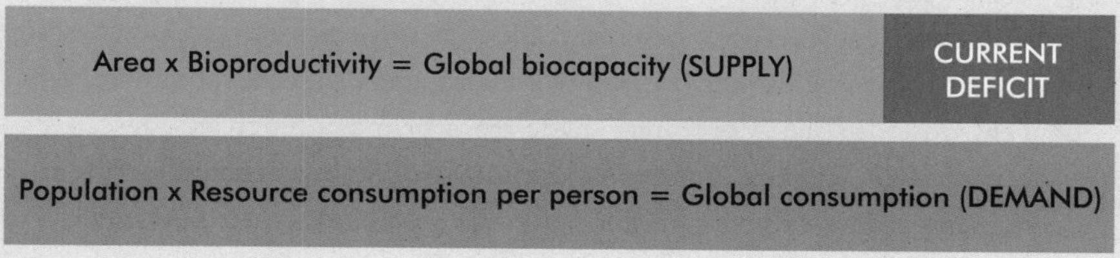

Figure 3:
BEST Futures [18]

Unsustainable growth rates

The Rule of 70

To find the doubling time in years of a quantity, divide 70 by its annual percentage growth rate.

At a growth rate of 1% doubling time is 70 years.
At a growth rate of 3% doubling time is 23 years.
At a growth rate of 7% doubling time is 10 years.
At a growth rate of 10% doubling time is 7 years.

Between 2000-2007 most industrial countries had growth rates of around 3% per year. Developing countries grew faster with the Indian and Chinese economies growing at average rates of 7% and 9% annually.

If these rates of growth were to continue, the global economy would more than quadruple in size by 2050.

However, humanity is already consuming 25% more resources than our planet is capable of producing each year. The global economy is now unsustainable. Economic growth based on the constant expansion of resource consumption is no longer possible.

Area x Bioproductivity = Global biocapacity (SUPPLY) | CURRENT DEFICIT

Population x Resource consumption per person = Global consumption (DEMAND)

From these figures, we can see that the effect of compound interest is to widen the gaps between the rich and poor countries. We can also see that not everyone on the planet consumes the same quantity of goods and services and/or is responsible for producing the same amount of pollution: while the average American consumed $101 per day in 2005 and is expected to consume $256 per day in 2050, the average Mozambiquan consumed $0.93 per day in 2005 and is expected to consume $0.21 per day in 2050. Moreover, statistics showing average consumption within countries hide the distribution of income within each country: within both the United States and Mozambique a few are much richer and many much poorer than the average.

This doesn't mean that all economic growth is bad — constant economic growth promises to benefit the majority of the world's population, including most of the people living in India and China. However, continued economic expansion will provide uneven benefits, with the richest people on the planet becoming much richer, and many of the poorest people becoming even poorer. In the following chapters we will look further at the structural links between global wealth and global poverty. But first we will examine the most important issue: whether the continued expansion of global consumption is even possible.

Available global resources

When we talk about global resources, we are referring to natural resources that are available or that could become available. For example, although there are enormous quantities of metals below the Earth's crust, they cannot be affordably accessed with today's technology. For water-stressed people in Jordan, the vast sheets of ice in the Antarctica might as well be on another planet. For resources to be useful, they must be available at a cost that justifies their acquisition.[19]

There are two basic types of resources: renewables and non-renewables. Resources are considered to be renewable if their quantities (stocks) can increase over time. Examples of renewable resources are fresh water, solar and wind energy, oxygen, wood, fish and products made from primary renewable resources (e.g. processed food or paper). Resources are considered non-renewable if their stocks decrease over time. Non-renewable resources are minerals, gas, oil, coal and products made from primary non-renewable resources (e.g. copper wire or plastics made from fossil fuels).

Supplies of all natural resources are limited. Quantities of non-renewables decrease as they are consumed, although some non-renewables, like metals, can be recycled. And while quantities of renewables can be increased there are upper limits to potential supplies. For example, although more trees can be planted there are only so many places where trees will grow. Quantities of renewables can also decrease through environmental degradation (e.g. through topsoil erosion or overfishing).

Not only are there limited resources on our finite planet, but most of the resources are not useful or available. For example, 2.5% of the world's water is fresh water, and of this only 1/3 is accessible since most fresh water is locked away in polar ice. The usable water is also unevenly distributed. As a result 40% of the world's population faced serious water shortages in the 1990s, and it is estimated that two-thirds of the world's population will face shortages by 2025.[20]

The earth has a surface area of 126 billion acres (51 billion hectares). Of the 90 billion acres (36.3 billion hectares) of sea, only the coastal areas are very productive. Of the 36 billion acres (14.7 billion hectares) of land, only 21 billion acres (8.3 billion hectares) is biologically productive as the rest is too cold, too dry or without the soils needed to support dense plant or animal life. Ecologists estimated that the Earth's total amount of biologically productive land and sea was equivalent to approximately 28 billion acres (11.2 billion hectares) in 2003.[21]

This limited quantity of biologically productive land and sea constitutes the renewable natural capital of our planet. It produces a sustainable flow of goods and services called natural income or interest. Since all the humans, animals and plants on Earth share and are supported by the annual global natural income, it is referred to as the planet's carrying capacity. It is critical for our survival that we know the size of this annual natural income and whether it is increasing or decreasing.

Figure 4: BEST Futures[22]

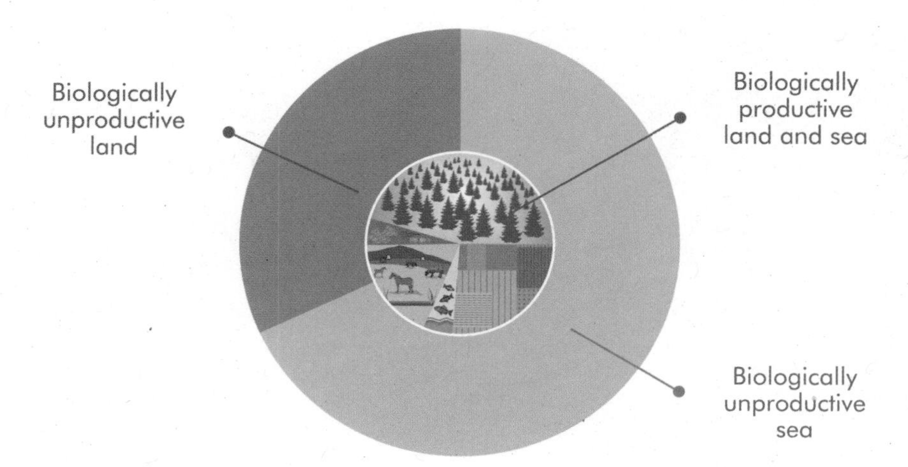

Natural capital is not a fixed quantity. The amount of natural capital can increase, for example when more farmland is brought into production or crop yields are improved. On the other hand, natural capital can decrease, for example when biological productivity is lost due to urbanization, desertification or pollution. If the amount of natural capital increases or decreases, the annual natural interest will also increase or decrease.

Figure 5: BEST Futures [23]

The general trend has been to slowly increase the planet's renewable natural capital. The total area of biologically productive land and sea increased from 22

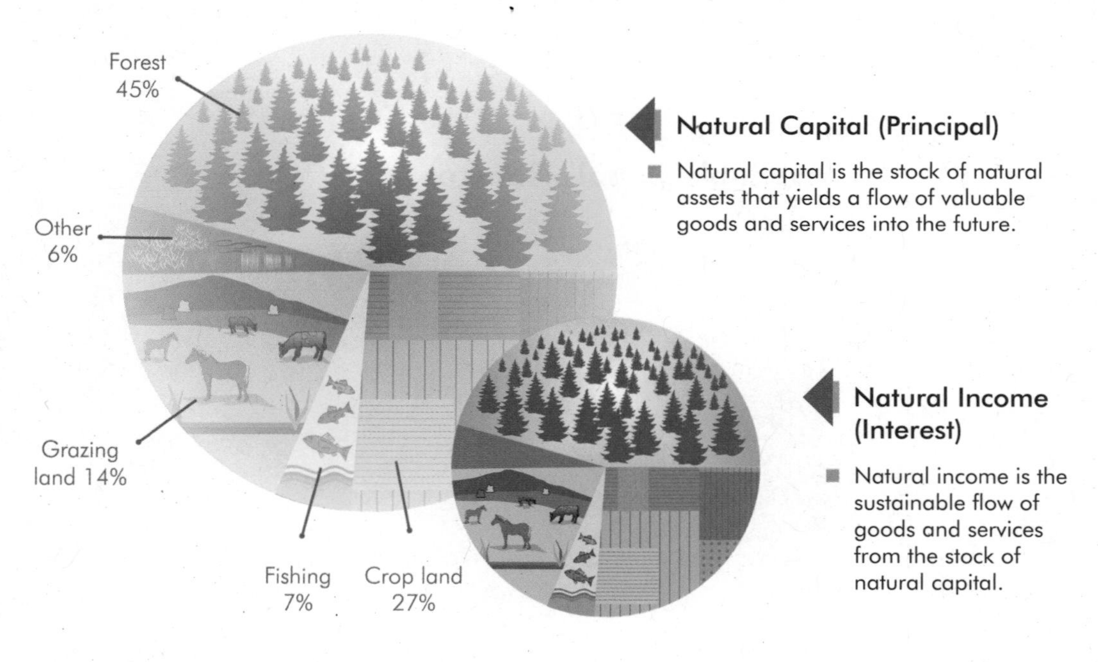

billion acres (9 billion hectares) in 1961 to 28 billion (11.2 billion hectares) in 2003, and is expected to further increase to 30 billion acres (12 billion hectares) by 2025. With careful management and technological improvements, it could further increase to 32 billion acres (13 billion hectares) by 2050 and to 35 billion acres (14 billion hectares) by the year 2100.

However, if the global economy continues with business as usual, the total quantity of biologically productive land and sea will only increase for another couple of decades. Environmental degradation will likely cause it to decrease to 27 billion acres (11 billion hectares) by 2050, and continue to decline after that. This disastrous scenario — in which a growing population will have to compete for declining resources — is predicted because the global consumption of resources is already exceeding the planet's carrying capacity.

In other words, humanity is living beyond its means. We are not only spending all our annual natural income, but each year spending more and more natural capital. This spendthrift lifestyle produces a vicious cycle in which the global natural income decreases each year as natural capital decreases, and the rate of decline accelerates. As everyone knows, the inevitable result of constantly spending more than you earn is bankruptcy. However, we are not just spending our natural income, but also the natural income that supports all the other species on the planet. This means that major ecosystems will begin to fail long before we consume every last renewable resource. Since human economies are completely dependent on their environments, collapsing ecosystems will collapse the global economy. Bankruptcy is not as far away as we might imagine.

We will examine the consequences of resource shortages and environmental degradation in the next chapter. Let's next look at the amount of natural income we each get to spend each year, and how much each of us is actually spending.

Ecological footprints

Ecological footprints are tools for measuring and analyzing humanity's demand on the biosphere. They are used to analyze the average annual natural resource consumption and waste output of countries, cities, organizations and individuals. An ecological footprint is the amount of natural capital required to produce a natural income large enough to meet biological expenses (resource consumption plus waste output). For example, the footprint of a country includes all the cropland, grazing land, forest, and fishing grounds required to produce the food, fiber and timber the country consumes, to absorb the pollution it discharges and to provide space for its infrastructure.

Although the total amount of global biologically productive land and sea was only 28 billion acres (11.2 billion hectares) in 2003, the global ecological footprint was 35 billion acres (14.1 billion hectares). This means that humanity spent more than it earned in 2003, and consumed and polluted the equivalent income of an extra 7 billion acres (2.9 billion hectares) of natural capital. Consuming natural capital involved deforestation, overfishing, overdrawing aquifers, etc., while producing more waste than the Earth could recycle involved polluting the air, water and soil.

This level of consumption is clearly unsustainable. If we wish to avoid environmental and economic bankruptcy, we will have to live within the planet's annual income. A sustainable average per capita footprint is calculated through dividing the total natural capital of 28 billion acres (11.2 billion hectares) by the total population (6.3 billion), which gives us a sustainable average per capita footprint (each person's *fair Earthshare*) of 4.4 acres (1.8 hectares). This was the average amount of productive land and sea that each person had in 2003 to live on without causing environmental degradation.

We can calculate the average per capita amount of overconsumption through dividing the global ecological footprint by the number of people in the world (6.3 billion). This gives us a per capita global footprint of 5.5 acres (2.2 hectares) in 2003. The difference between the average global per capita footprint and the fair Earthshare — 1.1 acres (0.4 hectares) — was the average amount that the environmental debt of every human on the planet increased in 2003. Of course this is only an average. Because billionaires and beggars have completely different levels of consumption, they have completely different environmental impacts.

Figure 6 shows that on average people in high income countries (e.g. in Europe, Canada and Japan) consume more than three times their fair share of renewable natural resources, while people in middle income countries (e.g. in China, Russia and Mexico) consume approximately their share of resources, and people in low income countries (e.g. in India, Nigeria and Viet Nam) are using less than half their fair share. Americans, with the world's largest economy, are using on the average more than five times their fair share.

In order to have been sustainable in 2003, the global economy would have had to reduce the average per capita global footprint by 20%.

A number of expert organizations focus on the problem of sustaining economic growth (e.g. the Wuppertal Institute in Germany, the Sustainable Technology Development program in the Netherlands, the Rocky Mountain Institute in the United States and the Natural Edge Project in Australia). They argue that global economic expansion can only continue if economic growth is decoupled from resource consumption (this is also referred to as separating economic growth from physical growth). This can be done through increasing the efficient use of resources. However, major gains in efficiency will need to be achieved quickly. In order to

Figure 6: BEST Futures [24]

Per capita ecological footprints

The Per Capita Ecological Footprint is a tool for measuring and analyzing the average annual natural resource consumption and waste output of individuals.

= **4.4** acres (1.8 ha.) — Each person's fair ecological footprint (Earthshare) in 2003 (The average amount of biologically productive land and sea available for each person on earth)

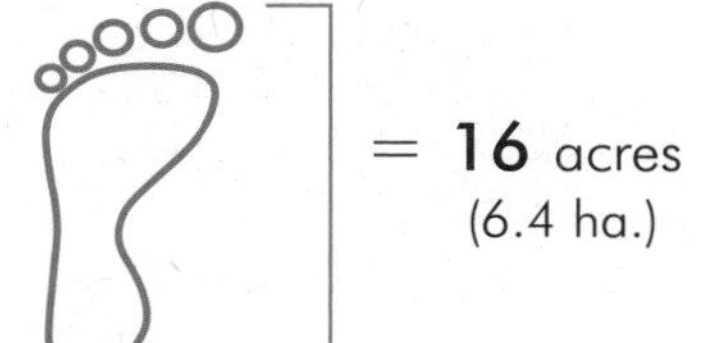

= **16** acres (6.4 ha.) — Each person's average footprint in high income countries (956 million people in 2003)

= **5** acres (1.9 ha.) — Each person's average footprint in middle income countries (3,012 million people in 2003)

= **2** acres (0.8 ha.) — Each person's average footprint in low income countries (2,303 million people in 2003)

The average American ecological footprint

= **4.4** acres
(1.8 ha.)

Each person's fair Earthshare in 2003

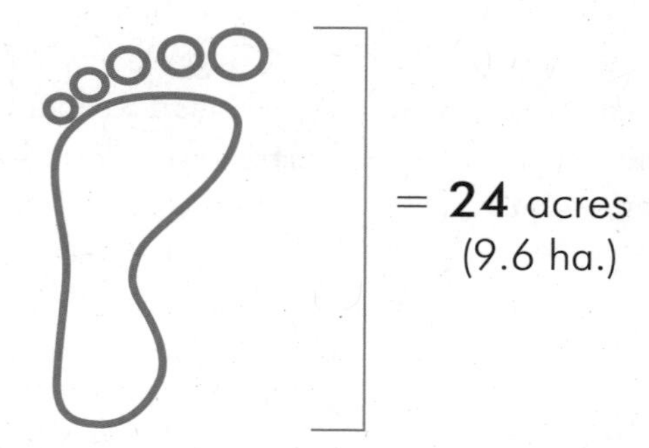

The average footprint of US citizens in 2003

Human economies will only survive over the long term if they are able to function within the carrying capacity of planet Earth.

The resources of 4 more planets would be needed for everyone in the world to live like Americans. The globalization of the American consumer society is not possible.

Figure 7:
BEST Futures [25]

sustain current rates of economic growth the use of resources (resource intensity) per unit of economic output will need to be reduced by 90% or more by 2050.[26]

Is this possible? Perhaps, but only if governments and businesses act quickly and begin making the massive investments required to dramatically increase economic efficiency. The problem is that most policy makers do not yet realize that there are biophysical limits to growth. As a result, rather than reduce the global consumption of either renewable or non-renewable resources, most political and

Sustainable ecological footprints

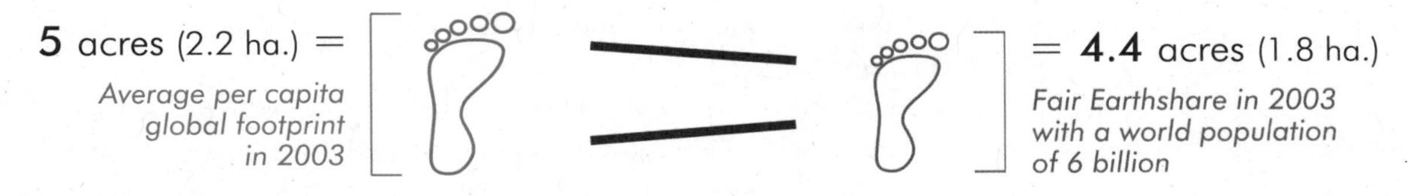

20% reduction needed in 2003

The global per capita footprint needs to be reduced now if human economies are to be sustainable.

5 acres (2.2 ha.) =
Average per capita global footprint in 2003

= 3 acres (1.2 ha.)
Estimated fair Earthshare in 2050 with a world population of 9 billion

40% reduction needed by 2050

By 2050, when there will be 9 billion people sharing the resources of the planet, we will have to reduce our average global per capita use of natural goods and services by 40% from 2003 levels.

Figure 8: BEST Futures [27]

---

business leaders are still trying to maximize economic growth with little regard for long-term sustainability. If global consumption continues to grow at current rates, the total global footprint will increase from 35 billion acres (14.1 billion hectares) to 60 billion acres (24.2 billion hectares) by 2050 — and more than quadruple the current annual environmental deficit.

Overshoot

In 1944 the US Coast Guard introduced 29 reindeer to St. Matthew Island in the Bering Sea. The original idea was to provide meat for American troops in the

Pacific, but after the Second World War ended, the island was abandoned to the reindeer. Conditions were ideal for the reindeer as there was plenty of lichens available for winter grazing and no predators. By 1957 the herd had grown to 1,350, and by 1963 to 6,000. Then the population suddenly crashed. A 1966 count found only 42 reindeer still alive.[28]

The reason for the die-off was that the exploding population had eaten all the lichens on the island. Unable to find anything to eat under the snow, most of the reindeer starved in the severe winter of 1963/64. This cycle of accelerating population growth, then expansion past sustainable limits, then die-off is frequently found in ecological systems.

Going too far and accidentally exceeding limits is called overshoot. *Limits to Growth* explained that: "The three causes of overshoot are always the same, at any scale from personal to planetary. First, there is growth, acceleration, rapid change. Second, there is some sort of limitation or barrier, beyond which the moving system may not safely go. Third, there is a delay or mistake in the perceptions and the responses that strive to keep the system within its limits. These three are sufficient and necessary to produce an overshoot."[29]

Overshoot does not have to result in a catastrophic collapse if problems are corrected in time. Unlike the reindeer on St. Matthew Island, humans can recognize the limits to growth, change our lifestyles, restore damaged ecosystems and create a sustainable economy. Or can we? Are we able to recognize the danger signs? Can we change our habits in time?

2 Growing Global Crises

Vital signs

As a former emergency paramedic I'm very aware that things are often not what they seem. The person sitting in the wrecked car may be perfectly all right, while the person walking away from the accident may be suffering from a serious head injury. You don't really know what is going on until you have checked the vital signs, conducted a thorough physical examination, discovered the history of the problem and listened to the patient's complaints. The point is not simply to list the problems, but to find out if a complaint is a new condition, if it is symptomatic of deeper issues, if the problems are interconnected and if the patient's overall health is stable, improving or rapidly deteriorating.

I learned this on my very first ambulance call. We were dispatched for a woman in labor, and I ran into the house nervously expecting to deliver my first baby. I went inside past three men sleeping on couches in the living room and found the woman locked in the bathroom. We had a confused discussion through the door — I didn't want her to give birth in the toilet — during which she told me that she wasn't due for another month and wasn't even having contractions. As I walked back through the living room I noticed that the three men were hardly breathing. On closer inspection I realized that they were about to go into respiratory arrest. I then yelled for my partners (I was still in training so there were three of us) and we worked on the heroin addicts and rushed them off to hospital. I never did see the pregnant woman.[1]

This chapter looks at some of the vital signs of the global economy and our planet. You will not find it pleasant reading — the global economy may look healthy, but it is actually critically ill. This information is not easy to accept — it is like a doctor telling you that your minor complaint is actually a life-threatening illness.

It is shocking to discover that the world we know is about to collapse. But if we don't face the problems they will only get worse. On the other hand, if we act now we can heal our sick planet with compassion and science. The first step is to carefully examine the problems. Once we know exactly what is going on we can devise a plan of treatment.

Figure 1: BEST Futures [2]

Energy slaves

The average person living in an industrialized country is supported by the work of 60 energy slaves.

Growing energy shortages

During the Stone Age our hunter-gatherer ancestors only had their own muscles for power and firewood for heat. Now most work is done by machines. Industrial economies run on energy. Energy is needed to power machines and transport, extract and process resources, provide heating, lighting and communications, produce food and pump water.

Although the majority of people don't live in developed countries, most people would like to. The world is industrializing because everyone wants a decent standard of living. But it will be an enormous challenge just to supply the world's basic energy needs. Since 1.6 billion people currently have no access to electricity, a new 1,000 megawatt power plant would have to be built every 48 hours for the next 50 years to bring global electrification up to the level of consumption Americans had in 1950.[3]

The increasing demand for energy comes from both the developing world, where more energy is required to meet basic needs, and from the developed world, where more energy is wanted to increase wealth. Current forecasts call for overall demand to increase 50% between 2007 and 2030. If present trends continue, most of this energy will have to come from three non-renewable fossil fuels — oil, coal and natural gas — which are forecast to provide 81% of the world's energy in 2030.[4]

The industrial system requires a constant flow of energy to function. If there is not enough energy available, energy prices will rise sharply and the global economy will begin to shut down. A critical question is whether enough affordable energy can be produced in the future to meet rising demand.

The size of this problem is illustrated in Figure 2.

Figure 2: Adapted from Limits to Growth: the 30-Year Update [5]

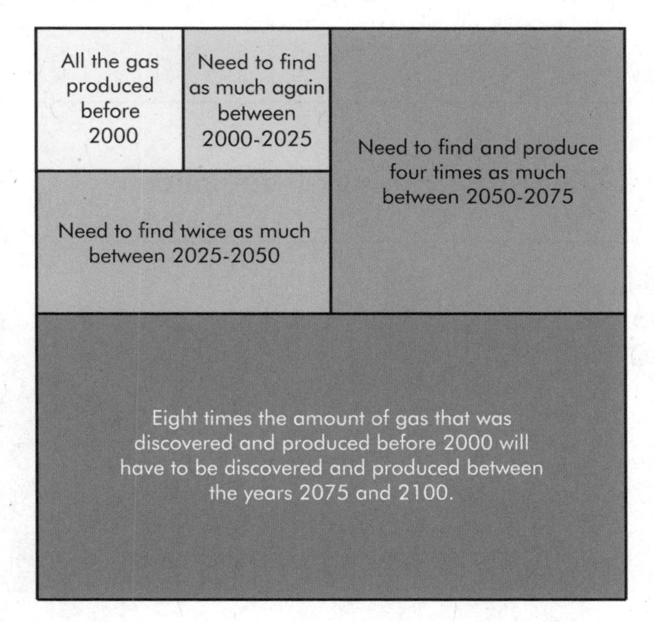

Since natural gas is a finite, non-renewable resource, it is obvious that global supplies must run out at some point. The only question is when. Given that gas discoveries and production have already begun to decline in many countries, we can be sure that it won't be long before severe shortages develop.

In order to understand global energy issues, let's take a closer look at future oil supplies. Oil is the most important energy source since it provides over 36% of all primary energy and is essential for food production and transport.[6] Oil prices (in US dollars) increased more than seven times from 2000 to 2008. The important question is whether this rapid rise in prices indicates that we are beginning to run out of oil.

Within the oil industry there are differing opinions about the amount of oil that can be produced in the future. Once again the debates are about limits to growth. On one side are the pessimists (including Chevron, Shell and Total), who believe that supplies of cheap, easily accessible oil are limited. On the other side are the optimists (including ExxonMobil and the OPEC countries), who believe that market forces and new technologies will guarantee adequate future supplies.

Figure 3: BEST Futures[7]

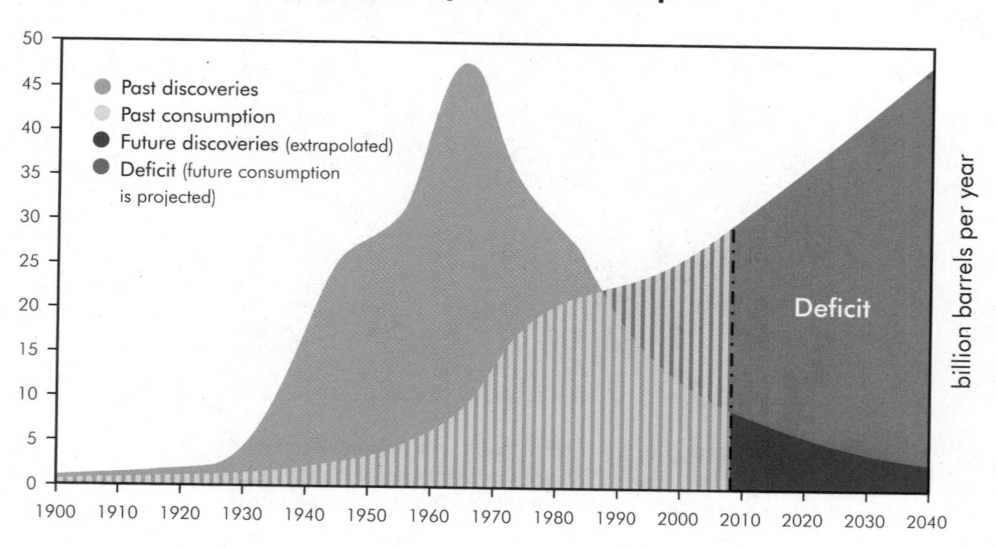

At current rates of growth, global demand is forecast to increase from around 86 million barrels of oil per day in 2007 to around 116 million barrels per day in 2030. While everyone agrees that there is still lots of oil in the ground — probably more oil than has already been consumed — there is disagreement over how much of it can be extracted at an affordable price.[8] Optimists believe that improving technology will enable new oil fields to be discovered and more oil to be extracted from existing fields.[9] Pessimists argue that global oil production will soon reach a maximum point (called peak oil), after which it will begin an irreversible decline.

The peak oil proponents give six reasons why global oil production cannot increase much further: (1) the oil in existing fields is declining at 4-6% per year;[10] (2) oil discoveries peaked in the 1960's; in the last few years the major oil companies have failed to discover oil as quickly as it is being consumed;[11] (3) while it may still be possible to produce more conventional oil in Saudi Arabia, Iraq and Iran, the point of maximum production has already been passed in the majority of oil-producing countries; (4) it is likely that many governments and corporations have been exaggerating the sizes of their existing reserves for political and financial reasons; (5) although huge quantities of oil are trapped in tar sands and oil shale, increasing demand cannot be met from non-conventional sources because high economic and environmental costs constrain production[12] and (6) oil prices have been rapidly rising. Their conclusions? Global energy shortages are just around the corner.

Although some significant initiatives are being made to increase the supply of energy from alternative sources,[14] no country has comprehensively addressed the problem of what to do in the event of oil shortages.[15] At present no cheap substitutes are available to replace oil in transport and in the production of fertilizers, pesticides, plastics and other products. Although natural gas can be substituted for oil in some areas, the global production of gas will peak soon after oil.

> The world is not running out of oil — at least not yet. What our society does face, and soon, is the end of the abundant and cheap oil.
>
> — Colin Campbell, founder of the Association for the Study of Peak Oil[13]

A major problem is that it will not be possible to rapidly make up shortfalls in gasoline and diesel supplies with current technologies. The production of biofuels is constrained by the need to use the world's limited amount of biologically productive land to grow food and fiber. While there are large reserves of coal in the world and synthetic oil can be produced from coal, this is currently an expensive and polluting process. Although electrical capacity can also be added through building more coal-fired or nuclear power stations, burning more coal will worsen global warming and nuclear energy has long-term issues related to safety, security and waste disposal.

There are many new technologies under development with the potential to produce cheap, clean energy not only from renewable sources (e.g. solar, wind, wave and tidal energy), but also from non-renewable sources (e.g. the plasma conversion of coal into hydrogen fuel for transport, nuclear fusion).[16] However, there is no indication that governments and businesses are planning to generate most of the world's energy from renewable and/or clean sources of energy in the next few decades.

In 2006 the International Energy Agency estimated that $20 trillion would have to be invested in all forms of energy production by 2030 to meet growing demand.[17] Decisions will have to be made about these huge investments in the near future because new technologies take years to develop, and major power installations and infrastructure take years to bring into production.

Given current trends and policies, the IEA expects most investments to be made in existing technologies, such as building more coal and gas fuelled power plants. These policies, which reflect the views of the majority of the world's political and business leaders, make two significant assumptions: first, that large new fields of oil and natural gas will be found over the next two decades; and second, that the use of fossil fuels will increase and greenhouse gas emissions (the cause of global warming) will continue to rise.

Because there are so many unknown variables, we really don't know what will happen. But it is quite possible that we are heading towards a catastrophic crisis. In May 2008 the International Energy Agency changed course and ordered an inquiry into whether the world is running out of oil.[18] What if the IEA has been wrong about future supplies and oil production begins to decline in the next few years? And what if the IEA has been right to predict that the increasing demand for energy will result in a rapid rise in greenhouse gas emissions between 2005 and 2030?

David Strahan, the author of *The Last Oil Shock,* has warned that "It is quite possible to run out of oil and pollute the planet to destruction simultaneously .... When oil production starts to fall, the economic impacts could well be devastating. Soaring crude prices could tip the world into a depression deeper than that of the 1930s, and collapsing stock markets cripple our ability to finance the expensive clean energy infrastructure we need. As the unemployment lines grow, the political will to tackle climate change may be sapped by the need to keep the lights burning as cheaply as possible."[19]

Strong words. But isn't he exaggerating a bit? Is polluting the planet to destruction a real risk?

Increasing climate change

Human economic activities are raising global temperatures through adding *greenhouse gases* that trap heat from the sun in the atmosphere.[20] These pollutants are primarily carbon dioxide, methane and nitrous oxide. Concentrations of carbon dioxide, which are higher than they have been for 650,000 years, are rising faster each year. This means that the rate of global warming is accelerating.

When climate scientists predict rising temperatures they are talking about long-term global trends. In the short term, weather cycles and regional variations can produce colder or hotter temperatures than average — for example the La Niña effect produced unusually cold weather in the Northern Hemisphere in the winter of 2007/8.[21] The Intergovernmental Panel on Climate Change (IPCC) forecasts that if current trends continue, average global temperatures will probably rise between 3.2°F-7.2°F (1.8°C - 4°C) by the end of the century. However, it is possible that temperatures will increase as little as 2°F (1.1°C) or as much as 11.5°F (6.4°C).[22]

Figure 4: BEST Futures. [23]

The shrinking Arctic ice cap

Summer Arctic ice in 1979

By 2003 summer Arctic ice had shrunk by 20%

The Arctic's summer sea ice may completely disappear by 2013.

The implications of increasing temperatures can be seen in Figure 5. Global warming adds energy to the atmosphere, causing weather patterns to change and extreme events to occur more frequently. Over the last 100 years, average global temperatures have risen by 1.4° F (0.8° C). Although this appears to be only a small increase, it has been enough to provoke major shifts in the Earth's climate. Glaciers are retreating, coral reefs are bleaching, deserts are advancing, storms are strengthening, rainforests are burning and polar ice is melting.[25]

Figure 5: BEST Futures[24]

Projected climate change impacts

Time	*Pre-industrial – 1980*	*2007 → global warming accelerates → 2100 (?)*
*Temperature**	0°F (0°C); +0.9°F (+0.5°C)	+1.4°F (+0.8°C); +1.8°F (+1°C); +3.6°F (+2°C); +5.4°F (+3°C); +7.2°F (+4°C); +9°F (+5°C); +11.5°F (+6.4°C)
Food		Extreme weather reduces global harvests Crop yields decrease by 10% for every 1.8°F (1°C) increase; widespread and increasing famine Mass starvation
Water		Melting glaciers; spreading deserts Perennial droughts in many areas including the US West, the Mediterranean, Southern Africa and Australia
Ecosystems	Healthy ecosystems	100+ species lost every day; 20% of coral reefs dead All coral reefs die; tropical rainforests dry out and become savannahs; oceans become acidic; 20% to 50% of all species die off 6th mass extinction; up to 90% of species die
Extreme weather events		Increasingly frequent and intense storms, floods, droughts, forest fires and heat waves
Global economy		Rising food prices; increasing storm damage and insurance costs Resource shortages increasingly cripple the global economy; tropical and sub-tropical regions become uninhabitable; sea levels rise and inundate coastal areas
Climate tipping points	Global temperatures stable (in equilibrium)	Melting arctic sea ice and glaciers; 50% reduction in the ability of soils and oceans to absorb CO_2 emissions Ocean algae die off; soils release carbon dioxide; frozen methane in tundra and oceans melts and escapes into the atmosphere; global warming becomes irreversible

*Increase in average global temperatures from pre-industrial times

It is easy to see that if an increase of less than 1.8°F (1°C) is already having serious impacts, then further increases are likely to have disastrous consequences. The IPCC estimates that if average global temperatures rise by more than 3.6°F (2°C), it will probably trigger rapid, major, and irreversible impacts, including the extinction of hundreds of thousands of species, the conversion of rainforests to dry savannah, the spread of deserts, increasing drought in dry areas of the planet, increasing precipitation and floods in wet areas, falling crop yields and rising sea levels.[26] The impacts of rising temperatures are explained in detail by Mark Lynas in his award-winning book *Six Degrees*.[27]

It is not possible to accurately calculate the impacts or costs of climate change since, for example, we can't put a value on the hundreds of thousands of species that will go extinct if temperatures rise by even a few degrees. The Stern Review on the economics of climate change concluded that rising temperatures "create risks of major disruption to economic and social activity, on a scale similar to those associated with the great wars and the economic depression of the first half of the 20th century."[29] At higher temperatures the consequences will be catastrophic. Much of the planet will become uninhabitable, and most of the species alive today will go extinct. It will be almost impossible to maintain advanced civilizations in these conditions.

Climate change is a result of the greatest market failure the world has seen. The evidence on the seriousness of the risks from inaction or delayed action is now overwhelming.

— Sir Nicholas Stern, academic and former Chief Economist of the World Bank[28]

The threat of runaway global warming is worrying an increasing number of scientists.[30] If the natural processes that keep the Earth's climate in equilibrium are seriously damaged, it may become impossible to prevent global temperatures from getting hotter year after year — even if all further greenhouse gas emissions are stopped. Some climate change tipping points have already been passed: for example, the Earth's ability to reflect sunlight is declining as the massive ice sheets that air-condition our planet melt, and the ability of oceans and soils to absorb carbon dioxide has been sharply reduced.

Global warming will not only have to be stopped, but also reversed to reduce temperatures back to a level at which biophysical processes can maintain an equilibrium — an equilibrium which was lost in the 1980s when average global temperatures rose higher than 0.9°F (0.5°C) above pre-industrial levels. If this isn't done quickly, global warming will trigger an irreversible destructive cycle in which a warming atmosphere and warming oceans will destroy the rainforests, ocean algae and other vital ecosystems that remove carbon from the air. This danger is explained in David Spratt's and Philip Sutton's book *Climate Code Red: the case for emergency action*.[31]

An indication that we have already passed a dangerous tipping point is that the permafrost in Artic regions has already begun to thaw and release increasing quantities of methane, a greenhouse gas that is more than 20 times stronger than carbon dioxide.[32] It is estimated that up to 10,000 billion tons (9071 billion tonnes) of carbon exists as frozen gas hydrates (methane plus water) in permafrost and under the world's oceans.[33] David Viner, a senior scientist at the Climatic Research Unit at the University of East Anglia, said "When you start messing around with these natural systems, you can end up in situations where it's unstoppable. There are no brakes you can apply. This is a big deal because you can't put the permafrost back once it's gone."[34]

Research on ocean sediments indicates that dramatic climate change can occur very quickly. Audrey Dallimore, a scientist with Natural Resources Canada, said that "Neolithic Chinese culture collapsed 4,000 years ago because the climate changed so fast the culture couldn't sustain itself. With natural climate change cycles, it appears there is no warning; there is no lead-up. Change was profound and happened in less than a decade .... Modern day atmospheric and ocean patterns suggest the same magnitude of climate change seen in the 4,000-year-old geologic evidence is happening now. It really is a 'sit up and listen change.' Something very different is happening."[35]

Catastrophic global warming is not an imaginary scenario. Of the five mass extinctions of life that have occurred in the history of our planet, four were caused by climate change. According to Gregory Ryskin, a professor of chemical engineering at Northwestern University, "explosive clouds of methane gas, initially trapped in stagnant bodies of water and suddenly released, could have killed off the majority of marine life and land animals and plants at the end of the Permian era" — a great extinction that wiped out 95% of the marine species and 70% of the land species that existed 250 million years ago.[36]

In 2006, the *Stern Review* estimated that emissions of carbon dioxide would need to drop more than 80% by 2050 to keep the increase in global temperatures under 3.6°F (2°C). In order to meet these targets many experts believe that industrialized nations will have to reduce emissions by up to 90% from present levels and developing nations such as China, India and Brazil will also have to make major reductions.[37] In their 2007 forecast, the International Energy Agency pointed out that, "Even if governments actually implement, as we assume, all the policies they are considering to curb energy imports and emissions, both would still rise through to 2030." In their best scenario, these additional atmospheric pollutants will increase average global temperatures by 5.4°F (3°C).[38] The report fails to mention that the current best case scenario will be enough to destroy one third of all the species on Earth as well as trigger runaway global warming.

The UN's *Human Development Report 2007/2008* was blunter: "There is now overwhelming scientific evidence that the world is moving towards the point at which irreversible ecological catastrophe becomes unavoidable .... There is a window of opportunity for avoiding the most damaging climate change impacts, but that window is closing: the world has less than a decade to change course."[39]

In reality we probably do not have even 10 years to stop catastrophic global warming — recent authoritative studies indicate that global warming is accelerating three times faster than the worst forecasts of the Intergovernmental Panel on Climate Change.[40] For example, American studies indicate that rather than sea levels rising by around 15.75" (40 cm) by 2100 as predicted by the IPCC, the true rise may be as much as 78" (2 meters) — an amount that will inundate islands and major coastal cities around the world. Dr. James Hansen, the director of NASA's Goddard Institute for Space Studies said, "[T]he Earth is getting perilously close to climate changes that could run out of control."[41]

This is serious stuff. The evidence not only indicates that climate change will do massive environmental and economic damage in the coming decades, but that the survival of most life on Earth is threatened by runaway global warming. So why have governments not declared climate change to be a global emergency?

The reason is that for decades well financed pro-growth lobby groups have successfully downplayed the dangers posed by climate change and blocked environmental initiatives.[42] The corporations and governments that have fought the hardest against limiting atmospheric pollution are the ones with the most to lose from programs that encourage conserving energy and using alternative fuels — for example ExxonMobil, Peabody Energy, General Motors and the government of Saudi Arabia.

Despite the enormous power and influence of the fossil-fuel industry, the scientific evidence on climate change has been overwhelming, and much of the public is now convinced of the dangers of global warming. However, most governments and businesses around the world still value political, military and economic expansion more than environmental conservation. As a result, much more research is being done on increasing industrial output than on reducing global warming, and on developing new weapons than on developing new sources of energy.[43]

Most economists agree that an enormous technological shift will be required to stop global warming and that this will only occur if polluters are taxed and the revenues used to subsidize the development and introduction of non-polluting alternatives. They agree that the best mechanism for this would be the establishment of an international market in which polluting industries would have to pay for each ton of carbon that they discharge into the atmosphere through buying emission rights.

The problem is that for this market to exist there will have to be a strict international agreement to limit global emissions, and most nations are still increasing rather than reducing emissions. Among the developed nations, the United States refused to sign the 1992 Kyoto Protocol on reducing greenhouse gas emissions, most European nations are polluting more rather than less,[44] and Canada (where emissions are now more than 25% above 1990 levels and rising) officially abandoned the goal of reducing emissions to below 1990 levels on the grounds that attempts to meet international commitments would ruin its economy.[45]

Harlan Watson, President Bush's chief climate negotiator, made the American position clear at a United Nations meeting in May, 2007. He said, "We don't believe targets and timetables are important, or a global cap and trade system. It's important not to jeopardize economic growth."[46]

Resistance to a tough and enforceable international agreement is not just coming from the rich countries. Most developing nations have to date opposed the idea of mandatory limits on the basis that they produce far less pollution per capita than industrialized nations. For example, while acknowledging that global warming will sharply reduce domestic grain production, the Chinese government's National Climate Assessment Report concluded that the country should focus on development before reducing greenhouse gas emissions.[47] Since China recently overtook the United States as the world's biggest producer of greenhouse gases, if China and other developing countries do not soon agree to mandatory limits, the chances of stopping runaway global warming will be nil.

The science is clear on the dangers of further global warming, on the measures needed to prevent it, and on the need for urgent action. In 2007 delegates from 187 countries agreed that "deep cuts in global emissions will be required" to prevent dangerous global warming. But the resulting Bali Action Plan contained no binding commitments.[48] In the words of Achim Steiner, the chief of the United Nation's Environment Program, "The missing link is universal political action."[49]

Awareness of the need to prevent climate change is growing, and most politicians and business leaders now talk about the need for sustainable development and environmental conservation. But few are willing to support conservation over economic expansion. Sometimes there is no need to choose between conservation and growth, since the cheapest and most sustainable way to achieve economic growth is through increasing efficiencies and reducing waste. However, in many cases real choices will have to be made, since economic growth beyond sustainable biophysical limits can only be achieved at the expense of the environment.

Will the world's leaders be prepared to put conservation above consumption? Will they be able to change the global economy's addiction to limitless quantitative growth in time to prevent catastrophic global warming? We'll examine these questions in the next chapters.

But first let's look at other urgent global issues.

Growing water shortages

Humans, like most terrestrial species, cannot survive without constant access to adequate supplies of clean fresh water. Many more people die from waterborne

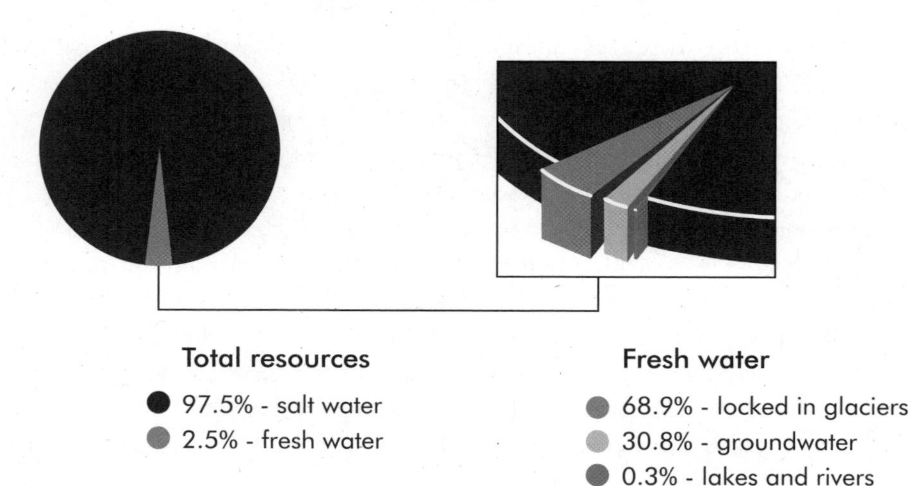

Figure 6: Adapted from a BBC News diagram [50]

diseases each year than are killed in wars.[51] The poor suffer the most as water scarcity often means having to carry water long distances, drinking contaminated water, paying high prices for it and starving when droughts wilt crops. In 2006, 1.1 billion people were still without clean drinking water, and 2.6 billion lacked access to adequate sanitation.

Water scarcity occurs not only because the amount of fresh water on Earth is limited, but also because very little is accessible in lakes and rivers. About a fifth of the water that is used worldwide is groundwater.

Again the key problem is one of biophysical limits to growth: while demand is constantly increasing, supplies are finite. Global water consumption increased approximately six times between 1900 and 2000 — twice as fast as population growth. The growing use of irrigation in agriculture has been responsible for most of this increase, and agriculture now accounts for 70% of global water use.[52] If the world continues to use current agricultural technologies to feed the world's increasing population, more water will need to be withdrawn each year from rivers and aquifers. Industrial and domestic demand is also constantly rising and additional supplies will be needed to support industrial growth and rising standards of living.

The United Nation's Millennium Ecosystem Assessment estimates that between 2000 and 2050, global water demand will increase by a minimum of 30% and a maximum of 85%.[53] But where will this extra water come from?

While the total amount of accessible fresh water is limited, supplies are also unevenly distributed. As a result, a third of the world's population now lives in water-stressed countries. This enormous problem is getting worse each day. The United Nations water agency has predicted that "By 2025, 1,800 million people will be living in countries or regions with absolute water scarcity, and two-thirds of the world population could be under stress conditions. The situation will be exacerbated as rapidly growing urban areas place heavy pressure on neighbouring water resources."[54]

The demands being made on existing fresh water supplies are already causing some of the world's largest rivers to be so overdrawn that they run dry for part of the year. Among these are the Yellow River, the Colorado and the Amu Darya. Many other rivers, like the Nile, the Ganges and the Indus, have had their flows greatly reduced. A combination of climate change and the diversion of streams for irrigation have also contributed to the decline or disappearance of many bodies of water including Lake Chad and the Aral Sea.

Climate change will increase water scarcity, as it will tend to make the wet areas in the world wetter and the dry areas dryer. The Intergovernmental Panel on Climate Change forecast that, "By mid-century, annual average river runoff and water availability are projected to increase by 10-40% at high latitudes and in some wet tropical areas, and decrease by 10-30% over some dry regions at mid-latitudes and in the dry tropics, some of which are presently water stressed areas .... Drought-affected areas will likely increase in extent. Heavy precipitation events, which are very likely to increase in frequency, will augment flood risk .... In the course of the century, water supplies stored in glaciers and snow cover are projected to decline, reducing water availability in regions supplied by meltwater from major mountain ranges, where more than one-sixth of the world population currently lives."[56] These areas include much of California and the American Midwest.[57]

> This is just a time bomb.
>
> — Carmen de Jong, hydrologist[55]

Another area of concern is the state of groundwater supplies. More than two billion people in the world rely on wells or springs for their domestic water supplies. In addition 40% of the world's food supply is grown on irrigated farmland.[58] Half of the world's population lives in countries where water is being withdrawn from aquifers faster than it is being recharged. As a result water tables are falling beneath many urban areas (e.g. Mexico City, Beijing and Bangkok), and in many of the most productive agricultural regions in the world (e.g. in China, India, Pakistan, Iran, Mexico and the United States). In the United States, the Ogallala aquifer below the Southern Great Plains is shrinking.[59] The Millennium Ecosystem Assessment report estimates that 15-35% of all irrigation withdrawals are currently unsustainable.

The quality of useable fresh water is also decreasing in many areas. Many of the world's rivers, lakes and aquifers are now badly contaminated by fertilizers, pesticides and industrial chemicals — for example, the Yangtze, China's longest river, is now so cancerous with pollution that experts fear it will become a dead river within five years.[60] Water problems will clearly worsen as global demand continues to rise and global supplies of usable water continue to decrease. If these problems are not dealt with in time, the result will be food shortages and famines, the collapse of major ecosystems, increasing disease and the forced migration of hundreds of millions of people in search of food and water.

Water shortages will be an enormous challenge in the coming decades. Not only do 2.6 billion people lack clean water and sanitation today, but another 3 billion people will need to share the world's supplies by 2050.[61] As well, the amount of water being drawn from many rivers and aquifers will need to be reduced in order to avoid the collapse of major ecosystems.

It is possible to add more supplies of clean fresh water through desalinating sea water, cleaning and recycling wastewater and pumping water from areas where there are surpluses to areas with shortages. However, all of these methods are expensive and energy intensive. A real danger is that global energy use and greenhouse gas emissions will rapidly increase as countries attempt to deal with growing water shortages. Cheaper and more sustainable approaches involve reducing water waste and improving the efficient use of water in agriculture and industry. The restoration of wetlands, the reclamation of deserts and reforestation will also improve water flows, filter pollution and improve regional climates.

International efforts have been making steady progress in providing more people in the world with clean water and sanitation. The question is whether governments will be able to continue these positive trends in the face of growing water crises. In 2000 the World Commission on Water estimated that an additional $100 billion a year will be needed to tackle global water scarcity. The impacts of climate change are likely to increase costs further. Will governments — and in particular the governments of the rich countries — be willing to come up with this extra money? At this point international development assistance for improving water quality and sanitation averages only $4.5 billion per year.[62]

The reality is that right now governments are more focused on economic growth than on the needs of either the poor or the environment. In the future water issues may have even more trouble attracting adequate funding. After all, water shortages are only one of many growing global problems.

Growing food shortages

The Millennium Ecosystem Assessment also predicted that world demand for agricultural products will increase by 70%-85% between 2000 and 2050. Is it possible for farmers to grow this much more food? Global food security is increasingly threatened by all the problems that we have discussed: increasing populations, looming energy shortfalls, rising temperatures and decreasing supplies of fresh water. Food production is also threatened by topsoil erosion, spreading deserts, pollution, overfishing and the loss of arable land to roads and cities.

Until recently the industrial world has always been able to produce more food than it consumes. However, just as the demand for energy and fresh water is beginning to exceed supply, the demand for food is also starting to exceed supply. The essential problem is the same: while there are no limits on potential demand, there are biophysical limits on potential production.

Like fresh water, food production can always be increased, but with rising costs. More food can be grown on marginal land, in the desert, in greenhouses or in factory vats, but only with expensive inputs of energy, water and other resources. The problem is not just whether farmers can find ways to almost double food production, but also whether they can produce huge quantities of extra food at prices people can afford.

While the poorest 854 million people were chronically hungry in 2006,[63] since then the rising price of food has greatly increased the number of malnourished people and is now threatening millions with starvation.[64] Their suffering is not caused by resource shortages but by economic inequality — although thousands of tons of fresh produce are shipped around the world each day to supermarkets in the rich world, at present the political will does not exist to ensure that the poorest of the poor have enough to eat.

There is enough food in the world right now for everyone on the planet to have enough to eat. The United Nations' Food and Agicultural Organization estimated that of the 2.34 billion tons (2.13 billion tonnes) of grain likely to be consumed in 2008, only 1.11 billion tons (1.01 billion tonnes) will be used to feed people.[65] This is because most grain is used for animal feed and to make industrial products; increasing amounts are now being used to make ethanol to fuel vehicles.

Figure 7: BEST Futures [66]

Grain consumption rises with meat consumption

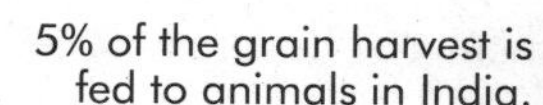

5% of the grain harvest is fed to animals in India.

60% of the grain harvest is fed to animals in the United States.

The resources do not exist for the world's population to copy meat-based Western diets.

The major single cause of food shortages is that rising global incomes are enabling 3 to 4 billion low-income consumers to diversify their diets and eat more meat, fish, milk and eggs. As consumers eat higher up the food chain, each person consumes more grain.

It takes approximately 2 pounds (900 grams) of grain to produce 1 pound (450 grams) of (farmed) fish or poultry, 4 pounds (1.8 kilograms) to produce 1 pound (450 grams) of pork, and 7 pounds (3.2 kilograms) of grain to produce 1 pound (450 grams) of beef.[68] Because most people in India cannot afford many animal products, the average Indian now consumes the equivalent of one-fifth as much grain as the average American.[69] The major cause of current food shortages is that there is not enough grain in the world for most people to eat meat based diets. As the growing middle classes of the world consume more meat, the price of grain rises and the poor go hungry.

> A silent tsunami which knows no borders is sweeping the world.
>
> — Josette Sheeran, Executive Director of the UN World Food Program[67]

But if a billion people are hungry today, will billions more be hungry within a few decades? This is quite possible because farmers will need to feed an extra three billion people by the year 2050. More people are being added to the world's population each year than live in the United Kingdom, and unless global meat consumption is sharply reduced, extra food will have to be grown to feed them all.[70] An increasing number of experts, like Britain's chief scientist Professor John Beddington, believe that the rapidly rising price of grains and other food staples is the first sign of a developing global food crisis.[71] However, at present most international agencies are not worried about long-term food shortages. They expect that rising prices will encourage farmers to produce much more food in the coming decades.

Let's take a closer look at the main areas where production is expected to increase, examining first the chances of significantly improving per hectare yields. In 1962 Paul Erhlich predicted that the world's growing population would soon run out farm of land and begin to starve. His grim predictions proved to be wrong because the rate of population growth slowed and new agricultural technologies greatly increased crop yields. But was he wrong to say that there are limits to the amount of food that can be produced on our finite planet? The Green Revolution introduced high yielding varieties of maize (corn), wheat and rice. These hybrids enabled cereal production to more than double in developing nations between 1961 and 1985. However, although global yields have continued to rise with better farming practices and the spreading use of industrial fertilizers and irrigation, the rate of increase is slowing down as farmers develop optimal mixes of seeds, inputs and agricultural practices. In fact per capita grain output peaked around 1985 and has been falling ever since.[72]

A major study has found that genetically modified (GM) soya and cotton actually produce smaller yields than conventional crops. Bob Watson, who directed the study for the International Assessment of Agricultural Science and Technology for Development, was asked if GM crops could solve world hunger: "The simple answer is no."[73] Lester Brown, the president of the Earth Policy Institute, has pointed out that although agricultural breakthroughs can improve harvests, the physical laws governing photosynthesis (the ability of plants to convert sunlight and carbon dioxide into sugars) place upper limits on the total amount of food that can be grown on our planet. This is because a limited amount of sunlight is available per hectare of farmland, and quantities of arable land and water are also limited.[74]

In other words, Ehrlich was right — the same limits to growth that apply to all renewable biological resources apply to the production of food. Of course it is technically possible to develop substitutes for crops grown on farm land, in the same way that farmed fish are being substituted for declining stocks of wild fish. In the future foods may be mass produced from various sources such as fungi and algae.[75] The problem is that substitutes often require more resource and energy inputs than the commodities they replace. High-yielding hybrid crops need more irrigation, fertilizer and pesticides than traditional varieties. Future alternative foods may require even larger inputs of raw resources and energy because they will not only need to be grown, but also processed in order to make them edible.

Although larger inputs of water, fertilizers and pesticides have produced rising global yields, the concern is that it will be increasingly difficult for the world's farmers to produce larger harvests. Triggering these fears is the fact that global harvests of wheat, corn, rice and the other grains that make up most of our diets are not increasing as quickly as consumption. The consumption of wheat and other coarse grains has outstripped production in all but one year since 2000, and the consumption of rice has outstripped production every year since 2001.[76]

Since farmers plant grains on nearly two thirds of the world's irrigated land, falling water tables are beginning to affect harvests. The combination of falling yields and growing demand is forcing more and more countries to import grain. As a result, international grain reserves are declining and are now at the lowest levels in decades. In China the annual wheat harvest has already dropped from a peak of 123 million tons (112 million tonnes) in 1997 to around 100 million tons

(91 million tonnes). Rice harvests are also declining, but more slowly. Although grain production is still rising in India, up to 25% of agriculture depends on over-pumping aquifers. Over the next few years millions of wells will run dry in India, Pakistan and many other major grain producers. As yields drop, more and more countries will be forced to import food, and declining global surpluses will turn into absolute shortages.[77]

Since it takes 1,000 tons (907 tonnes) of water to produce one ton (0.9 tonnes) of wheat, and 2,000 tons (1814 tonnes) of water to produce one ton (0.9 tonnes) of rice, the export and import of one million tons (0.9 million tonnes) of grains is equivalent to the export and import of one or two billion tons (0.9 or 1.8 billion tonnes) of water. In the near future, countries like Australia that are facing increasing water shortages may have to restrict food exports in order to conserve domestic supplies.

Global wheat and rice production is failing to keep up with rising consumption. The International Rice Research Institute has warned that increasing shortages of rice threaten the food security of 3 billion people. They expect that prices will continue to rise as land used for producing rice and irrigation water is lost to industrialization and urbanization.[78] Although the production of corn still exceeds demand, demand may soon exceed supply since the use of crops for fuel is rapidly increasing. In the United States, where biofuels are produced primarily from corn, construction is starting on a new ethanol plant every three days. Since US annual corn exports account for 25% of the world's total grain exports, the increasing use of corn to fuel vehicles means that less and less corn will be available in the coming years for export.[79]

As the gap between production and consumption has narrowed, the cost of food has begun to rise. In 2007 global food prices rose on average by 40%, driven by a combination of rising demand, floods and droughts caused by climate change, the rising cost of fertilizers and other oil-based inputs and the increasing diversion of corn and other edible crops to the production of bio-fuels.[80] Rising prices are not only making food unaffordable for increasing numbers of the world's poor but also putting the cost of fertilizer and other farm inputs out of the reach of many poor farmers. This is forcing farmers in many countries to plant fewer crops, despite the promise of higher prices for future produce.[81]

The evidence indicates that it will be difficult to significantly improve yields through introducing new technology or through irrigating more land. But will it be possible to increase global harvests by bringing more land under cultivation?

The ability of farmers to increase food production is not only being restricted by falling water tables but also by soil erosion and pollution.[82] For example, overgrazing by cattle, sheep and goats in northern and western China has stripped off protective vegetation. The result is massive dust storms that remove millions of tons of topsoil and quickly convert productive rangeland into desert. The Gobi desert has been rapidly advancing and is now only 150 miles (240 kilometers) from Beijing.

Desertification — the expansion of desert areas, caused by environmental degradation and climate change — is a huge and growing problem. In Nigeria deserts are taking over 877,000 acres (351,000 hectares) of land each year.[83] Deserts now occupy a third of the planet — more than the surface of China, Canada and Brazil combined. The United Nations has warned of "a creeping catastrophe" — two-thirds of Africa's arable land may disappear by 2025, along with one-third of Asia's and one-fifth of South America's.[84]

Farm land is also disappearing due to contamination from the excessive use of fertilizers, polluted water, heavy metals and pollution from factory farming.[85] Air pollution is affecting farming as well as human health — a vast haze (the Asian Brown Cloud) made up of ash, acids, aerosols and other particles now stretches across South Asia.

While these trends are alarming, the worst threats to future food security come from climate change. The International Panel on Climate Change forecasts that while a low temperature rise of 1.8°F- 3.6°F (1°C-2°C) will improve grain yields at higher latitudes, it will reduce grain yields closer to the tropics. At higher temperatures yields will fall almost everywhere.[86] The Chinese government agrees with their findings. Their own scientists predict worsening droughts and extreme weather will reduce the production

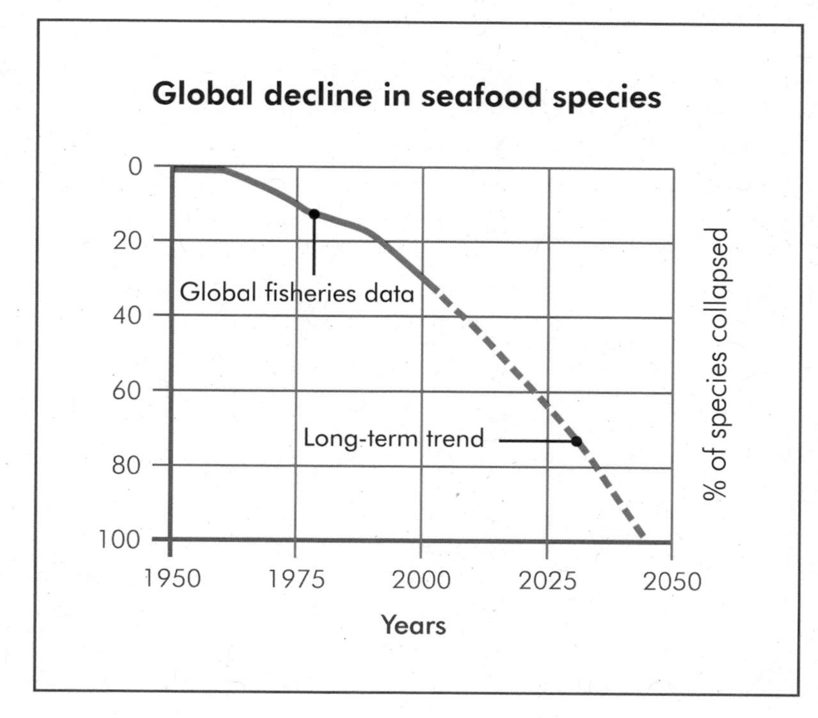

Figure 8: Adapted from a BBC chart[87]

of rice, corn and wheat in China by 10% by 2030, and by up to 37% during the second half of the century.[88]

Many experts think that even these grim forecasts underestimate the severity of the coming food crisis. Gwynne Dyer, an international political analyst, states that a rise in global temperatures of even 3.6°F (2°C) will cause massive crop failures in most major food-producing regions of the world. Because the world will not have any food reserves, millions of people will starve.[89]

The future will not see any increase in the catches of wild fish. The reality is that one-third of all ocean fish stocks have already collapsed (i.e. to below 10% of their original yield) and more are collapsing each year. Marine biologists now predict that if present trends continue all the rest of the world's fisheries will be gone by 2050.[90] This will be a tragedy not only for the environment, but also for many coastal communities around the world that rely on fishing for both their food and their livelihoods.

Fortunately, aquaculture is likely to produce increasing amounts of protein in developing nations. Although farming carnivorous species such as salmon does not produce more food (five kilograms of fishmeal are required to produce one kilogram of salmon), raising herbivorous species such as carp, tilapia and catfish in farm ponds is extremely efficient.[91] In China and India fish polyculture is growing rapidly. Worldwide, farmed fish may soon produce more animal protein than cattle ranching.

Despite the positive outlook of aquaculture, it may prove difficult to significantly increase overall food production. While better agriculture techniques will tend to improve yields in many parts of the world, this is likely to be countered by declining yields resulting from a combination of increasingly extreme and unpredictable weather events, falling water tables, increasing temperatures, increasing pollution and increasing desertification.

The situation is serious but not hopeless. In his book *Feeding the World,* Vaclav Smil argued that too much attention has been focused on increasing production and not enough on eliminating waste in production, storage, processing and consumption. He is optimistic that it should be possible to avoid global food shortages through a combination of methods, including eliminating waste, proper environmental protection, reducing meat consumption, planting wasteland and "well-proven economic and technical fixes."[92] Some of these well-proven fixes are using drip irrigation to reduce water use, or preventing the spread of deserts through ending overgrazing and planting trees and grasses. Of course, these actions cost money, but the cost of solving the problems will be much less than the cost of doing nothing.

An example is the problem of producing food in Africa. If soil conditions continue to worsen, Africa may be able to feed just 25% of its population by 2025. Since a major cause of desertification and soil degradation is caused by stripping the soil of stabilizers and nutrients (trees, crop residues and animal dung) for use as cooking fuels, much of the decline could be prevented by providing people with alternative ways of cooking. Dr. Rattan Lal of Ohio State University has estimated that for an annual cost of about $2 billion — the cost of 20 jet fighters — soil improvements could produce an extra 22-33 million tons (20-30 million tonnes) of food per year, enough to feed the number of people being added annually in developing countries.[93]

Zafar Adeel, director of the United Nations University's International Network on Water, Environment and Health, pointed out that "Addressing desertification is one of the most effective ways to deal with climate change — you get the biggest bang for your buck. Unfortunately, there is a lack of understanding on the part of policy-makers." In 2007 international contributors cut funding for the UN's land-degradation convention by 15%.[94]

Can the world avoid future food shortages? Do our governments have the vision and the political will to solve the growing threats of global warming, water shortages, pollution, desertification and other related problems? Not yet.

Accelerating rates of extinction

An even worse threat than food shortages is the mass extinction of life on Earth. After examining over 40,000 species, the 10,000 scientists of the World Conservation Union have concluded that over 40% of all species are presently at risk. They estimated that these include one in four mammals, one in eight birds, one in three amphibians, half of all reptiles, half of all insects, one in three conifers and three out of four flowering plants.[96] German Environment Minister Sigmar Gabriel estimated that up to 150 species are now disappearing each day — one every 10 minutes.[97]

No one knows exactly how many or which species disappear each day, because no one knows how many different species exist. Estimates range from 1.5 million (the number of already classified species) to 100 million.[98] Scientists are still discovering new types of birds, mammals and fish every year, and systematic efforts are only beginning to be made to catalogue the myriad varieties of plants, insects,

> This we know: the earth does not belong to man; man belongs to the earth .... All things are connected. Whatever befalls the earth befalls the sons of the earth. Man did not weave the web of life: he is merely a strand in it. Whatever he does to the web, he does to himself.
>
> — Ts'ial-la-kum, Chief Seattle (1754-1866)[95]

worms, fungi and microbes. Any one of these may have a critical role in preserving the ecological balance of a major ecosystem or may hold the secret to curing cancer or to extending our life spans.[99] Many unknown species are hidden in complex ecosystems such as rainforests, mangrove swamps and coral reefs, and when their habitats are destroyed they disappear without trace, taking their secrets with them.

Although we don't know the exact rate of extinction, we do know that the rate of loss is accelerating. The evolutionary process has always meant that on average every year one species in a million disappear, while a slightly higher number of new species emerge to replace them. The rate of extinction is now somewhere between 100 and 10,000 times the normal rate.

All the species on the planet have to live on the world's limited renewable biological resources. As humans increasingly consume and pollute these resources, less and less is left over for all other species. Under the pressure of disappearing habitats, climate change, pollution, invasive species and relentless overharvesting, their populations are rapidly declining. Between 1970 and 2003 the populations of terrestrial species declined by an average of 31%, marine species declined by 23%, and freshwater species declined by 28%.[100]

Global warming has already severely damaged and disrupted ecosystems.[101] An international study has concluded that climate change alone will probably cause the extinction of between 15% and 35% of all species studied by 2050, if global temperatures rise as expected by over 3.6°F (2°C).[102] The United Nations agency responsible for preserving biodiversity warned that "Unless action is taken now, by 2100, two thirds of the Earth's remaining species are likely to be extinct."[103] In the less likely but possible event that temperatures rise as high as 11.5°F (6.4°C) by the end of the century, up to 90% of species will die off.

Climate change accelerates species extinction in a number of ways. Up to half the world's species live in tropical rainforests, and already more than half the rainforests have been cut down for timber or to clear land for agriculture. At current rates of deforestation, another 40% of rainforests in the Amazon alone will disappear by 2050.[104] With global warming many of the remaining areas will dry out and either become grasslands, or be regularly swept by forest fires.

Not only does climate change accelerate species loss, but deforestation accelerates climate change. The burning and cutting of forests produce a quarter of all greenhouse gases. Every 24 hours, deforestation releases as much carbon dioxide into the atmosphere as 8 million people flying from London to New York.[105]

Coral reefs, the home of hundreds of thousands of marine species, are also in danger since they begin to die when ocean temperatures rise by a few degrees. When the corals die, the reefs often collapse into rubble, depriving fish of food and shelter. A combination of rising temperatures, pollution and destructive fishing techniques (trawling, poisons and dynamiting) has already killed 20% of the world's coral reefs and degraded many more.[106] As carbon dioxide levels rise, the oceans are also becoming more acidic, making it increasingly difficult for corals to build their skeletons and for other species to build their shells.[107] Almost all coral species will be killed by temperature increases of above 3.6°F (2°C), along with krill and other species of zooplankton crucial to the marine food web.

Pollution is another growing problem, not only because the waste from industry, factory farming, cities and landfills is killing off entire ecosystems — forests, fields, lakes, rivers and coastal fisheries — but also because toxic chemicals become more concentrated as they travel up the food chain. In low concentrations chemicals are affecting the social and mating behaviors of many species, and in high concentrations they are lowering sperm counts as well as causing birth defects, reproductive problems and cancers.[108] Health problems due to ingested poisons and pollutants are increasingly showing up in whales, bears, eagles, dolphins and other top predators — including humans.[109]

Apart from the destruction of habitat from deforestation, desertification, pollution and the spread of farms, cities and roads, the plants and animals of the world face the almost insurmountable problem of having to migrate to avoid rising temperatures and changing environments. Studies show that species have been moving towards the poles over the last 50 years at the rate of about 4 miles (6.4 kilometers) per decade. This is not fast enough, since areas with the same temperatures (isotherms) have been moving towards the poles at the rate of 35 miles (56 kilometers) per decade and will soon be moving polewards at the rate of 70 miles (112 kilometers) per decade. Because interconnected ecosystems of trees, birds, insects and other species cannot migrate that quickly, many of them will find themselves trapped in strange environments for which they are not adapted. Mountain species are especially vulnerable to climate change as they have no place to go. This process will help drive up to 50% of the world's species to extinction.[110]

Scientists are calling this the sixth mass extinction event in the history of our planet. Although mass extinctions have occurred before (the last great extinction occurred 65 million years ago), this will be the first time that humans will have caused a disaster of this magnitude.[111] And it will be an unimaginable tragedy — after most of the existing mammals, birds, flowers, fish and other living things have disappeared, it will take tens of millions of years for new species to evolve and take their place.

Life in its present forms only exists because of the existence of a complex bio-physical equilibrium.[113] Every extinction upsets this equilibrium and weakens the web of life that supports human societies. If we lose many of the other life forms on

Figure 9: BEST Futures [112]

Spreading pollution

Between 50,000 and 100,000 synthetic chemicals are being manufactured. Most have never been tested for safety. Three new chemicals enter the market every day. Air pollution alone causes 5% of all deaths worldwide.

Most of Asia is now permanently covered by a thick brown haze. Up to 30% of our food is contaminated by hazardous chemicals.

Pregnant American woman

One in six American women have enough mercury in their wombs to put their children at risk for autism, blindness, cognitive impairment, heart, liver and kidney disease.

Polar bears

The survival of many species is threatened as toxic chemicals spread and are concentrated in the food chain.

Air pollution

Air pollution in China causes over 500,000 premature deaths each year.

Earth, it will not only be a huge spiritual loss, aesthetic loss and recreational loss, but it will also do irreparable damage to our economies. We cannot survive without the ecosystem services that other species provide us: services such as climate regulation, oxygen, clean water, food, waste recycling, building materials, crop pollination, agricultural nutrients, bioenergy and medicines.[114] Sigmar Gabriel pointed out that "biological diversity constitutes the indispensable foundation for our lives and for global economic development."[115] Not only is 40% of world trade based on biological products or processes — without the countless ecological services of plants, insects, microbes and other species, human societies cannot exist. To destroy the biodiversity of our planet is to self-destruct.

According to the Millennium Ecosystem Assessment report, approximately 60% of all ecosystem services have already been degraded.[116] Further degradation is likely to continue under even the most optimistic scenarios in which governments, businesses and civil society immediately act to prevent further global warming and preserve the environment. The situation will be much worse if constructive action is not taken in time and critical ecosystems are allowed to collapse.

Again the cost of taking action is much less than the cost of inaction. The areas where species are threatened with extinction are well known and can be protected through creating terrestrial or marine parks where humans are not allowed to hunt, fish or otherwise harm the local ecosystem. For example, the Earth Policy Institute suggested that marine reserves should be created where fishing is prohibited altogether: "A global network of marine reserves protecting up to 30 percent of the world's oceans would cost around $13 billion — far less than the subsidies that currently promote overfishing. Such a network would also create some 1 million new jobs and bolster the number of fish that can be caught in nearby waters."[117]

Organizations like the Alliance for Zero Extinction are trying to keep the most threatened species from going extinct by creating protected areas. The problem is that it is not enough to create a park: the park must also be protected from illegal logging, hunting and fishing, and from encroaching farms. This means that the people who live in the area need to be provided with alternative ways to earn their livings, and the parks must be properly patrolled and guarded.

However, in many parts of the world loggers and poachers have better funding and more weapons than park wardens. For example, Rachmat Witoelar, Indonesia's environment minister, said that illegal logging was ravaging 37 of his country's 41 national parks. At current rates of destruction 98% of the forests on the island of Sumatra and Borneo may be gone by 2022, and with them thousands of species.[118]

Saving the biodiversity of the world is still not a priority for most governments. In 2007, member states cut the budget of the Convention on International Trade in Endangered Species (CITES), the international organization responsible for listing and protecting species in danger of extinction. Jochen Flasbarth, the leader of the European Union delegation, commented, "If you look for the real problems of biological diversity around the world, it's clear that they lie in the forests and the marine environment. And as soon as you interfere in these areas you are confronted with huge economic interests."[119]

Growing crises

There are many good things happening in the world — positive global trends are the focus of the second part of this book. For example, the Africa Rice Center in Sierra Leone has developed a hybrid variety of rice that is increasing yields in Africa by up to 50% without the application of expensive fertilizers;[120] and the Norwegian government is planning to keep rare crops from going extinct by preserving seeds from around the world in a vault dug into a frozen mountainside.[121]

However, despite thousands of constructive programs, the overall situation is steadily deteriorating. The rate of global warming is accelerating. Energy and food prices are rising quickly as world demand begins to outstrip production.[122] Water shortages are worsening. Deforestation, desertification and pollution are spreading, and species are being lost at an ever-increasing rate. Given these facts, shouldn't everyone be worried?

These problems are all interconnected. As water tables drop, more oil is needed to drill the wells deeper and pump the water further. As the price of oil rises, the increasing cost of fertilizers, pesticides, and tractor and transport fuel drives up the price of food. As more fossil fuels are burned, global temperatures rise, deserts spread, and the amount of arable land decreases. As forests are cut down to make way for new farmland, biodiversity decreases and the ecological services that they provide decline. And so on in a dangerous and deadly cycle of increasing destruction.

These problems also all seem to be intractable — despite endless international conferences and initiatives, each year they are getting worse. It is not as though we don't know about the problems — scientists have been issuing warnings for 50 years — or that we don't know the solutions. We do. Humanity has never been so wealthy or possessed so much knowledge. It is not that difficult to feed a starving child, to preserve the habitat of a rare species, to stop pollution or soil erosion. The needs are there, the technologies exist, and the resources exist — but the will to solve the problems is missing. Why?

One answer is that political, business and union leaders do not want to accept that there are limits to growth. The collapse of cod stocks off the eastern coast of Canada is an example.

For over four hundred years the Grand Banks was the most productive fishery in the world. However, in the 1960's large companies brought in large trawlers that more than doubled the annual catch. Government scientists warned that these huge harvests were unsustainable, and in the early 1970's they were proven right: the size of the catch plummeted. Politicians then found themselves caught between the scientists, who wanted low quotas set to preserve the cod stocks, and the local

Figure 10: Millennium Ecosystem Assessment [123]

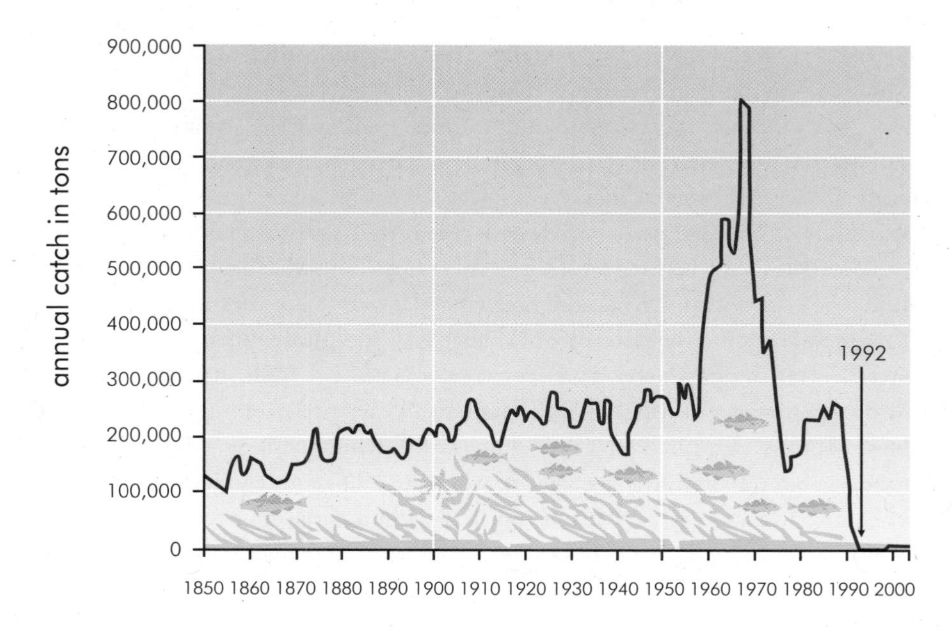

communities and the fishing industry, who wanted to keep higher quotas in order to preserve jobs and profits.

> You used to be able to fish right in the port. Now the only thing you can catch here is water.
>
> — Sall Samba, an octopus fisher in Mauritania and father of six[124]

The fact that fishers were still able to catch large quantities of fish persuaded the politicians that large catches were still sustainable, and they refused to lower quotas to the levels recommended by the experts. In reality, the only reason there were still sizable harvests was that better technology was enabling trawlers to track down and catch the last remaining schools of cod. In the late 1980's fishers were shocked to discover that the fish were gone. Since then, despite the closure of the fishery, the cod stocks have never recovered.

Have governments around the world learned a lesson from this ecological and economic disaster? Do they now realize that there are limits to the size of biological resources, that overexploiting resources has consequences, that scientific warnings should be taken seriously and that ecological collapses can happen suddenly with irreversible results? Apparently not. Boris Worm, the leader of an international team researching the worldwide decline in ocean fisheries, said "The way we use the oceans is that we hope and assume there will always be another species to exploit after we've completely gone through the last one."[125]

Scientists are now warning that every major fishery in the world will collapse unless catches are reduced and protected marine parks are established to allow threatened species to recover. However, to date they are having little influence on politicians, fishing communities and business groups who don't believe — or don't want to believe — that there are strict biological limits. For example, scientists are now warning that the North Sea cod stocks are about to follow the Grand Banks cod into terminal decline, but so far their requests for a moratorium on further fishing have been refused. Dr. Worm was amazed by the response: "It's very irrational. You have scientific consensus and nothing moves. It's a sad example; and what happened in Canada should be such a warning, because now it's collapsed it's not coming back."[126]

A major factor in the resistance to change is the political influence of vested interests. The fishing industry is a clear example. Scientists agree that trawlers are a major threat to ocean fisheries since they drag heavy nets and huge rollers along the sea bottom that catch rare, slow-growing fish while crushing corals and other marine habitat. Not only is deep-sea trawling incredibly destructive, it is not even economically viable — it could not continue without large government subsidies. However, the political power of the fishing industry in many countries is enough to keep subsidies coming and block efforts to ban trawling.[127]

This is not the only case of governments subsidizing all the wrong things. Around the world governments spend much more money subsidizing polluting industries than clean industries; much more on expanding unsustainable types of agriculture than on developing sustainable alternatives; much more on treating illnesses than on supporting health; and much more on waging war than on building peace. Although 15 million children die annually from polluted water (many more than the 3,000 people who died in the terrorist attack of September 11, 2001), providing clean water for the world's poor is not given anywhere near the same priority or funding as warfare.[128] Politicians still fail to see the links between environmental destruction, poverty and conflict. Although governments are beginning to address some of the urgent global issues, the critical question is whether they are capable of acting in time to prevent major disasters.

Rich world, poor vision

In the year 2000, all the governments in the world agreed to address global problems through setting eight Millennium Development Goals (MDGs) — targets for reducing world poverty and hunger, for improving health and education, for providing women with equal opportunities and for protecting the environment.

In 2005 the United Nations *Human Development Report* stated, "For many of the MDGs the jury is now in, with the evidence that a 'trickle down' approach to reducing disparities and maintaining overall progress will not work. The MDGs set quantifiable targets that lend themselves to policy responses rooted in technical and financial terms. Ultimately, however, the real barriers to progress are social and political. They are rooted in unequal access to resources and distribution of power within and among countries."[129]

Sufficient resources exist to solve all the world's major problems, but the global system has been designed to increase productivity and profits, not to redistribute wealth or protect the environment.

The Earth Policy Institute estimates that the worst social and environmental problems can be solved by increasing existing spending by approximately $190 billion dollars per year.[130] The additional costs of preventing global warming and developing non-polluting energy sources can be paid for through removing subsidies from fossil fuels and placing a carbon tax on greenhouse gas emissions.

This amount is sufficient to meet the world's most urgent needs. However, these are minimum goals: completely eliminating malnutrition, providing basic sanitation for everyone in the world, eliminating malaria and other deadly diseases, etc. will cost much more. But since the world's annual GDP is now larger than $48 trillion US dollars ($48,000,000,000,000), even if an additional $480 billion per year (or more than twice as much as the Earth Policy Institute has estimated) is needed to resolve the worst social and environmental problems on the planet, these expenditures would still only require the reallocation of 1% of the global budget.[132]

It is still possible for the world to avoid disaster. In the coming decades good leadership and good luck may combine to provide clean energy, clean water, good harvests

Figure 11: BEST Futures [131]

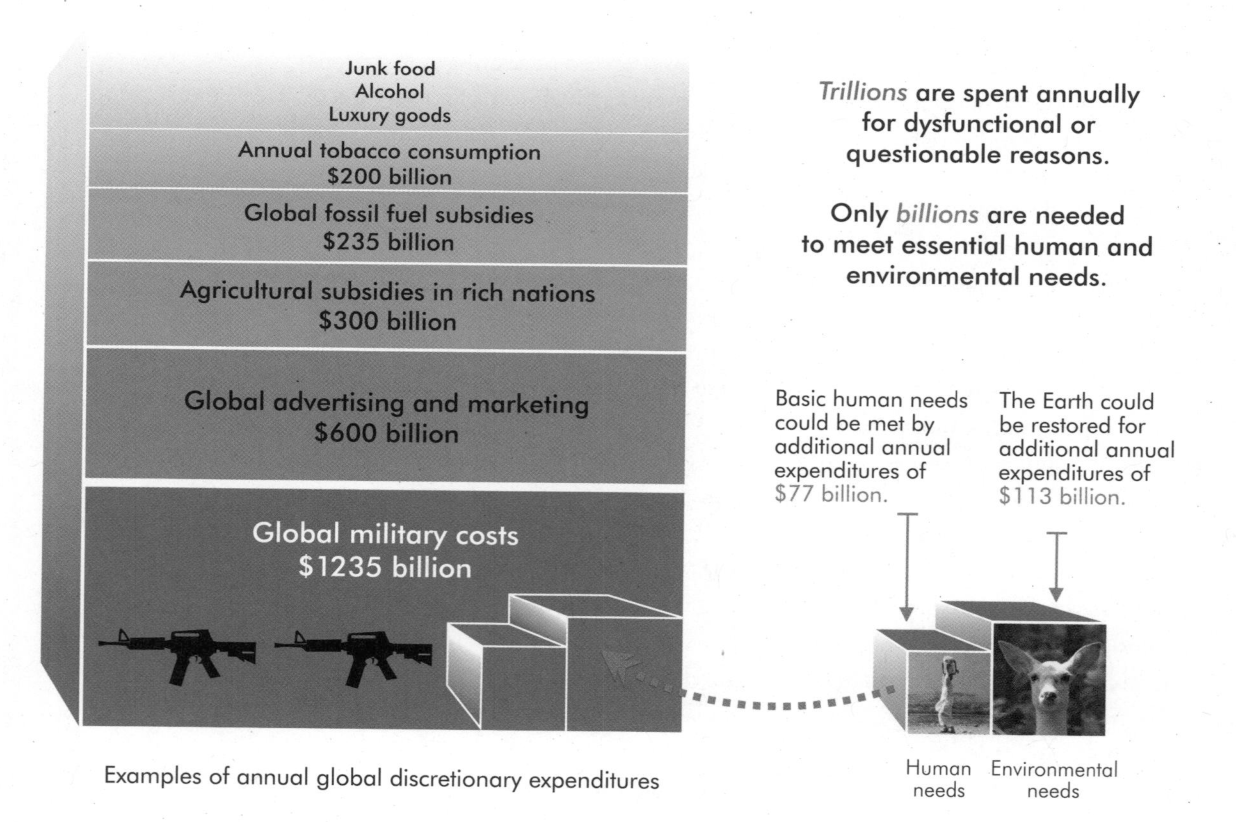

and responsible environmental conservation. However, at present few leaders even believe that there are environmental limits to growth. Almost every government and business is still making plans to increase consumption. If present trends continue, the rest of our lives and the lives of our children will be dominated by growing global crises.

Spaceship Earth

Is it good enough to hope that everything will turn out fine? Would you take the people you love on a trip if you knew that you might run out of fuel, water, food and shelter? At what point would you call off the trip and make new plans: if the risks were only 1%, or if they were 10%? Would you still take the trip if you knew that the risk of disaster was more than 60%? [133]

The reality is that you are already taking a journey with all your family and all your friends. We are all on Spaceship Earth, and although we don't know exactly how great the risks are, the crew is warning that the life-support systems are breaking down.

But the officers aren't listening. Like the officers of the *Titanic*, their main concern is making the ship go faster. Whenever the lookouts and engineers call in warnings, they brush them away. There is nothing to worry about; the ship is indestructible; there are no limits.

So what will you do? This is not a hypothetical question.

3 The Unsustainable Global Culture

The fundamental problem

Insanity is repeating the same behavior and expecting a different result.

— Anon[1]

WHY IS IT SO DIFFICULT for politicians, business people, economists and others to accept that there are environmental limits to growth? Why are they deaf to the warnings of scientists and blind to the approaching crises? The answer is that they, like us, are products of our cultures and our times. In every age most people have shared the values of their society. Today we live in capitalist industrial societies which confidently proclaim that market forces plus technology will solve every problem. Most policy makers support these values, not only because it is in their personal interests to defend the system, but also because they passionately believe that capitalism is the best system that has ever existed. And in many ways they are right.

For two and a half centuries industrialization has improved the living standards of most people in the world — in general people now live longer and are better fed, better housed and better educated than ever before. Capitalism (the dominant type of industrial system) has helped to overthrow the dictatorial rule of kings and give individuals more rights and freedoms. Although industrialization has brought misery to many (especially where it has been imposed by force), in general it has brought tremendous benefits to our species.[2]

This is not to say that the industrial system is wonderful — the environment is being destroyed and the world is full of cruelty and suffering. But it is as meaningless to say that the industrial system is bad as it is to label Stone Age societies good or bad. For all its faults and crimes industrial civilization has been a necessary stage in the evolution of our species. From an evolutionary standpoint the fundamental problem with the dominant world system is not that it is immoral, but that it is unsustainable.

And the reason why industrial capitalism is unsustainable is not that it uses industrial processes (which we need to raise living standards) or that it relies on market forces to set prices (an important economic tool) or even that it is based on self-interest and competition (these are part of our biological makeup). The problem is that it is organized by a belief system that does not recognize the need for limits — limits on environmental exploitation, limits on economic competition and limits on social inequality. Industrial capitalism is like a car that has an accelerator but no brakes.

Greeds not needs

Although the global system creates extremes of rich and poor, powerful and powerless, the expanding global economy has been gradually lifting most people out of poverty. This is a slow and uneven process — at current rates of progress it will take Sub-Saharan Africa 160 years to reduce child mortality by two-thirds and hunger in South Asia won't be halved until sometime in the next century.[3] Nevertheless, if the global economy were to continue to expand, most of the world's hunger and preventable diseases would eventually be eliminated. The problem is that because the global economy cannot continue to expand indefinitely, in the future poverty is not likely to decrease but increase.

The global system will collapse because it will go ecologically bankrupt. Bankruptcy happens when a person, a business or a global economy no longer has sufficient income to cover expenses. Whether or not a bankruptcy will occur is a practical economic question. But why a bankruptcy will occur is a moral question. The destruction of our children's future and millions of other species is the biggest moral issue humanity has ever faced.

While the global system cannot be labelled bad or evil, we can say that it is amoral because its primary concern is profits and not the well-being of humans and animals. It is also accurate to say that it is environmentally and socially destructive and that it is irrational because it is in the process of committing suicide.

This is why we can't separate the issue of sustainability from ethics. Because worldviews and values motivate and organize social structures and economic processes, all economic and environmental outcomes are culturally determined. The reason why the global system will inevitably collapse is because it is organized by destructive values. The consumer culture drives limitless economic expansion through encouraging limitless consumption. It works through encouraging people to pursue insatiable desires rather than meet essential needs, and while there are enough resources on our planet to meet everyone's needs, there will never be enough for everyone's greeds.[4]

Socially structured behaviors

The world system is not about justice or kindness, but about power and wealth: it is designed to reward success, not morality. As a result, being rich or being poor has little to do with being a good or a bad person. Some of the richest people on Earth (e.g. Bill Gates, George Soros and Warren Buffet) are deeply concerned about the state of the planet; they are donating much of their wealth to heal and educate the disadvantaged. Other rich people have little integrity and do not care what damage they do in their pursuit of further wealth.

> The test of our progress is not whether we add more to the abundance of those who have much; it is whether we provide enough for those who have too little.
>
> — US President Franklin D. Roosevelt (1882-1945)[5]

Social systems exist because they are able to maintain and reproduce themselves. Although they have cultures, they don't have feelings. The industrial system doesn't care whether a capitalist produces lifejackets or landmines anymore than the agrarian system cared whether a king built convents or castles. However, social systems shape social behaviors. A business cannot function without financial capital, just as a feudal lord could not rule without armed retainers. Both industrial and agrarian systems are heirarchical and predatory in nature: in agrarian societies successful aristocrats became more powerful through eliminating their opponents through conquest or marriage, while in industrial societies successful companies grow larger by eliminating their opponents through competition or merger.

Whatever a king's personal values might have been, every agrarian kingdom was a military dictatorship, which meant that every king had to be prepared to fight and kill opponents or risk losing not only his throne but his life. In the same way, whatever the values of their owners might be, companies must be able to successfully compete and make profits in order to survive, which means that any business that makes the welfare of its employees and the public more important than its bottom line risks commercial failure. In feudal societies young aristocrats were taught warfare from an early age. Today the children of the rich and aspiring rich go to business schools to learn how to compete and maximize profits. In our society most successful business leaders have hyper-competitive personalities and values: in a 2002 survey, 82% of top US executives admitted cheating in golf.[6]

In every age people have to support the values of their society and work within its rules if they wish to prosper. Although personal ethics are still important — since all social events are ultimately shaped by the decisions of individuals — if we wish to understand long-term problems such as the existence of chronic war, poverty, crime, disease and environmental destruction, we need to examine the cultures and social structures that produce these behaviors.

Fear and addiction

The world system is dominated by competitive, capitalist values in which success is defined by the ability of individuals, companies and nations to accumulate wealth and power. The capitalist system works in ways similar to compound interest: through investing capital (cash and other financial assets), people with money make more money, and the more money they have, the more money they are able to make. For example, in 2006 the fortunes of Britain's 1,000 richest people (including 68 billionaires) grew by 20%, with the wealth of the richest person increasing by more than £4 billion.[7]

Everyone needs to consume resources in order to meet their needs. But our economic system is not designed to ensure that everyone has enough for a decent life. Instead it promotes conspicuous consumption and the limitless accumulation of wealth, which has little to do with meeting needs or even comforts. After essential economic requirements are met (for food, shelter, clothing, health care, education, transportation, etc), people consume in order to meet other emotional and intellectual needs like recreation and creative expression. After these needs are satisfied, further consumption is not about needs, but greeds — for luxury, status and power.[8] Being wealthy will not let you eat more, sleep more, learn more, exercise more, or be more comfortable. But it will allow you to influence and control people, politicians and policies.

Of course there are real advantages to being rich instead of poor. A study of civil servants in the United Kingdom found that a lifetime on a low wage physically ages a person eight years earlier than high earners. In the United States, where there is more violent crime and no universal health insurance, the situation is even more extreme: men in the top 5% of the income distribution live about 25% longer than men in the bottom 5%.[9] This is due to interrelated causes; low income earners have more accidents, suffer more stress and are more likely to commit suicide. Highly paid and educated people also tend to eat better, live in safer neighborhoods, enjoy more holidays and have access to better health care.

But there are other major factors influencing levels of stress and happiness besides incomes: self-esteem, physical and emotional security, community support and respect and opportunities for self-expression. While everyone needs to feel loved and respected, a person does not need to be wealthy to be valued by their family, friends, co-workers and community. Moreover, even financial status is a relative concept.[10] People in rich countries need to buy much bigger houses to gain the same level of status as people in poor countries. Americans who earn low incomes often have standards of living as high as or higher than African professionals. But doctors in Lagos are much more likely to feel satisfied with their lives than janitors in Los Angeles.

In every country advertising and movies glamorize being rich and powerful. Consuming is equated with beauty, youthful energy, desirability, happiness and

Figure 1: D. Myers[11]

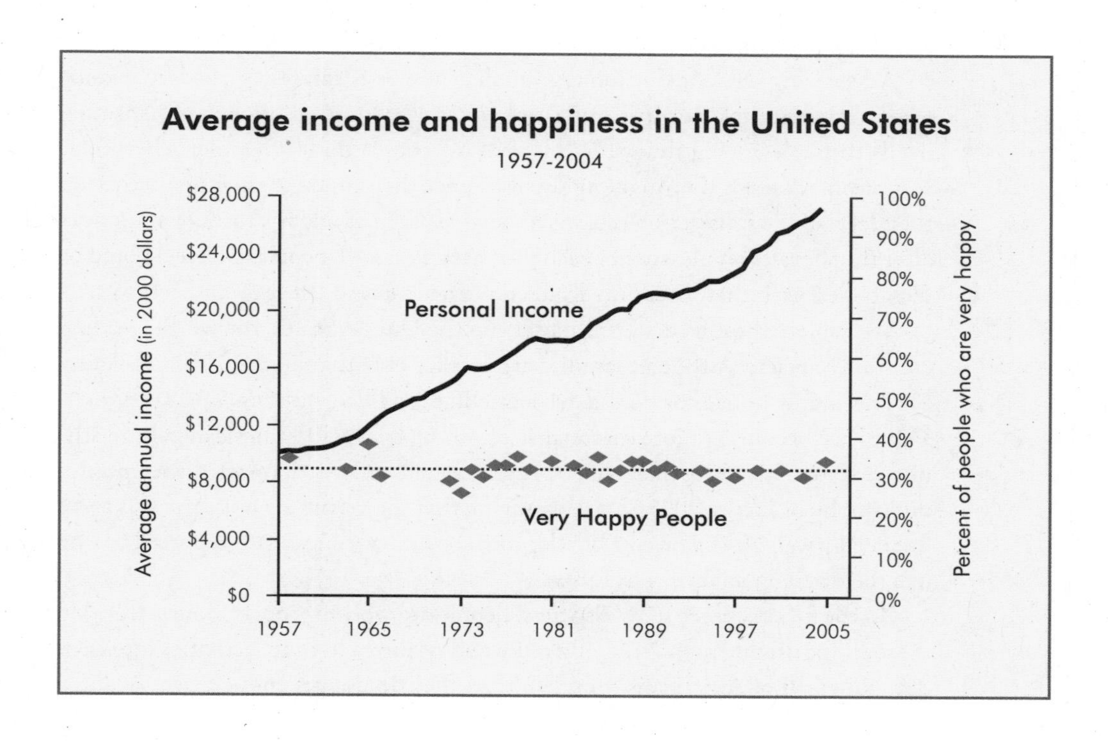

personal freedom. These are destructive illusions. Most people will never be able to live in luxury because there are not enough resources on the planet to maintain current levels of consumption, let alone enough to allow everyone in the world to shop till they drop. After all, over half the world's population are struggling to survive on less than $5 per day;[12] while they may see ads and dream (or rage) about the consumer society, there is little chance that they will ever be able to have an affluent lifestyle.

Moreover, many advertisements are lying. The studies have been done, and they agree with traditional wisdom: after basic needs are met, money can't buy happiness.

Although average incomes in Britain and the United States have tripled since the 1950s, on the whole the British and Americans are less happy now than then.[13] In 1973, 86% of people surveyed in Britain said that they were satisfied with their standard of living, while in 2006 85% were satisfied.[14] Economists have found that a person in Britain who earns $24,000 per year and spends time each day with friends and relatives is on average just as happy as a person earning $230,000 per year who almost never sees his or her family and friends.[15]

Advertising sells the illusion that qualitative needs can be satisfied by quantitative consumption. But it is impossible to satisfy most emotional and spiritual needs through buying things. While well-off people do feel happier when their income increases or they make a purchase, once they adjust to their new situation their level of life satisfaction returns to normal. Consuming is a form of addiction: after the short-term pleasure of each purchase wears off, people soon feel bored or dissatisfied with their lives and again consume in a vain effort to find comfort.

We can see the effects of this pattern in the United States, the world's leading consumer society. Although families are smaller now than in the 1950s, the average one-family house has doubled in size and been filled with more possessions.[16] Values are becoming more materialistic: although 79% of American college students stated in 1970 that their primary goal was to develop a meaningful philosophy of life, by 2005 75% were saying that their primary objective was to be financially well off.[17] The cost of adopting consumer values? 81% reported feeling like they were in an *existential vacuum*.[18]

As the US becomes a 24/7 society, people are working longer hours, working more shifts, driving more, studying more and trying to fit more activities into each day. Now half of Americans complain than they do not get enough sleep,[19] and a quarter suffer from acute anxiety, depression and other mental illnesses each year.[20] As lifestyles simultaneously speed up and become more sedentary, people are cooking less, eating more fast foods, exercising less and becoming fatter and sicker. One in eight adult New Yorkers now has diabetes and the numbers are growing each day.[21] Some health experts warn that the average child in the developed world may not live as long as his or her parents.[22] These problems are not confined to the United States. In Canada one in three adults say that they are workaholics, half say that they do not spend enough time with family and friends, and four in ten feel that they do not have time to have fun.[23]

Consumer advertising often sells not only illusions but fear. There is a flip side of the message that happiness comes from being rich and looking beautiful — the implication that anyone who does not fit this image is a loser. People are under pressure from the time they are children to buy the right brands and live consumer lifestyles. This causes tremendous suffering since it is not possible for most people to be rich, to stay young or to fit into the fashion industry's image of slim and beautiful. For example, three quarters of young Australian women with healthy weights want to weigh less.[24] The media's portrayal of young women as sex objects also means that many girls in the developed world are confused about their roles, lack confidence and are prone to depression and eating disorders.[25] One in four teenage girls in the United States has a sexually-transmitted disease.[26]

The consumer culture focuses on having instead of being, on ownership instead of relationship, on external appearances instead of internal well-being. This promotes selfishness, competitiveness and insecurity, and weakens families and communities. Some indicators of unhealthy relationship skills and alienation are high levels of mental illness, high divorce rates, high rates of drug addiction and high crime rates. Over the past 50 years these problems have worsened in almost every developed country.

Real and artificial scarcity

On balance the majority of people in developed countries are happy, healthy members of functional families. But this doesn't mean that we live in a healthy culture. To take an extreme example, although most people living in Nazi Germany were decent, emotionally healthy people, the psychopathic values of the society drove it to war and destruction. We can see that the modern global system with its consumer culture is also dysfunctional because of the existence of deep and intractable social problems.

Luxury expenditures

Product	Annual expenditure	Social or economic goal	Additional annual investment needed
 Makeup	$18 billion	*Reproductive health care for all women*	*$7 billion*
 Christmas presents for pets in the US	$5 billion	*Universal literacy*	*$4 billion*
 Perfumes	$15 billion	*Elimination of hunger and malnutrition*	*$19 billion*
 Ocean cruises	$14 billion	*Clean drinking water for all*	*$10 billion*
 Ice cream in Europe	$11 billion	*Immunizing every child*	*$1.3 billion*

Figure 2: BEST Futures [27]

Industrial civilization is not only environmentally unsustainable but socially unsustainable: it creates emotional scarcity in the developed world and material scarcity in the developing world. In rich countries advertising creates emotional insecurity through constantly suggesting that people should be something other than who they are. The consumer culture creates false needs for power, status and wealth instead of satisfying real needs for meaning, community and health. It creates the illusion of scarcity in the rich world, where people try to satisfy their emotional and spiritual needs through consuming things, and real scarcity in the poor world, where the resources do not exist to meet basic human needs for food, shelter, health and education.

In the rich world communities and families are weak, and mental illnesses and addictions are widespread. In the poor world there is not enough food, clean water, sanitation, housing, health care or education. Globally, rich nations are plagued by obesity while poor nations are plagued by hunger; more research is done on developing cosmetics for the rich than on discovering cures for the diseases of the poor; and many of the best scientists on the planet waste their lives developing new weapons for countries that already have the power to destroy all life on the planet.

Structural inequality and structural violence

Supporters of capitalism say that there is nothing wrong with inequality. In a capitalist economy, a few people have to be allowed to accumulate and invest assets in order for the wealth of the whole society to increase. Conservatives argue that as the rich get richer, their wealth will trickle down and create new jobs for the poor. In their view, taxing the rich in order to provide social programs for the poor does more harm than good because it slows the process of wealth creation (the rate of economic growth) that benefits everyone. Let's examine some of the effects of inequality.

Figure 3: BEST Futures [28]

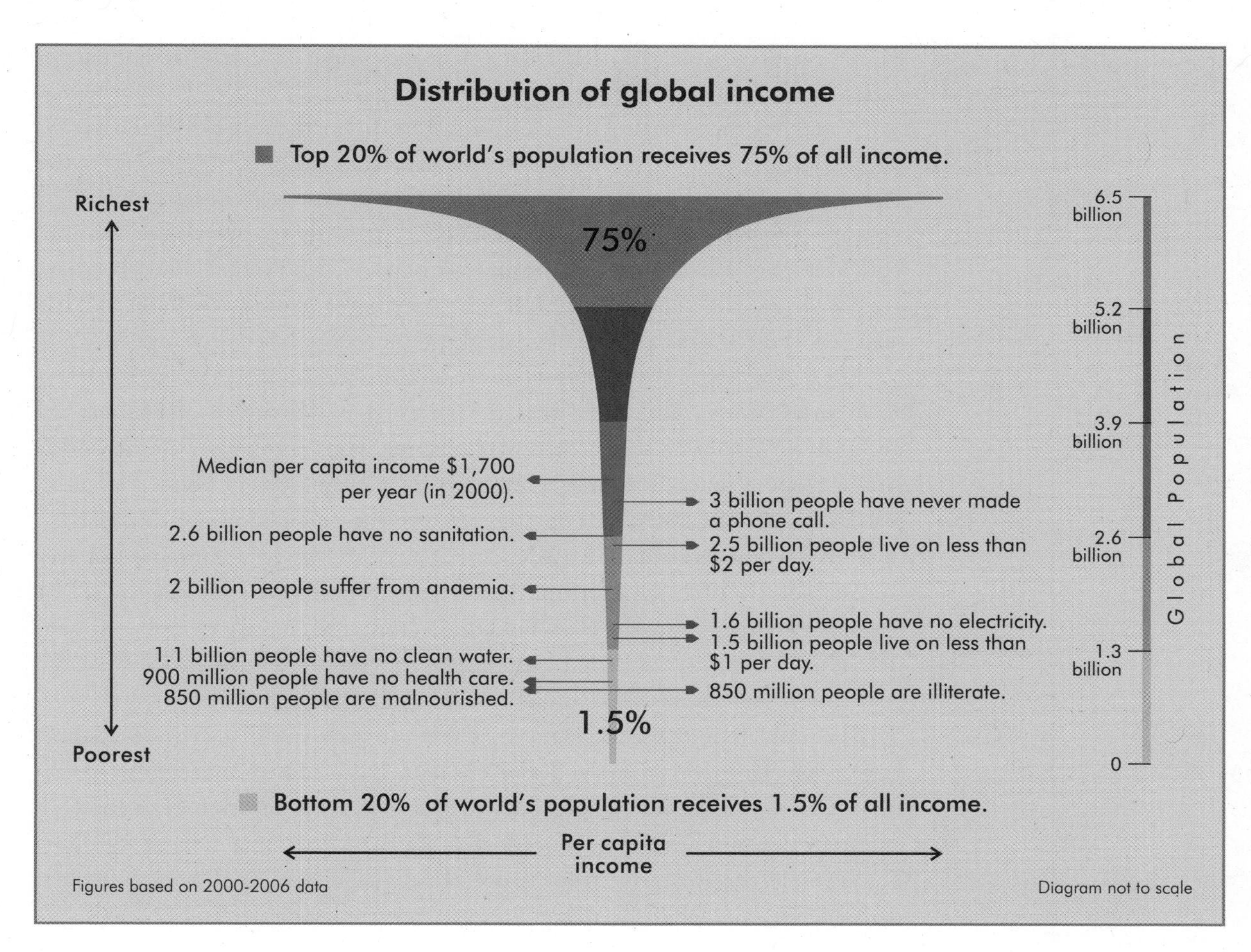

> There are people in the world so hungry, that God cannot appear to them except in the form of bread.
>
> — Mahatma Gandhi, Indian political and spiritual leader (1869-1948)[29]

Branko Milanovic, the World Bank's lead economic researcher, points out that since the industrial revolution the economic gap between the rich and poor countries in the world has steadily widened. This is the result of the process explained in Chapter 1 — in a capitalist system economic growth benefits the rich more than the poor. Although average incomes in many developing countries like India and China have been rapidly rising, so has inequality within those countries. For example, 300 million people in China are still struggling to survive on less than $1 per day.[30] As a result, economic inequality among the world's people remains at historically high levels.[31]

In 1998 a little over 40% of the world population lived on incomes of less than $1,000 per person per year, and only 10% of the people on Earth had annual per capita incomes of over $9,600.[32]

We can see the impact of inequality in Figure 3. The richest 20% of the world's population enjoy 75% of the total global income, while the poorest 20% have to live on only 1.5%.[33] Of course many changes have taken place over the last decade and the data appearing in Figures 3 and 4 is no longer completely accurate.[34] For example, both average global incomes and the number of people who are malnourished have increased, and the number of people who have never made a telephone call has declined. But the basic picture of great global inequality has not changed.

The gap becomes even wider when we look at wealth instead of income.

Figure 4 shows what people own (the net worth of individuals. At the very top of this human pyramid are a thousand billionaires who control many of the world's largest corporations. At the very bottom are tens of millions of bonded laborers, sex slaves and child slaves who do not even own their own bodies or control their own lives.[35] This is structural inequality: because the world is composed of very unequal social structures a person's opportunities in life are largely determined by the country they live in, the class and ethnic group they belong to and their gender. This is not only an issue of unfairness: a structurally unequal system is a structurally violent system.

The problem in the world is not that some are rich, but that many are poor.[36] Poverty blocks opportunities and shortens lives. If you come from a wealthy country or family, you are much more likely to receive a good education, get a good job and to be rich and healthy all your life than if you come from a poor background. For example, the average life expectancy of people living in the United Kingdom is around 79 years, while the average life expectancy of people living in Mozambique

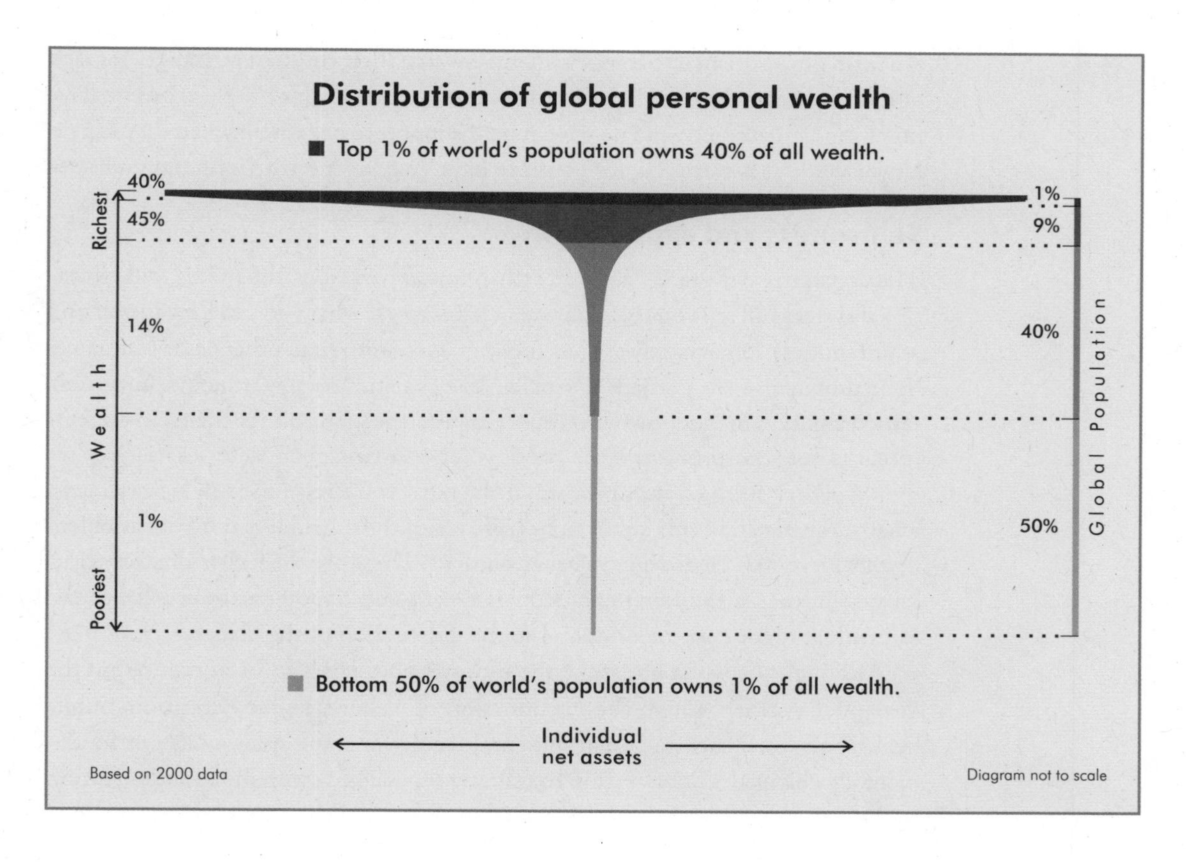

Figure 4: BEST Futures [37]

is around 41 years.[38] Extreme poverty is also extreme violence. More than 10 million children die each year before their fifth birthday — one every three seconds. Almost all childhood deaths are preventable. The United Nation's *Human Development Report* stated, "For every child who dies, millions more will fall sick or miss school, trapped in a vicious circle that links poor health in childhood to poverty in adulthood. Like the 500,000 women who die each year of pregnancy related causes, more than 98% of children who die each year live in poor countries."[39] Poverty doesn't just kill, it also blinds and cripples: chronic malnutrition stunts the physical and mental development of millions of children.

A quarter of the world's population — some 1.5 billion people — survive on less than $1 a day.[40] Another 1 billion people live on $1–$2 a day. Extreme poverty means that 40% of the world's people face constant threats to their survival. Poverty is a form of terrorism to parents who must watch their children go hungry for lack of food or become sick for lack of medicine. It is terrifying for elderly people to have no savings, no pensions and no government support to help them when they can no longer work. It is terrifying for youth to have no education, no jobs and no hope.

Selfishness and exploitation

The consumer culture is based on selfishness: advertising, the media and educational systems all tell you that your main job is to get what you can for yourself and your family. If some people are less lucky or successful than others, it is ultimately their problem — or perhaps a problem for charities or governments. Your only responsibility is for your own life: other people are responsible for theirs. In our culture it is not your problem if the goods you buy are made in factories that exploit their workers. It is not your problem if the meat you eat is raised in factory farms where animals and birds suffer in horrific conditions. And it is not your problem if your lifestyle is destroying the environment. These are the values of a very brutal system that is indifferent to people's health and happiness, to cruelty, to the extermination of entire species and to the destruction of the planet.

Although the worst abuses of the industrial system have been reduced in the developed world through the introduction of welfare, public education, public health and workplace and environmental regulations, the system has not fundamentally changed. Globalization has simply transferred sweatshops and pollution to the poor world. Most of the products consumed in the rich world are made in developing countries by workers who barely make a living wage, have no rights, no job security, no health care, no holidays and no pensions. One reason why living standards are high in the rich world is because living standards are low in the poor world.

Poor families often cannot afford to send their children to school because they need their children to help in the fields or earn a wage. An estimated 218 million children between 5 and 17 work in developing countries.[41] Extreme poverty and hardship brutalizes people and cheapens lives. In many countries in the developing world poor children are rented out as servants, forced to work in slave-like conditions making carpets, matches or other handcrafts, and are all too frequently sold into the sex trade. A comprehensive 2007 study by the government of India discovered that abuse is common: 69% of Indian children have been physically abused and 53% sexually abused.[42]

The values and structures of the global system produce violence in many forms. There is the physical violence of poverty that comes in the form of hunger, disease, brutal work, frequent accidents and early deaths. And there is the mental violence that comes in the form of discrimination, denied opportunities, financial insecurity and lost hope. Although there is less poverty in wealthy countries, physical and mental violence are still present in the form of preventable diseases, addictions, abuse and mental illnesses. Around the world one-third of all women report having been physically and/or sexually abused at some time in their lives.[43] Poverty and abuse also force millions of young people to engage in prostitution. Since prostitutes are frequently subjected to physical and sexual assault, they suffer higher rates of post-traumatic stress disorder than combat soldiers.[44]

Because the global system is based on selfishness and exploitation, it is permeated with greed, indifference, fear and violence. Fear, hatred, anger, conflict and violence are the natural products of a system where wealth is transferred through interest payments from the poor to the rich, large fortunes are mostly unearned (can anyone earn millions of dollars a year by their own work?) and desperate people are constantly reminded by global media that the lucky few live in luxury. The poor fear destitution and envy the rich; the rich fear that the poor will take away their wealth. In order to maintain such extreme inequality walls have to be built between the rich and the poor. These walls come in many forms: in the regulations that keep poor people from emigrating to rich countries in search of work; in the guarded and gated communities of the rich; in the unequal trade and financial agreements forced on poor countries and workers by rich countries and corporations.[45]

We live in a global system of apartheid, in which whole countries are ghettos for the poor. If you have a passport from a rich country it is easy to travel, and if you are wealthy you can live almost anywhere. But if you have a passport from a poor country you need visas to travel. No one wants you if you are poor. Everywhere in the world security guards, police, customs officials and soldiers separate the poor from the rich; everywhere doors are opened for people with money and closed in the faces of the penniless.

Hierarchy and discrimination

The industrial system is a hierarchical system with many gradations from rich and powerful to poor and powerless. While the major divisions in the world are based on class and country — your opportunities in life are largely determined by whether you were born into a rich or poor family in a rich or poor country — they are also determined by your gender, your ethnic background and the color of your skin. We live in a world that is ruled by men, and primarily by rich, white North American or European men. If you are a healthy, intelligent and hard working white male from a rich and well connected family in a developed country, it will probably be relatively easy for you to get a good education and a good job. But if you are an equally healthy, intelligent and hard working black or aboriginal female from a poor family in the developing world, it will probably be very difficult for you to get a good education or find a good job.

Of course every woman isn't doomed to a life of hardship. In a world where money and power go hand in hand, women who live in rich countries or who come from rich families in poor countries often have more economic power and more educational opportunities than most of the men living in developing countries. And there are always unusually talented or lucky (and untalented or unlucky) people: some women from disadvantaged backgrounds become rich and successful, while some men from rich families end up in poverty. However, these are exceptions. In general, the cultural and structural biases of the world system discriminate in favour of men and against women.

Although the status of women has improved greatly over the last hundred years, there is still no country in which women share power equally with men. In 2006 only 17% of the parliamentarians, 10% of government ministers and 10% of the mayors in the world were female.[46] Women are underrepresented in positions of authority in the developed world as well as the developing world: for example, although half the British workforce is female, only one in four managers, 9% of the judiciary and 10% of senior police officers are women.[47]

This lack of power is also economic: 70% of the people living in extreme poverty in the world are women and children. Although women do more than half of the world's work, two-thirds of this work is unpaid.[48] Women are not paid for the work they do at home such as cooking, cleaning, raising children, looking after the elderly, carrying water, fetching firewood and gardening. When they work outside the home they are often given inferior jobs or paid less than men for doing the same work: for these reasons Japanese women earn 51% as much as men.[49] Because much of women's work involves caring for people instead of competing, and working for love instead of for cash, much of the work women do is not even counted as part of the global economy. As a consequence most of the women in the world have little financial independence and are unable to secure loans to go to school or set up their own businesses.

The fact that many women earn little money and have little power means that their status is lower than men. In some countries female children are not just devalued, but disposable. In South Asia there are now 100 million more males than females. The 100 million *missing women* have died from the combination of poverty and gender discrimination. Poor Indian families often view girls as economic liabilities because they must be given dowries when they marry, and poor Chinese families want boys who can earn higher incomes. As a result many baby girls are either aborted or allowed to die from malnutrition or disease.[50]

Cultures of violence

Fear and violence permeates the global culture. It is used to sell movies and win political support. In almost every country violent acts are frequently shown on the news, on television, in movies, and in video games. In the United States 60% of prime-time TV shows contain violence, despite repeated warnings from psychologists that the more children watch violent shows, the more likely they are to become violent abusers as adults.[52]

> If you came home and you found a strange man … teaching your kids to punch each other, or trying to sell them all kinds of products, you'd kick him right out of the house, but here you are; you come in and the TV is on, and you don't think twice about it.
>
> — Jerome Singer,
> Professor Emeritus of Psychology,
> Yale University[51]

Violent behaviors are learned. Allan Johnson pointed out, "It is difficult to believe that such widespread violence [that girls twelve years old in the United States now stand a twenty to thirty percent chance of being violently sexually assaulted in their lifetimes] is the responsibility of a small lunatic fringe of psychopathic men …. The numbers reiterate a reality that American women have lived with for years: sexual violence against women is part of the fabric of American life."[53]

Not all societies are equally violent. For example, the US murder rate is 10 times that of Britain. Although conflict is a normal part of everyone's experience, violent conflict is not inevitable. Conflict develops when people believe that they have incompatible objectives. Violence occurs when people believe that it is acceptable to resolve conflict through force. Because both objective and subjective factors contribute to conflicts, violence is more likely to occur when people's real needs are not being met, and more likely to occur when people live in communities with violent traditions.

Cultural factors that contribute towards violence include feelings of superiority; the devaluation of others; authoritarian traditions; a monolithic culture; a

The causes of genocide in Rwanda

Genocide in Rwanda resulted from the potent combination of ecological crises (overpopulation and resource shortages), economic crises (hunger and inequality), political crises (civil war), a history of interethnic violence and fascist leadership. Mass murder occurred in regions where food energy per person per day was less than 1,500 calories, while there was less violence in regions where consumption was over 1,500 calories.

Figure 5: BEST Futures [54]

dogmatic ideology and a history of responding aggressively to conflict. On the other hand, cultural factors that contribute towards peace include diversity, tolerance, equality, democracy, the rule of law and a history of peacefully resolving conflicts. Europeans have reduced the likelihood of wars between their countries through promoting tolerance, democracy, interdependence and pluralism.

War and domination

At the core of global culture is war. This is easy to see: the nations of the world spend more than 1.2 trillion US dollars per year ($1,200,000,000,000) on studying war, developing new weapons, arming and training military forces and waging war,[55] while only spending a small fraction of that amount on studying peace, eliminating the causes of conflicts, educating people to peacefully resolve conflicts and actively intervening in conflicts in order to create, maintain and build peace.

For example, relatively little money is given to the United Nations, the global organization tasked with promoting world peace and international development. The governments of the world only provide the United Nations with an annual core budget of 5 billion dollars, a peacekeeping budget of 7 billion dollars and an annual international development budget of 15 billion dollars.[56]

By the close of the Cold War NATO and the Warsaw Pact had reached agreements on dismantling 20,000 nuclear warheads and scrapping 52,000 tanks, warplanes and artillery guns.[57] Many people hoped that the collapse of the Soviet Union in 1991 would usher in a period of world peace. Since then the total number of wars and conflicts has declined by 40%, and the number of democratic governments in the world has increased.[58] Unfortunately, although global military expenditures did decrease through the 1990's, none of the major powers took definitive steps to eliminate weapons of mass destruction. Following the election of the Bush administration in 2000, the United States effectively ended further disarmament negotiations, declaring the right to develop new weapon systems, the right to pre-emptively attack countries it was not at war with and the right to use weapons of mass destruction in warfare.[59]

At present the international arms race is accelerating, lead by the United States, which currently spends more on warfare than the rest of the world combined. For example, the US is developing space-based laser weapons;[60] Russia has successfully tested a new intercontinental ballistic missile capable of penetrating American defences;[61] China is building a fleet of nuclear ballistic missile submarines capable of launching pre-emptive strikes;[62] India is developing long-range intercontinental ballistic missiles and a blue water navy;[63] and the United Kingdom is upgrading its nuclear submarines.[64]

Most analysts agree that the major military threats in the coming decades will occur within countries rather than between countries and involve asymmetrical warfare against urban guerrillas or terrorists. So why are the great powers spending hundreds of billions of dollars developing weapon systems that are designed to fight conventional wars? In reality the arms race has little to do with real defence needs; it has everything to do with projecting national power, influencing regional and global events and accessing and controlling foreign resources.

We need to ask why we are still fighting wars in the 21st century. Our species knows all about the terror, waste, misery and ultimate futility of war. Historians and social scientists have studied both war and peace and written thousands of books on both subjects. We now know that identity and resource conflicts occur when people believe that their needs are not being met or are being threatened. We now know that people compete and fight over material goods when they fear material scarcity, and people compete and fight over religious, ethnic and national issues when they fear the loss of cultural identities. And we know how to negotiate a comprehensive global disarmament treaty — all that has to be done is to expand current agreements and verification procedures.[65] So why aren't our governments investing in peace rather than war? And why are people throughout the world still following leaders who prefer international competition to international cooperation, conflict to compromise and mutual destruction to mutual disarmament?

The answer is that the industrial culture values competition, violence and domination more than cooperation, peace and equality. Since the dawn of the agrarian age five thousand years ago humans have been living in militarized societies and engaging in large-scale organized warfare. Although our social structures have evolved from tribes, to kingdoms, to nations, the causes of war have not been eliminated. We still live in a world where people live in fear of other individuals, groups and nations: fear of being denied access to resources, fear of being attacked and dominated and fear of having their communities and identities destroyed. The industrial culture tends to provoke conflicts because it is based on a win/lose paradigm that combines competition, exploitation and violence. It believes that life is a constant battle to dominate and control nature and to dominate and control other humans. It teaches us that individuals, corporations and countries succeed through defeating their opponents with superior intelligence, technology, wealth and force.

Hitler rose to power on the anger of Germans who blamed the economic hardships of the 1920s and 1930s on the humiliating terms of surrender that the victorious Allies forced Germany to sign at the end of the First World War. Following the Second World War, the Western powers were afraid of the spread of communism, so instead of punishing Germany and Japan they supported their economic and political development. This approach was successful in eliminating the causes of conflict — with more freedom, education and prosperity than ever before, most Germans and Japanese are content with their lives and strongly anti-war.

Unfortunately, Western economic aid to Western Europe and Japan was not extended to the rest of the world. Instead, for 45 years the West and the Soviet Bloc fought proxy wars in Latin America, Africa and Asia. The proxy wars were fought primarily through military means, with economic aid only seen as a means of securing short-term political support. Both sides armed supporters in developing countries, assassinated opponents, subverted and overthrew popular governments, installed military dictatorships, provoked wars and wrecked economies. These actions have left a legacy of authoritarian governments, corruption,[66] poverty, war and failed states. There are now around 640 million small arms and light weapons in circulation in the world — one for every 10 people.[67] Many countries (e.g. Columbia, the Congo, Somalia and Afghanistan) have more machine guns than tractors.

However, it is impossible to win people's support through brute force. The failure of this approach is shown by the defeat of the two industrial superpowers by poorly armed guerrillas in Vietnam and Afghanistan. The American and Soviet military disasters were the result of ignoring the real grievances and real needs of local populations and trying to rule through corrupt local officials and military force. As more and more people were killed and arrested, anger and hatred against the occupying forces grew along with support for guerrilla movements.

We can see how deeply many Americans believe in the use of force by the fact that the US government claims the right to attack and occupy countries it is not at war with and to assassinate non-Americans in covert operations. Because of this belief, instead of giving help to the Iraqis who were fighting to free their country from dictatorship, the United States invaded and occupied Iraq. The results have been similar to the tragedies of Vietnam and Afghanistan: while most Iraqis originally welcomed the Americans as liberators, within a few years the majority were supporting various insurgent groups. Rather than reducing terrorism, the invasion of Iraq has increased it. Omar al-Baghdadi, an al-Qaeda leader in Iraq, said, "From the military point of view, one of the [enemy] devils was right in saying that if Afghanistan was a school of terror, then Iraq is a university of terrorism ... The largest batch of soldiers for jihad ... in the history of Iraq are graduating and they have the highest level of competence in the world."[68]

Most people in the world believe in the right of self-defence and support wars fought to liberate people from oppression. Millions volunteered to fight against fascism in the Second World War. But most people also oppose aggressive wars. If rich Western countries were to help reduce poverty and exploitation in the Muslim world, Islamic extremists would find themselves isolated.[69] In an interview with a western reporter a militant leader in Pakistan, Abdul Rashid Ghazi, said, "How much have you spent on the war on terror — trillions of dollars. If you had spent this on helping to develop Pakistan and Afghanistan, we would have loved you and never attacked you."[70] At present, Western military interventions in the Middle East are perceived as aggressive rather than defensive, and oppressive rather than liberating. They have increased Muslim poverty and anger, with the result that extremist groups are gaining recruits.[71]

The same dysfunctional approach has been used in the US *war on drugs*. In Afghanistan and Columbia, the only source of income for many poor farmers is growing crops used to produce heroin, cocaine and marijuana. American policy is to destroy illegal crops without compensation, leaving people hungry and angry. Efforts to suppress the drug trade without dealing with the root problems of poverty and corruption has resulted in an increase in the cultivation of illegal crops, an increase in support for anti-government guerrillas, and an increase in the power of local war lords and international criminal syndicates.[72] The war on drugs has been equally counter-productive within the United States: the effect of punishing rather than treating drug addicts has been to vastly expand the prison population — with 2 million people behind bars, the US now has more people in jail than any other country. By the end of 2000 there were more Afro-American men in prison than in university.[73]

The United Nations' 2005 *Human Development Report* pointed out that "Nine of the 10 countries ranked at the bottom in the human development index (HDI) have experienced violent conflict at some point since 1990 .... While there is no automatic link between poverty and civil conflict, violent outcomes are more likely in societies marked by deep polarization, weak institutions and chronic poverty .... Indeed, the 'war against terror' will never be won unless human security is extended and strengthened. Today's security strategies suffer from an overdeveloped military response to collective security threats and an underdeveloped human security response."[74]

The violent cultures and structures of the industrial system produce endless cycles of war and terrorism, crime and punishment. The alternative is to reduce poverty, inequality, fear and alienation through helping people meet their real material and social needs. Unfortunately, the global economy is not designed to reduce inequality, but to increase it. Although chronic violence and wars will only end if there is a global culture that promotes compassion, love and cooperation, at present the global culture is based on selfishness, fear and competition.

Cultural destruction

As industrial civilization expands, it consumes and degrades not only natural resources, but also other civilizations and cultures. When it comes in contact with traditional agrarian or tribal societies, the force and attraction of its superior power and wealth begins to break down the economies, values and social institutions of the older societies.

Rapid urbanization in the developing world has been accompanied by soaring rates of poverty, crime and addiction. While industrial civilization has provided billions of people with increasing amounts of personal freedom, the price has often been the loss of community and meaning. The *clash of civilizations* between the West and the Islamic world is not a war between democracy and dictatorship, but the reaction of traditionalists against what they perceive as an attack by the consumer society on their values and way of life — a conflict described by Benjamin Barber as Jihad vs. McWorld.[75]

Of the 6000 languages existing in the world today, one-half are not being taught to children and will be extinct in a generation. Less than 600 languages are stable: this number may decline by 2100.[76] Every language and culture has taken thousands

Figure 6: BEST Futures[77]

Mayan nobles drinking chocolate in 1051 C.E.

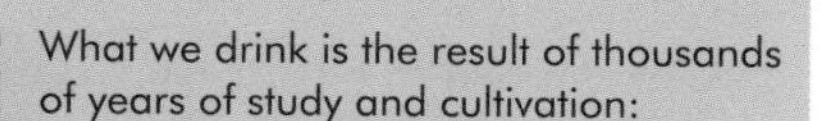

of years to evolve and is both a unique perspective on reality and a priceless museum of human experience.

One hundred cultures and thousands of species are now disappearing every year. This is an irreplaceable loss of human and genetic knowledge. The destruction of cultural and genetic diversity not only lowers the quality of our lives but threatens our ability to survive. These mass extinctions are weakening the resilience of our natural and social systems and leaving them more vulnerable to crises and collapse.

How much is enough?

> Simplicity doesn't mean to live in misery and poverty. You have what you need, and you don't want to have what you don't need.
>
> — Charan Singh, mystic (1916-1990)[78]

The industrial system has improved living standards for much of the world's population, but at the cost of ruined cultures, communities and ecosystems. We now have more but belong less. Many people feel alienated from their families, from their work, from nature and spirituality. Because emotional, intellectual and spiritual needs can only be satisfied through actions that develop community, creativity and faith, a life of consuming will always leave people feeling empty and incomplete.[79] And people who are alienated from themselves and each other will never feel satisfied and will never stop consuming.

In our materialistic age the majority of people place more faith in money than in God. As a result, criticizing the accumulation of wealth has become a modern form of heresy. Any politician, businessperson or teacher who speaks against the consumer society is putting their career at risk. All the material resources and scientific knowledge needed to resolve the major problems on the planet have been available for decades, but the will to change the political and economic priorities of society has not. As a result, increasing global wealth has been accompanied by increasing global poverty. Although many leaders have good intentions, their efforts to implement change are constrained by the existing system, whose worldview, values and structures oppose the development of new priorities such as the reduction of consumption and the redistribution of wealth.[80]

We will not be able to avoid the environmental collapse of our world and the economic collapse of our civilization by technological means alone. As long as the world system is organized by values that promote materialism and violence, global consumption will continue to increase and the environment will continue to degrade. In order to preserve the environment we need not only better technologies but also better values.

Satish Kumar, the editor of *Resurgence* magazine, says that in addition to efficiency we need sufficiency.[81] Sufficiency is learning to be satisfied with enough; it is taking care of real needs rather than false greeds. The consumer culture has no concept of enough: millionaires want to be billionaires, and billionaires want to own even more. In reality no one needs to be a millionaire in order to be happy. In fact consuming more than we need is not only immoral in a hungry world, but it is also the road to environmental destruction, emotional and spiritual poverty, war and economic ruin. We would all be far happier living in a peaceful world without poverty, pollution, disease, crime and war. But a peaceful and sustainable world will only be possible if we learn about limits — if we learn to live within the planet's biophysical limits; if we learn to share the planet's limited resources; if we learn to give more and take less and if we learn to be satisfied with enough.

Figure 7: BEST Futures

Wind turbines at sunrise

New technologies

While technological advances will improve efficiencies and reduce waste, they will not change the societal values and structures that promote limitless consumption and growing inequality.

Technological advances can postpone environmental collapse, but only social advances can transform an unsustainable consumer society into a sustainable conserver society.

Technological fixes cannot solve social problems.

4 The Need for a New Model

The historical expansion of the industrial system

MOST POLITICAL AND BUSINESS LEADERS RECOGNIZE that the world has serious problems. However, because they believe that the global system is structurally sound, their response to growing global crises is to make economic and technical adjustments (e.g. to reduce interest rates or lower CO_2 emissions). While these policies may slow the speed at which crises develop, ultimately they will not work because the global system has an environmentally unsustainable design. In order to understand why the industrial system is doomed to collapse, it is necessary to examine how and why this particular design developed.

The industrial system was not always unsustainable. The Industrial Revolution began around 1750 C.E., at a time when the global population was $1/8^{th}$ the size that it is now and when the average person used a fraction of the resources that are consumed today. Figures 1-6 illustrate the historical expansion of the global footprint (the total human consumption of the planet's annual natural income).

In 1650, human populations and environmental impacts were limited. At that time, most people in the world lived in small communities and made a living from farming, herding and/or hunting. Energy came from renewable sources: human and animal labor plus wood for heat and wind and water for power. Few travelled far from home and most of the world was unexplored wilderness.

The Industrial Revolution, which began around 1750, applied science to increasing production. Major scientific and technological breakthroughs improved productivity and created new products. The harnessing of fossil fuels gave birth to increasingly powerful machines that transformed production, transportation and warfare. Populations began to grow more rapidly as better food, sanitation and health care reduced mortality and increased longevity. At the same time rising productivity enabled per capita consumption to gradually increase.

Figure 1:

BEST Futures [1]

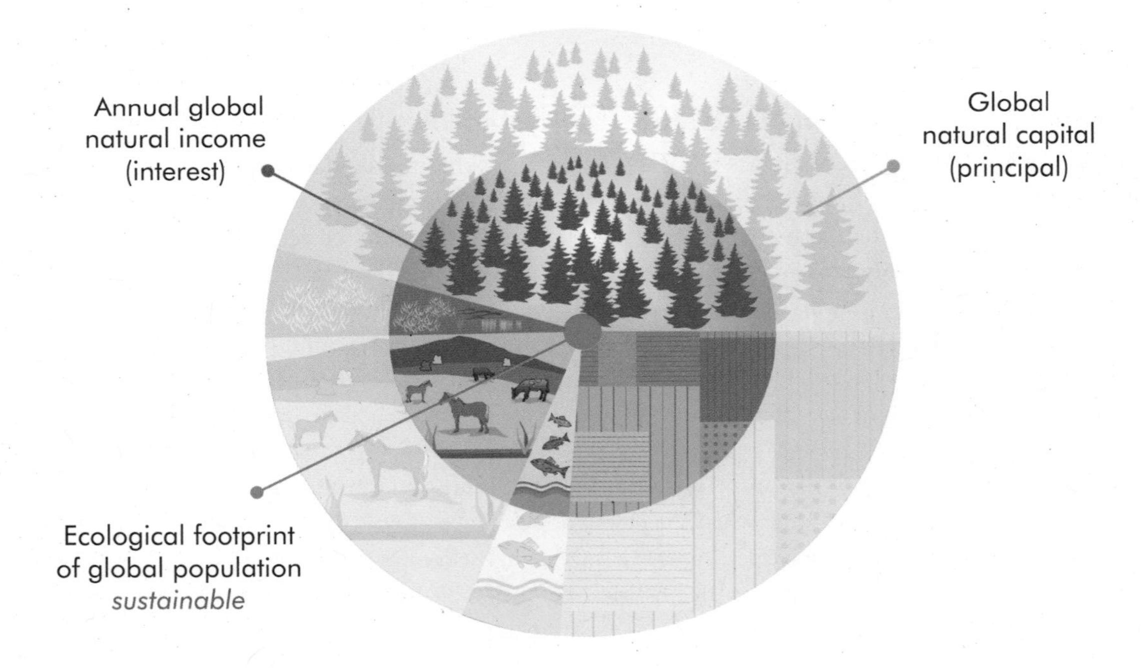

However, industrialization was a slow and uneven process. By 1850, only a few industries in Europe and North America were mechanized. The vast majority of people in the world still made a subsistence living from farming or made traditional handcrafts in the cities. Although industrialization was creating a wealthy class of factory owners, their employees worked and lived in terrible conditions. Nevertheless, industrial products such as tools, textiles and ceramics were beginning to change patterns of production and consumption around the world.

Figure 2: BEST Futures

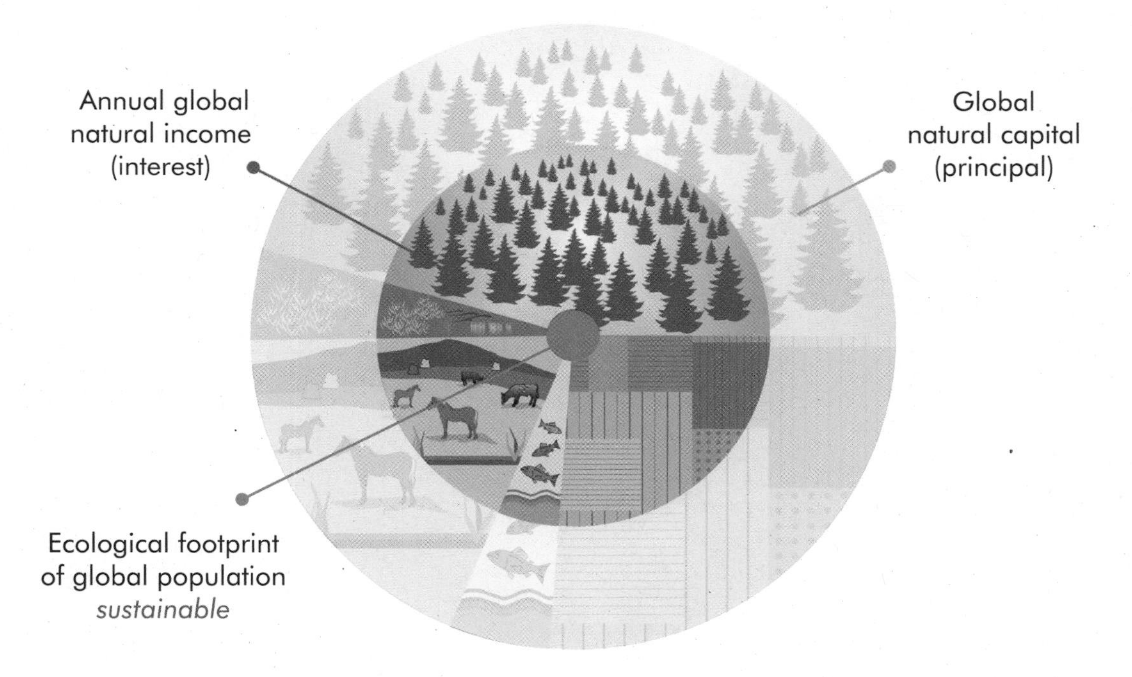

The process of industrialization involved a constant search for new markets and new sources of raw materials as well as continuous technological, commercial, financial and military expansion. Before the Industrial Revolution, Europe was less wealthy than many of the territories it was exploiting. By mechanizing, European industry was able to out-compete traditional industries in other nations. The gap which developed grew wider over time.

During the 20th century three major industrial systems vied for power. While

Figure 3:
BEST Futures

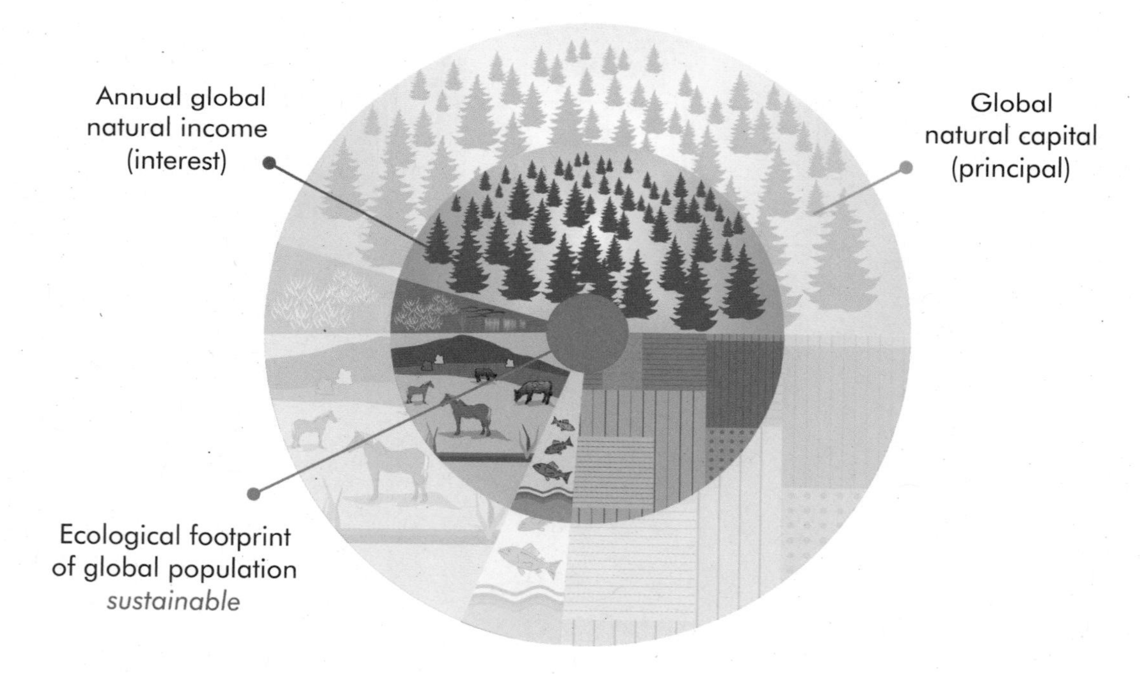

socialism promised more social rights (e.g. jobs, education and health care), capitalism promised more individual rights (e.g. property rights, and freedom of travel and expression) and fascism promised increased national pride and power. By the end of the century both fascism and state socialism had failed, and capitalism, the most politically and economically open industrial system, dominated the global economy.[2]

Capitalism was successful because its pluralistic, relatively autonomous institutions made it more dynamic and adaptable than its authoritarian and bureaucratic

Figure 4: BEST Futures

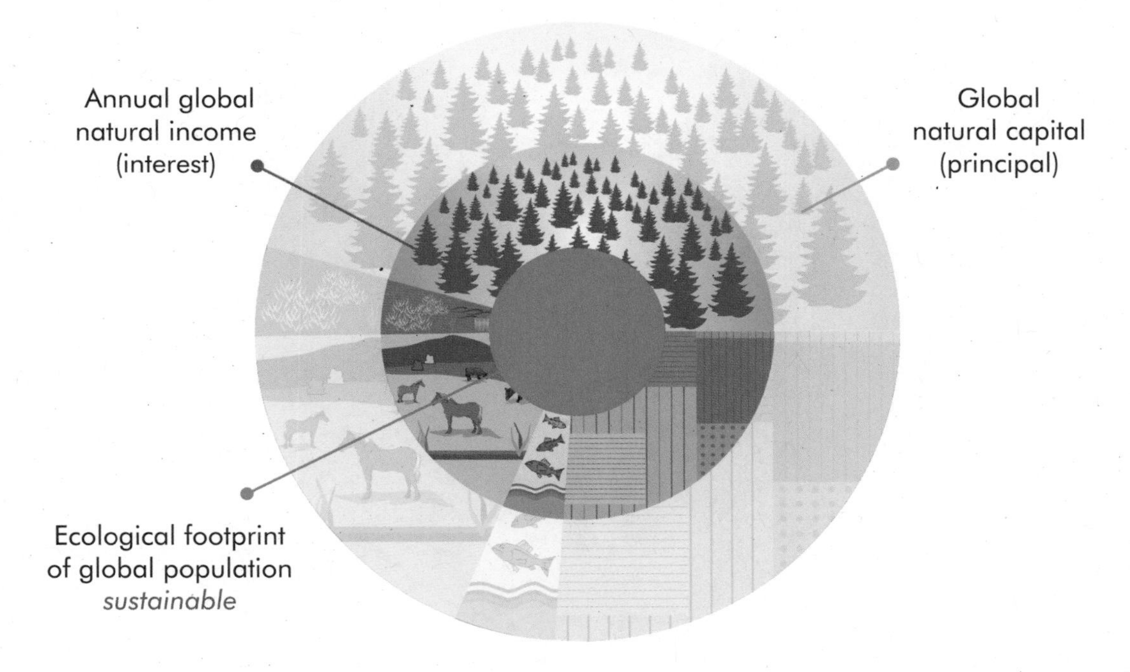

opponents. For example, Western capitalist democracies prevented working-class revolutions through accepting some of their opponents' key demands. Now every developed country has public education, old age pensions, minimum wages and (with the exception of the United States) universal health care. Although these socialist reforms were initially resisted by conservative elites, they reduced social tensions, created educated work forces and increased productivity and wages. Rising wages in turn increased consumption and stimulated growth.

Figure 5: BEST Futures[3]

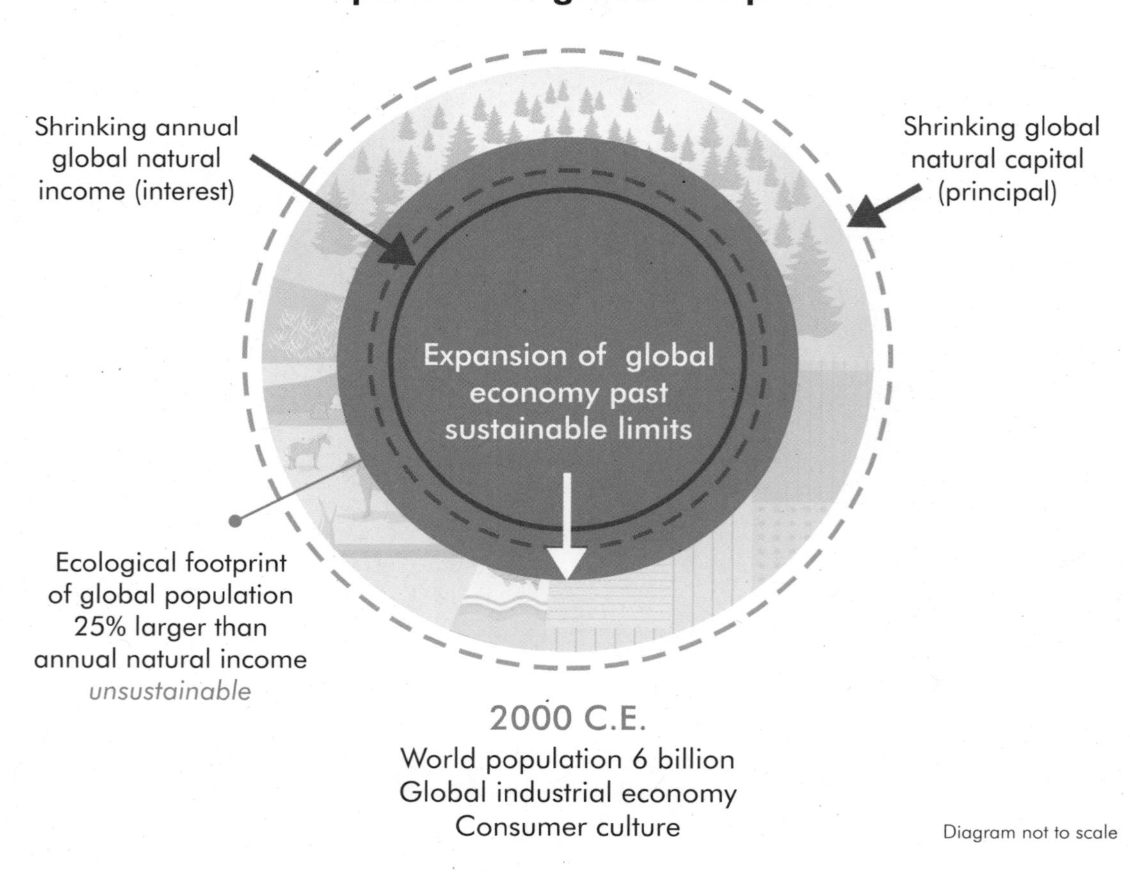

Over the 20th century the global population increased four times while the consumption of global goods and services increased approximately 20 times.[4] However, since most people still lived in rural areas in developing countries, most of the increase in consumption took place in the industrialized countries of the North.[5] While the global population is projected to increase by 50% between 2000 and 2050, world consumption is expected to increase by 400% due to the rapid industrialization of developing countries and the increasing wealth of the developed world.

Figure 6: BEST Futures [6]

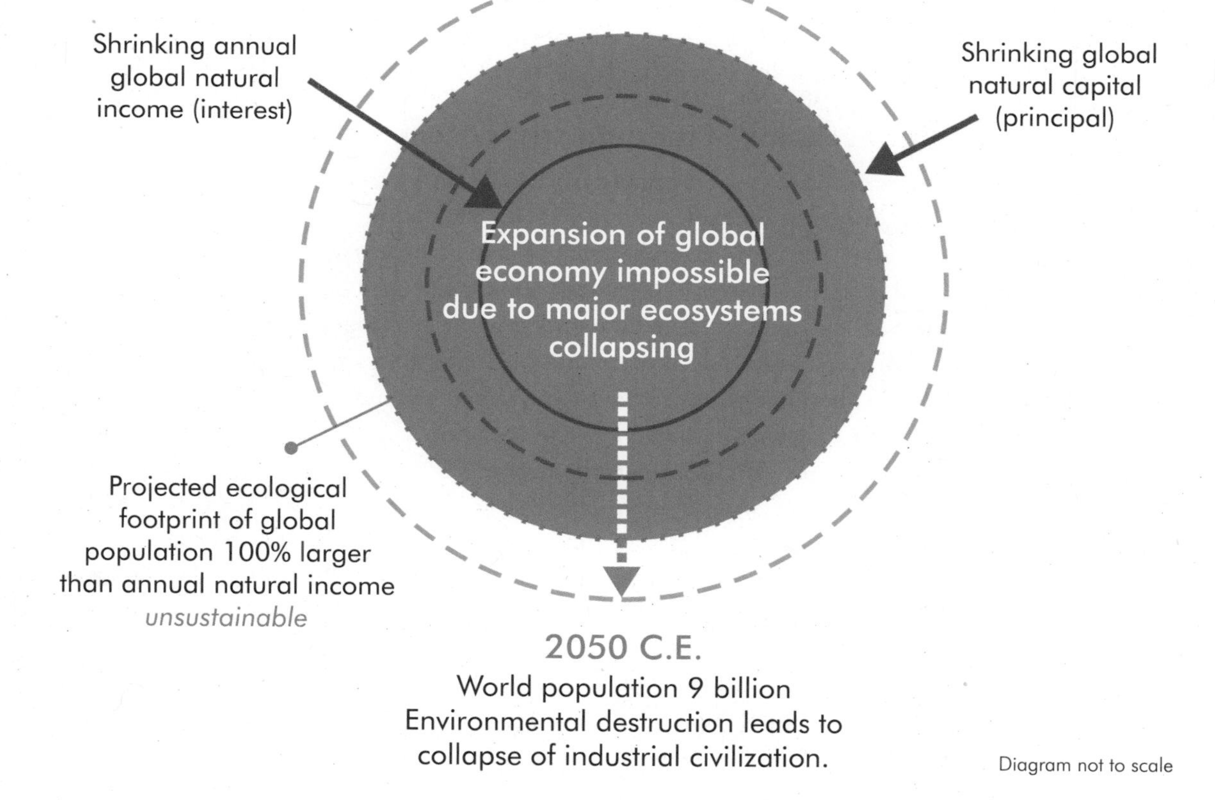

Ecologists estimate that human economies were sustainable until the 1980s. The global economy became unsustainable when the annual consumption of renewable resources and production of waste began to exceed the carrying capacity of the planet. At this point humanity began to spend not only the annual global natural income, but also the planet's natural capital. By 2000 world consumption was approximately 25% more than the global natural income; if current growth continues, it is estimated that by 2050 the annual deficit will be around 100%. This means that humanity will consume the equivalent of two planets worth of natural income each year.[7] However, it is unlikely that this level of consumption will ever be reached due to growing resource shortages and the collapse of major ecosystems.

The industrial system is doomed because its views, values, social structures and technologies are designed for constant physical growth, which means that it cannot reduce its consumption of natural resources without becoming an entirely different type of societal system. Because the Earth's carrying capacity has been passed, the global system has begun to fail.

The evolution of the industrial system

Human societies are organized by their cultures. The worldview and values of a society give it direction and organize the institutions that govern day to day activities.[8] Our world is on the brink of disaster because the international political and economic system is organized around the unsustainable worldview of the Industrial Age. It will help us to understand why the industrial worldview cannot solve the global problems of the 21st century if we examine its origins.

Five hundred years ago the largest, richest and most technologically advanced agrarian kingdoms in the world were in Asia. So why did the Industrial Revolution begin in Great Britain? In part the answer is favorable geography: Britain had fertile soils, abundant coal resources and navigable rivers and coastlines that facilitated transport and trade. But the critical difference between Britain and countries like China was not geography but culture: the British developed ideas and social structures that supported scientific innovation, social mobility and the expansion of commerce. The Industrial Revolution was not just a technological revolution, but also an economic and political revolution. The views, values and social institutions of the Agricultural Age had to be completely transformed in order for a more open, creative and productive economy to develop.[9]

This transformational process had three main components: the replacement of a religious worldview with a secular, rationalist worldview; the replacement of rigid, caste-based social structures with more open, pluralistic institutions and the replacement of handcrafts with industrial production. While these three elements had different origins, they became immensely powerful when they began to interact and integrate in 18th century Europe. At that point, they acquired the power to change agricultural societies into industrial societies and to replace the rule of kings with the rule of commerce.

The origins of the industrial worldview

Our present rationalist view of reality has its origins in Ancient Greece. Greek philosophers and scientists were the first to emphasize the importance of reason, measurement and value. The combination of dualistic analysis (dividing things into their constituent parts) and quantitative measurement became the basis of Western science. The analytical worldview, which separated subject and object, the knower and the known, began a profound conceptual parting of the ways between Eastern and Western philosophies.

Greek methods were later adopted by the Romans, who applied them to practical matters such as road building and military engineering. Following the collapse of the Western Roman Empire, classical knowledge was retained in the Byzantine world and then transmitted by Muslim scholars back to Western Europe. Its revival stimulated a return to a human focus (expressed in the more naturalist art of the Renaissance) as well as interest in science and exploration. Classical theory was linked with contemporary empirical evidence: the scientific method was further developed and the world began to be mapped.

The development of industrial social structures

However, dualistic ideas by themselves could not transform the static social structures of agrarian societies. These changes were made possible by the concurrent rise of Protestantism and capitalism. The Commercial Revolution began with voyages of discovery in the 15th and 16th centuries, which enabled European powers to establish colonies and build international trading networks. Following the Spanish conquest of the Aztec and Incan empires, a flood of gold and silver stimulated European commerce. With the rapid expansion of trade, the power and independence of the merchants steadily increased.

The Protestant Reformation was the catalyst that enabled the development of more open and dynamic institutions. In 1517 Martin Luther challenged not only corruption within the Roman Catholic Church but also its authority in religious matters. In asserting that a person's salvation depended on faith alone, Protestantism applied Renaissance individualism to the religious sphere. The views of Luther and other religious dissenters spread rapidly due to the recently invented printing press.

The Protestant revolt won the secular support of many rulers who wanted political and economic independence from the church. It was also supported by urban capitalists and others who wanted freedom from feudal and religious restrictions on trade and social mobility. While the split within Christendom led to decades of religious persecutions and warfare across Europe, it also ended the religious monopoly of European Catholicism and stimulated the growth of national monarchies. The Treaty of Westphalia ended the 30 Years War in 1648. By declaring the territories of princes independent of any supra-national authority, this act symbolized the emergence of the modern European nation-state system and the end of the feudal order.

The spread of ideas of religious liberty and the separation of church and state supported the development of pluralist institutions and representative government, and the spread of puritan values and the Calvinist work ethic supported the rise of capitalism. The idea of popular sovereignty, maintained by constitutional law, developed first in the Netherlands and England, which were commercial maritime states with strong urban middle classes. In England civil wars were fought between absolutist kings allied with the established church and religious dissenters allied with Parliament. After the Glorious Revolution of 1688, Parliament dictated the conditions under which British sovereigns would rule.

The 17th century also saw the beginning of the Age of Reason. The European world of ideas was profoundly affected by the rapid development of science. Galileo, Descartes and Newton described a universe that resembled a giant mechanism with measurable properties such as size, shape, quantity and motion. While it had definable parts and behaviors, this universe was without feelings or attributes such as justice, dignity, goodness or worth. This new view of reality totally transformed humanity's relationship to the world: now nature was no longer a living organism to be respected and nurtured but a collection of objects to be manipulated and mastered. This rationalist perspective supported the rise of a secular society: since God had created a universe that ran on immutable natural laws, there was no place for miracles, and faith must give way to reason.[10] In this paradigm, time and space were independent and absolute properties of an infinite nature. Similar concepts were also developing in political and economic theory. The state was also held to be independent, with absolute control over its territory. In the same way economic space was associated with absolute economic rights. Economists (imitating physicists) posited the existence of a natural mechanism that automatically regulated economic society — the impersonal market.

Not all 18th-century Enlightenment thinkers shared these mechanistic political and economic views. Many humanists began to believe that utopian civilizations were possible. Voltaire was convinced that reason would bring about reform, while Condorcet depicted history as a march toward equality and justice. Revolutionaries would later take up the humanists' criticisms of injustice and use them to overthrow monarchical absolutism. Monarchical absolutism and its economic counterpart, mercantilism, were dealt lethal blows in the American and French Revolutions (1776 and 1789). Pluralist political and economic institutions were developed that supported the replacement of a hereditary aristocracy by a propertied meritocracy. Although European autocracy did not end in the 18th century, the seeds of liberalism, constitutionalism and democracy were firmly established.

The development of industrial technologies

The third requirement for the evolution of a new type of societal system — along with the emergence of a new view of reality and more functional social institutions — is the development of more productive technologies. The Industrial Revolution marks the application of the rationalist scientific method to technology and production. This was a paradigm shift in the way humans understood and manipulated the natural world. It accelerated the invention of machines and technologies capable of processing more energy and resources, but also the emergence of a more complex and open type of societal system. Organized around the rationalist worldview, the industrial system was composed of congruent individualistic values, pluralistic societal structures, machine technologies and competitive economic processes.

Each new level of societal organization utilizes more energy. While agrarian civilizations are primarily powered by harnessing renewable energy (humans, animals, wind and water), industrial civilizations are primarily powered by burning stored energy (initially wood and coal, and now oil and gas). The Industrial Age began with the invention of steam engines. Railways and steamships allowed more resources to be accessed through greatly extending two-dimensional environmental control (e.g. opening up the interiors of continents). Mechanization contributed to rapidly increasing populations, urbanization, restructured societal relationships and rising living standards.

As social and economic relationships changed, new social conflicts emerged. Complex industrial economies require well-educated and mobile workforces. The printing press helped lay the foundations for public education and the mass society. But while mechanical publishing stimulated commerce and spread the values of rationalism and individualism, it also increased nationalism and raised aspirations for higher living standards and social change.

Industrial production methods increased exploitation and alienation as well as wealth. Although both Adam Smith and Karl Marx were rationalists who believed in the immutable laws of the capitalist system, Smith emphasized the role of the individual and laissez-faire competition, while Marx emphasized the role of the collectivity and class conflict. Ideological conflicts over the direction of industrialization convulsed the 19th and 20th centuries. Although socialist and labor movements succeeded in improving health and living standards in the industrialized Western world, individualist values and heirarchical social structures remained dominant.

> Our enormously productive economy ... demands that we make consumption our way of life, that we convert the buying and use of goods into rituals, that we seek our spiritual satisfaction, our ego satisfaction, in consumption ... we need things consumed, burned up, replaced, and discarded at an ever-accelerating rate.
>
> — Victor Lebow, retail analyst, 1955[12]

The Enlightenment, which reshaped the classical worldview into the West's modern view of reality, was responsible for numerous positive contributions: liberal democracies; the ideals of equality, freedom, and justice; modern physics, chemistry, biology and medicine; public education and health; the end of institutionalized slavery and the rise of feminism. On the other hand, this modernity must be held accountable for the diminution of religious awe and wonder; the destruction of the environment; the loss of community; the replacement of quality by quantity and the commodification of life summarized in Max Weber's famous phrase: "the disenchantment of the world."[11]

Designed for constant growth

The industrial system is ecologically unsustainable because it is designed for continuous and limitless growth. This design reflects its worldview, which can be described as both rationalist (its philosophical methodology) and expansionist (its economic and political values).

Industrial societies are much more dynamic and creative than their agrarian predecessors because they encourage constant economic development. Agrarian kingdoms believed in order and cyclical continuity — unchanging divine commandments, the divine mandate of kings and inherited and unchangeable social status. Life revolved around the unchanging cycles of agricultural production. In contrast, industrial societies believe in the concept of progress: people should try to improve their lives through hard work and creativity, and companies and countries need to constantly grow and innovate in order to survive.

In agrarian societies the economy was primarily regulated by social relationships — by a person's rights and obligations to others as determined by their social position. Markets were secondary and existed essentially to facilitate the exchange of goods. In a world where faith and duty were the highest values, commerce had low status; many Christian and Islamic societies believed that it was immoral to charge interest on loans. In contrast, because the core value of the modern consumer society is getting rich, capitalist economies are regulated through financial transactions made by individuals and corporations based on their access to financial capital (their wealth). Because the primary purpose of markets is not to exchange goods, but to increase capital,[13] the global economy has evolved into a financial system designed to support constant capital expansion.

Capital expands through people borrowing and investing money. The money is created by banks in the form of interest-bearing debts. Loans are predicated on the assumption that the borrowers will be able to pay back the original loan with interest because they have or will be able to generate more wealth than they have borrowed. This process creates the financial need for constant economic expansion.[14]

An obsolete model

The worldview and design of the industrial system creates problems that it can't fix. Not only is the worldview expansionist, but it is also rationalist and reductionist: it views reality as being made up of discrete objects rather than interrelated systems. This has created a world system that is made up of independent individuals, corporations and nations, each of whom acts in their own self-interest with little regard and no responsibility for the common good. As a result local and global economies convert natural capital into manufactured and financial capital without taking into account environmental or social costs.

The consequences can be seen everywhere: because individuals and corporations are more interested in short-term profits than long-term social benefits, productive farmland is sold to commercial developers, ancient forests are clear-cut and skilled jobs outsourced overseas. These failures are structural: they are the result of a nation-state system that makes governments accountable only to their own citizens, and of a corporate system that makes executives accountable only to their shareholders. This creates what is called the tragedy of the commons: because no government or corporation is responsible for common resources or collective problems: the air is polluted, stocks of wild fish are destroyed and millions of refugees left stateless.

Governments of the world find it difficult to take action on global warming, global poverty or world peace because national interests often conflict with global interests, and no one is responsible for the planet. The only organization with global responsibility is the United Nations, but it has no independent authority — all the UN's funds come from national governments.

Corporations are even more restricted than governments in their ability to act in the common interest. By law they must take whatever actions most benefit the financial interests of their shareholders. Law professor Joel Bakan pointed out that their relentless pursuit of economic self-interest so often disregards the social, economic and environmental interests of others that it can be described as psychopathic. For example, General Electric, one of the world's most respected corporations, was found in major breaches of the law 42 times between 1990 and 2001 for fraud, deception and violations of pollution and safety regulations.[16]

> Did you know that the worldwide food shortage that threatens up to five hundred million children could be alleviated at the cost of only one day, only ONE day, of modern warfare?
>
> — Peter Ustinov, actor, writer and director (1921-2004)[15]

We live in an increasingly interconnected world. The carbon dioxide emissions of one country affect every other country. Economic crises in one country immediately affect global stock markets. Wars in one region affect the security of countries on the other side of the globe. However, although we have a global system, it is not responsible or accountable to the people of the world. Its worldview, values and social structures are not designed to ensure the welfare of either the environment or the majority of the world's population. These failures are reflected in the lack of confidence most people have in global political leadership.[17]

The industrial system is incapable of solving problems of war, poverty and environmental destruction because its competitive views, values and structures support the domination of nature by humans as well as humans by other humans. Jeff Vail has commented, "The process of evolution within a system dominated by competing hierarchies demands that one set of goals consume all others: continuous growth, expansion, and increased domination. Any corporation or nation that pursues a more human-oriented goal will soon find itself squeezed out of existence for not following the simple rules of natural selection."[18]

The expansionist worldview and institutions of the Industrial Age were designed for the loosely connected local and national economies of a world with few people and many resources, not for the interconnected global economy of a world with many people and scarce resources. Constant, unregulated expansion is viable in an empty world, but dysfunctional and destructive in a full world. Because the global economy is no longer environmentally sustainable, and because corporations and national governments cannot manage a global system, the industrial model is now obsolete.

5 Cascading Crises and System Failure

Why earlier civilizations have collapsed

Most people cannot imagine that the powerful nations that they live in could suddenly collapse. But societies are dynamic living systems that are constantly rebalancing (equilibrating) in response to both internal changes and changes in their social and natural environments. Major changes create stresses within the society whether they are constructive (like the introduction of new ideas or technologies) or destructive (like a plague or an invasion). If a society fails to adapt to destructive changes such as declines in the production of food or energy, the stresses are likely to escalate into uncontrollable crises. Unless something happens to reverse the process, the society will then begin to disintegrate.

At some point almost every society has to deal with major changes and crises, which is why most of the societies and civilizations that have ever existed no longer exist. Some, like the Chinese Empire, declined slowly; some, like the Soviet Union, collapsed suddenly; some, like Britain transforming from feudal monarchy into democratic industrial nation, evolved.

In some cases, the process of collapse was so catastrophic that a complex civilization reverted to a simpler type of societal system. A well-known example is the fall of the Western Roman Empire. With its collapse, government, law and order came to an end, roads and aquifers were neglected and commerce was abandoned. As cities and countryside sank into poverty and chaos, populations declined, knowledge was lost and isolated parts of Europe regressed to the Stone Age.

Around the world we can visit the ruins of once powerful civilizations that failed because they either ran out of resources or were destroyed by invaders. But we can't visit the ruins of thousands of other ancient societies: because Stone Age peoples built their homes out of wood and skins and left no written records, almost nothing remains to tell us that they ever existed. Because we can't see 99% of human history, we don't realize how many societies have disappeared or how much human views, values and institutions have changed over time.

Of course not all societies suddenly disappear. Most economic and political changes do not cause societies to either collapse or evolve. For example, the lives of peasants were often unaffected by the overthrow of kings and the installation of new ruling families. Some powerful cultures (e.g. ancient Egypt and China) were even able to survive repeated invasions, famines and civil wars. Despite extreme social tension and massive destruction, they were able to adapt to changing conditions, integrate invaders and new ideas and continue functioning with the same basic culture, institutions and economy as before.

So when we talk about collapse, we need to distinguish between the collapse of a political system and the catastrophic collapse of an entire societal system. While a political system can collapse without destroying a society's basic social and economic structures, a catastrophic collapse typically involves the irreversible destruction of a society's economy and institutions and a rapid reduction in population size and density. For example, while the Allied victory in the Second World War destroyed the fascist governments of Italy, Germany and Japan, it did not destroy their industrial economies. As a result all three countries were able to quickly recover from the war. In contrast, after Angkor (the capital of the Khmer Empire) degraded its surrounding environment, the city became unsupportable and was permanently abandoned.[1]

The causes of collapse

In his book *Collapse*, Jared Diamond examined why societies succeed or fail. He identified five main causes of collapse: environmental damage, climate change, hostile neighbors, the loss of support of friendly neighbors and the failure of leaders to develop constructive responses to growing problems.[2] While every situation is unique, a frequent cause of collapse has been a combination of environmental problems and resource shortages. These have weakened a society and provoked internal unrest or civil war, at which point outsiders have been able to attack and destroy it.

The underlying problem is usually resource shortages: a society runs out of adequate supplies of water, food, wood and other essential materials. Villages and cities are normally established in places with plentiful resources, such as fertile farmland and pastures, rivers, forests and good hunting or fishing. As they prosper, the population increases and uses more resources, eventually farming all the land close to the settlement, cutting down all the nearby trees for fuel and building materials and killing off all the local wild game. As a result, people have to go further and further in search of agricultural land, wood and other materials.

Over time three problems begin to emerge. First, the land that has been intensively farmed, grazed or cleared of trees progressively degrades. Nutrients are lost and erosion sets in. As a result, yields gradually decline. Second, since the best farmland and other resources are used first, resources that are accessed later tend to be of increasingly lower quality. And third, the cost of getting the new resources rises with distance. At some point the cost of bringing food or wood from far away is more than they are worth, and the society runs out of ways to affordably access new resources.

Crises begin when the growing population starts to experience resource shortages due to the combination of declining yields and the lack of economical new resources. In Diamond's historical study, societies facing resource shortages have often either fought wars with neighbors or fought civil wars in attempts to secure access to scarce supplies of land, water and other resources. Climate change can then be the event that triggers catastrophic collapse: it exacerbates existing resource shortages with crop failures. A current example of this destructive dynamic is the conflict in Darfur, where spreading deserts and deforestation is a major cause of the conflict over water and land between African farmers and Arab nomads.[3]

In *The Collapse of Complex Societies*, the American anthropologist Joseph Tainter explained that agricultural and industrial civilizations are limited in their ability to expand not only by the increasing cost of acquiring new resources but also by rising maintenance costs. For example, as the Roman Empire expanded it required new layers of bureaucracy, longer lines of communication and additional regional governments and garrisons. The more it grew, the more complex and expensive it became to maintain. The combination of diminishing agricultural yields from exhausted soils, the increasing cost of acquiring new resources and the increasing cost of maintaining the Empire eventually resulted in a declining population, economic bankruptcy, chaos and collapse. Tainter concluded that complex societies are vulnerable to collapse because of the economic law of diminishing returns: while they must expand and develop more complex structures in order to acquire new resources, as they expand the costs of obtaining additional resources tends to rise. At the point where costs begin to outweigh benefits, the society becomes unsustainable.[4]

Of course the law of diminishing returns only applies if technology remains the same. More efficient technologies can reduce the costs of acquiring resources that are further away or of lower quality. Here industrial societies have an enormous advantage over earlier social systems: where innovation used to be a random and rare occurrence, now new technologies and economic processes are consciously developed and continuously introduced.

Does this mean that industrial societies do not have to worry about collapse? Unfortunately not. There is no escaping the problem of biophysical limits to growth. The world system today is being challenged by many of the same issues that destroyed ancient empires, with the difference that the challenges are global rather than regional in scale. The quality of new resources is constantly diminishing, and more and more resources are needed to support the growing populations and infrastructure of our increasingly complex system. Our industrial economy can continue to expand as long as we can still find sufficient quantities of affordable energy and other resources. But as soon as we run out of affordable supplies, the global system will be in trouble. New technologies can allow us to utilize lower quality resources and to use resources more efficiently, but they cannot create new materials out of nothing. They can postpone the collapse of an ever-expanding economy, but they cannot prevent it.

The loss of resilience

The ability of a social system to cope with stresses depends not only on their size and severity, but also on a society's resilience. Societies are resilient (able to manage shocks) to the extent that they are internally functional, externally relevant and rich in both resources and creative diversity (adaptive capacity). They are vulnerable to collapse to the extent that they are internally dysfunctional, poorly adapted to their environments, poor in resources and unable to adapt to change.

A growing danger is the global loss of resilience. In the past, collapses were localized since few goods were traded between civilizations, the disintegration of one society often had few economic effects on others. In contrast, because every country in the world is now part of a global economy, every regional crisis sends shock waves around the world. The increasing vulnerability of the global system is shown by how rapidly financial crises are now able to spread. In 1997 for example, Western investors lost confidence in the Thai economy and began to withdraw money. A domino effect rapidly spread currency speculation from Thailand to Indonesia and Korea, and then to Russia and Latin America. By 1998 one-third of the globe was in recession.[5] While the 1997 crisis was largely confined to weaker economies, a major shock to the international economy — such as a panicked sell-off of US dollars — could quickly produce a global depression.

The globalization of consumer society is creating a brittle monoculture that is simultaneously accelerating unsustainable economic expansion and creating an ever more uniform and tightly integrated global economy. Industrial civilization globalizes inequality and concentrates power at the expense of local autonomy, community and diversity. As the many varieties of human civilizations and societies become undifferentiated parts of an expanding global monoculture, the system loses checks and balances. The result is an increasingly closely connected but unstable world system.

The danger is that the number and complexity of large crises is increasing at the same time as the global system is progressively losing its ecological, economic and social resilience. If existing political and economic institutions do not have the reserves, flexibility and back-up systems to enable them to rapidly adapt to resource shortages and failing ecosystems, local crises may quickly become uncontrollable global crises.

Parameters and perfect storms

Like plants and animals, societies are living systems that cannot survive without constant flows of energy and resources from their environments. In order to get these inputs, every society must have coordinated, functioning social institutions. A society will experience stress if there are disruptions in the flow of resources for any reason: external problems acquiring resources or internal problems processing or distributing them. This means that crises can have environmental, economic or social causes and can occur anywhere in the system.

It is easier for societies to manage problems that are temporary and/or local in nature than problems that are sustained and/or generalized. For example, it is easier for a society to manage a single failed harvest than long-term climate change, a work stoppage in one factory than a national strike, a border dispute than all-out warfare. The work of Jack Goldstone, a sociologist at George Mason University, indicates that societies are more likely to break down when they face multiple converging stresses. He wrote, "Massive state breakdown is likely to occur only when there are simultaneously high levels of distress and conflict at several levels of society — in the state, among elites, and in the populace."[6] While converging stresses can result from disparate developments (e.g. a harvest failure occurring at the same time as invasion), they are often caused by cascading crises (e.g. a harvest failure causes famine which then triggers a rebellion). As crises interact with each other the problems multiply and become more difficult to manage.

The large number of interacting problems facing humanity in the coming decades increases the probability of major crises. Enormous threats are posed by climate change, energy shortages, water scarcity, food shortages, loss of biodiversity, growing economic inequality, increasing global financial instability and conflicts over scarce resources. Other growing threats also exist, such as pandemics and the proliferation of nuclear weapons.[7] While any of these issues will be extremely difficult to manage by itself, in combination they will be unmanageable. For example, until recently UN food estimates have assumed that both the weather and energy prices will remain relatively stable for the foreseeable future.[8] But what will happen to agricultural production if the costs of irrigation, fertilizers and transportation continue to rise due to declining oil supplies? What will happen if this problem is compounded by other factors such as climate change? And what will be the political consequences in China, India and other countries if these interacting crises produce a deadly combination: a global depression, inflation, increasing food shortages and growing unemployment? Scenarios such as these are the recipe for the type of perfect storm that could cause the catastrophic collapse of the world system.[9]

Global crises could start almost anywhere. Since societal systems have complex and chaotic dynamics,[10] it is not possible to make precise predictions about the future. Nevertheless, it is possible to define system parameters — the operating conditions and resources that a society must have to survive. The biological Law of the Minimum (Liebig's Law) states that the population of any species is limited

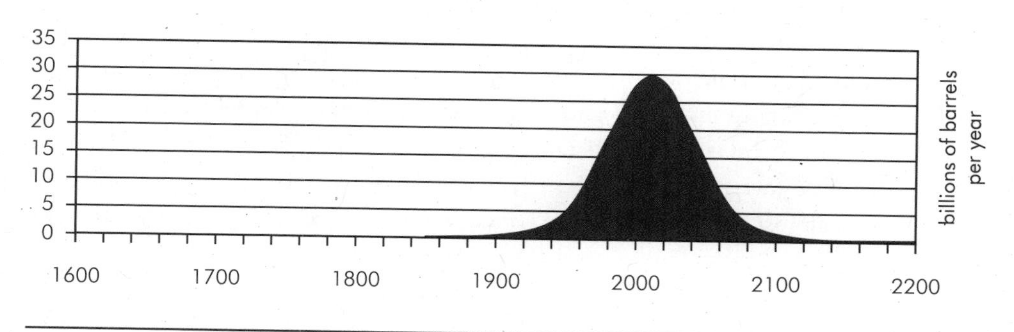

Figure 1: BEST Futures [11]

by the necessity (water, food, suitable climate) in least supply. This means that it will only take a shortage of one irreplaceable global resource to trigger a global crisis. Although we are not yet in a position to predict which resource will run out when, we do know that resource shortages are inevitable because the industrial system requires constantly increasing quantities of energy, water, metals, fiber, grains and other critical resources in order to function. We also know that the availability of many resources is declining due to overexploitation, pollution and climate change.

Many experts believe that oil production will peak and begin to decline sometime between 2008 and 2015.[12] While the first 150 years of industrialization were powered by wood and coal, the rapid expansion of the industrial system after 1900 closely corresponds with the rapid expansion of oil production. Since oil is the most important source of energy on the planet today, when production begins to decline, the global economy is likely to go into a severe depression. This chart indicates how vulnerable our industrial civilization is to resource shortages. We could create similar charts to show how rapidly other essential resources such as groundwater, topsoil, wood, fish, natural gas, lead, zinc or copper are being depleted, and how rapidly the planet's air, earth and water are being polluted.

In terms of the long existence of humans on earth, the Industrial Age is only a brief, passing phenomena. Non-renewable fossil fuels made mechanization possible. Mechanization has allowed our species to expand and consume both renewable and non-renewable resources at an unsustainable rate. Now the resources are almost running out, major ecosystems are failing, and industrial civilization is about to collapse.

The consequences of system failure

We don't have to look to the ancient past to see the consequences of societal collapse — the world is full of failed and failing states. In the last few decades many societies have collapsed, with consequences ranging from war, genocide and ethnic cleansing to civil wars and economic ruin. Some examples are Cambodia, Rwanda, Yugoslavia, Afghanistan, Iraq, Somalia and the Soviet Union. In almost every case government services such as law and order, public health care and education almost disappeared, living standards sharply declined, mortality rates rose and criminal gangs took control of entire regions.

While failed states are a major source of conflict, terrorism and drugs, they do not destabilize the global economy.[13] The real danger to the survival of our species does not come from the collapse of individual nations, but from the collapse of major ecosystems and the global economy as a whole. The global economy was able to recover from the Great Depression of the 1930s because major ecosystems were intact and the world was still full of undiscovered resources. But if the next great depression is caused by a combination of climate change and resource shortages, the world may not have the ecological or economic resilience to recover.

Attempts may be made to try to avert a global depression through intensifying the exploitation of already degraded ecosystems. If this is done, a vicious cycle of environmental, economic and social destruction will ensue. Environmental crises will rapidly escalate, triggering a cascade of uncontrollable economic and political crises. At some point interacting crises would converge and create a *perfect storm* that causes the catastrophic collapse of the global system. As both international and national institutions begin to fail, wars will break out over scarce resources. The fall of the Roman Empire would be repeated on a global scale, but with wars fought with weapons of mass destruction.

As ecosystems, economies and social institutions progressively collapse, human populations will sharply decline due to starvation, disease and warfare, and cities will be abandoned. Survivors will have to learn how to use primitive technologies to eke out livings in devastated environments. Social and biophysical systems may be damaged to the point where it becomes impossible to support advanced civilizations on Earth.

The industrial system, which has been able to manage crises and changes for over two hundred years, is becoming increasingly unable to cope with interacting environmental, economic and social crises. As the global economy begins to fail,

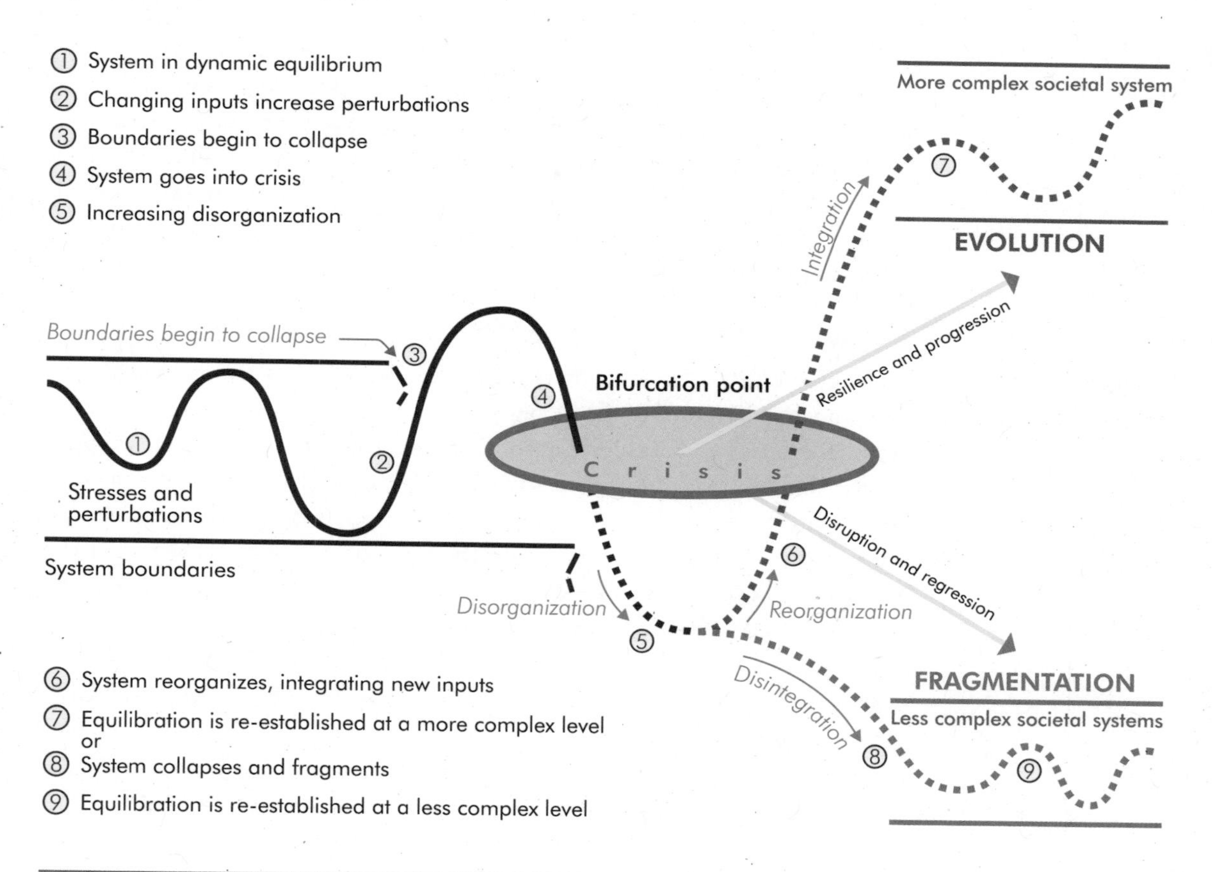

Figure 2: BEST Futures

it will become more disorganized and dysfunctional. However, at this stage disaster is not inevitable. Two (but only two) future outcomes are possible: the global system will either continue to disintegrate to the point where it suffers irreversible damage and collapse, or it will reorganize itself into a new type of sustainable system.

Although most of the human societies and civilizations that ever existed have disappeared, not every one has suffered catastrophic collapse. Some met the challenges of changing conditions by developing new and more environmentally relevant worldviews, technologies and institutions. The weakening of the existing system is not only a time of great danger, but a time of great opportunity.

We are now entering a time of increasing global crises that can only result in either the catastrophic collapse of our unsustainable industrial system or its transformation into a sustainable planetary civilization. We are already well into the first part of this process — growing global crises. The question is no longer whether our unsustainable system will eventually collapse; the question now is whether humanity has the time and ability to avert disaster through creating a sustainable planetary civilization.

Why we can't see the dangers

> The world is sleepwalking its way to the edge of catastrophe .... The Doomsday Clock of climate change is ticking ever faster towards midnight.
>
> – Charles, Prince of Wales[14]

On January 17, 2007, the Bulletin of the Atomic Scientists (BAS) announced that it was moving the hands of the Doomsday Clock forward to five minutes to midnight. "Reflecting global failures to solve the problems posed by nuclear weapons and the climate crisis, the decision by the BAS Board of Directors was made in consultation with the Bulletin's Board of Sponsors, which includes 18 Nobel Laureates. In a statement supporting the decision to move the hand of the Doomsday Clock, the Board focused on two major sources of catastrophe: the perils of 27,000 nuclear weapons, 2000 of them ready to launch within minutes; and the destruction of human habitats from climate change."[15]

It all seems a little unreal. Although all the evidence indicates that our world is speeding towards disaster, most people, including most global leaders, are unaware of the urgency of the situation. It is not as if we can leave the problems to future generations. The climate scientists warn that we only have 10 years to turn around global warming; many experts warn that oil production could peak any time; global grain reserves may disappear within a few years. At a time when all of humanity needs to be mobilized to avert catastrophe, our leaders are asleep at the controls.

There are many reasons for the ignorance and apathy. Part of the blame lies with vested interests that do not want to stop making weapons or toxic chemicals, cutting down tropical forests, polluting the air or exploiting child labor. We cannot afford to ignore the existence of these powerful and dangerous groups because they have been actively working for decades to block meaningful environmental and social reforms. But although these groups are influential, they are still in the minority, even among business, military and political elites. The only reason that their lies have worked is because their interests converge with the traditional worldview and values of the industrial system and its consumer culture. When leaders, seem unaware of the existence of major global problems, people want to believe them. It is very difficult for most people to accept that the civilization we live in is about to collapse, despite its vast wealth, its glittering technology and its millions of experts.

Since most experts share the industrial worldview, they also believe that reports of growing crises are greatly exaggerated. In addition, because most scientists, most economists, most educators and most journalists work within institutions that reward conformity and punish dissent, they tend to avoid taking critical positions. As a result they often ignore or minimize the danger of future crises when they teach, speak or write in the media. With all these factors converging to present a positive image of the consumer society, it's no wonder that the majority of people are only beginning to realize that the planet is in serious trouble.[16]

However, there are other reasons for the slow pace of reform besides the resistance of interests, inertia and ignorance. More people are not actively working to save the planet because a complete picture of what is happening is only beginning to emerge, and events are developing too quickly for people to easily understand. Thomas Homer-Dixon discussed the problem of cultural acceleration in *The Ingenuity Gap*.[17] He argued that existing institutions and ways of thinking are no longer able to cope with either the pace of global change or the complexity of current events.

Another explanation comes from Albert Bartlett, Emeritus Professor of Physics at the University of Colorado. He believes that people have difficulty accepting that the global system is approaching its maximum limits of physical growth because they do not understand the principle of compound interest. In order to illustrate the difficulty of imagining the consequences of constant growth he compared the global economy to bacteria growing in a bottle.

> Bacteria grow by doubling. One bacterium divides to become two, the two divide to become 4, the 4 become 8, 16 and so on. Suppose we had bacteria that doubled in number this way every minute. Suppose we put one of these bacteria into an empty bottle at 11:00 in the morning, and then observe that the bottle is full at 12:00 noon. There's our case of just ordinary steady growth: it has a doubling time of one minute, it's in the finite environment of one bottle.

> I want to ask you three questions. Number one: at what time was the bottle half full? Well, would you believe 11:59, one minute before 12:00? Because they double in number every minute.
>
> And the second question: if you were an average bacterium in that bottle, at what time would you first realise you were running of space? Well, let's just look at the last minutes in the bottle. At 12:00 noon, it's full; one minute before, it's half full; 2 minutes before, it's a quarter full; then an 1/8th; then a 1/16th. Let me ask you, at 5 minutes before 12:00, when the bottle is only 3% full and is 97% open space just yearning for development, how many of you would realise there's a problem?[18]

The Doomsday Clock of the nuclear scientists may be set at five minutes to midnight, but the planet is no longer mostly empty — it is almost completely full. Because we are quickly running out of resources, time is not on our side. But we have no choice — if we are to avoid catastrophe we must not only alert humanity to the growing dangers, but also quickly transform the entire global economy, political system and culture.

Is this possible? Can we prevent major ecosystems from collapsing in the next decade? Can we rapidly transform our destructive global system into a constructive system? It is possible. The accelerating pace of change works both ways. Not only is there increasing destruction, but also growing awareness and more and more constructive innovation. New views, values, social structures and technologies are emerging all over the planet. The process of transformation has already begun.

Part

2

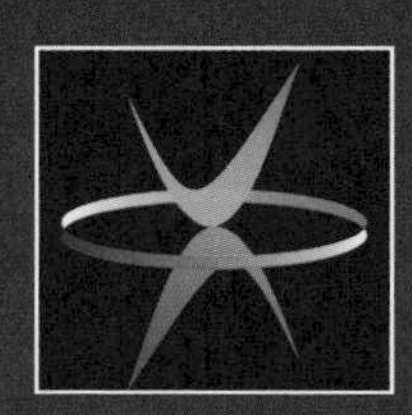

Transformation

— the emerging trend

This is the largest social movement in all of human history .... coherent, organic, self-organized congregations involving tens of millions of people dedicated to change .... If you look at the science that describes what is happening on earth today and aren't pessimistic, you don't have the correct data. If you meet the people in this unnamed movement and aren't optimistic, you haven't got a heart.

— Paul Hawken, *Blessed Unrest*[1]

6 Sustaining Development or Developing Sustainability?

What is sustainable development?

IT IS EASY TO DESCRIBE THE PROBLEM FACING HUMANITY — because our global system is unsustainable, it is heading towards environmental, economic and social collapse. The solution is equally straightforward — in order to avoid catastrophic collapse, our unsustainable system must be transformed into a sustainable system.

However, many people do not have a clear understanding of either the problem or the solution because the term *sustainable development* has been corrupted to mean sustaining material growth, rather than developing ecological sustainability.[1] So we need to define *sustainable* and *unsustainable*. Put simply, something is sustainable if it can persist over time and unsustainable if it can't.[2] In this book we examine whether the global industrial system is sustainable — if it can persist (survive) for the foreseeable future.

The real bottom line is not financial profits but survival, and our survival is utterly dependent on our environments. The air we breathe, the water we drink and the food we eat all come from our environments. The energy and raw materials used by our economies come from our environments. As a result, the long-term viability of human societies depends on the long-term viability of the biophysical systems that support them. We know that the present industrial system is unsustainable because it is progressively degrading major ecosystems, and history and science tell us that any human society that destroys its environment cannot survive for long.

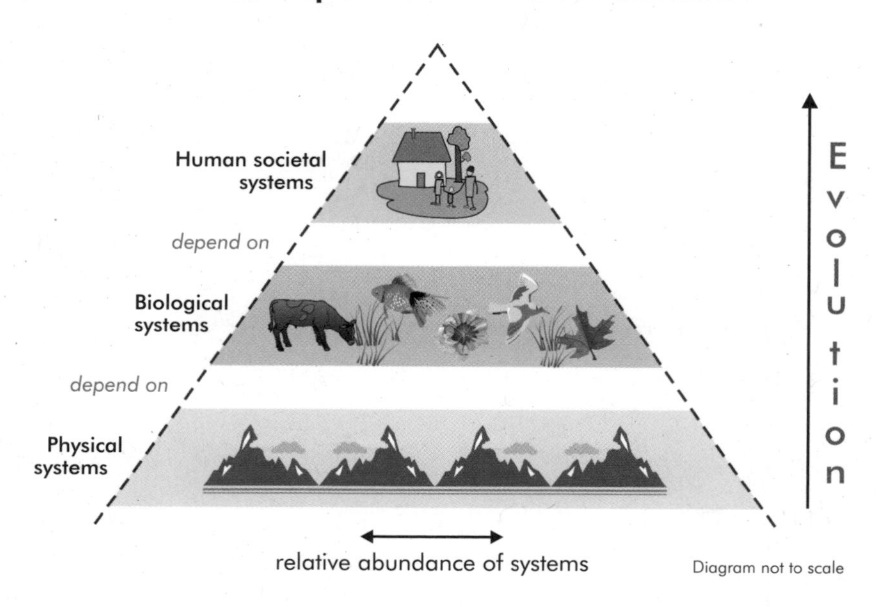

Figure 1:
Adapted from E. Laszlo [3]

In 1987, the Brundtland Report defined sustainability as the ability to meet the present needs of humanity without compromising the ability of future generations to meet their own needs.[4] Sustainable development has also been defined as improving the quality of human life while living within the carrying capacity of supporting ecosystems.[5] Every living system (plants, animals and societies) is sustainable to the extent that it is able to meet its essential needs on an ongoing basis. This is a key concept: in order to function, a living system must be able to satisfy its essential needs. Once we grasp the relationship between sustainability and needs, we can begin to clearly understand societal and environmental interrelationships and dynamics. When living biological or social systems are able to meet their essential needs, they usually have sufficient resilience (adaptability) to withstand normal environmental stresses and to reorganize in healthy ways in response to changing conditions. When they can't meet their essential needs, they weaken and become increasingly dysfunctional and incapable of managing stress.

Since all living systems are members of wider communities and ecosystems, essential needs are more than minimal physical inputs. For example, human needs are more than material needs for food, shelter and safety: they are also emotional, intellectual and spiritual needs — for meaning and belonging, and for relationships with both human communities and nature. Because living systems can only survive if they are simultaneously individually healthy and members of healthy communities and ecosystems (i.e. healthy parts of healthy wholes), it is accurate to say that the essential needs of all biological and social systems are for health and wholeness.[6]

It is important to remember that sustainable does not mean fixed or static, since open systems are constantly adjusting in response to their environments. However, every living system is only able to operate and change within a limited range. Because all systems have internal structures and external boundaries, failure occurs when the parameters of the system are exceeded (e.g. if there is not enough water or if there is too much heat). Open living systems are sustainable as long as they are sufficiently healthy to manage and adapt to change, either by adjusting in order to continue functioning within existing parameters, or by evolving in order to function within new parameters.[7] They become unsustainable at the point when further changes will cause system failure.

The sustainability of ecosystems and social systems can be predicted using objective data alone once we discover the critical thresholds at which system failure occurs. Examples of such data include whether populations of keystone species are stable or declining or if supplies of critical societal resources are increasing or decreasing. For this reason we can say that ultimately sustainability is not by itself a moral question, but a question of functionality within parameters. However, this does not mean that sustainability is a separate issue from ethics. Because worldviews and values motivate and organize social structures and economic processes, the economic and environmental outcomes of human activities are culturally determined. The present global system is unsustainable because it is driven by unethical, destructive values that do not respect the lives and well-being of either humans or other species. This means that we cannot create a sustainable world without developing ethical, constructive values.

How we spend out time, energy and money involves moral choices — what we do, what we buy, what we say and how we vote. When our actions support competitive, consumer values we are choosing a path that leads to the extinction of most life on Earth and the destruction of human civilizations. When our actions support cooperative, conserver values we are choosing a path that leads to the preservation of life and the future of humanity.

What are biophysical needs?

Biophysical systems are made up of many interacting and interdependent components. Because ecosystems continually process energy and information from their surrounding environments, they must constantly adjust (equilibrate) in response to changing conditions. Individual species risk extinction when they lose critical habitat and genetic diversity, and with these the ability to adapt to environmental stressors. Not only are healthy species genetically diverse, but healthy ecosystems are composed of a wide variety of interdependent species. Diversity increases a system's resilience, which is its ability to absorb shocks and adapt to changes. Larger and more complex ecosystems that are not tightly entwined are more resilient than smaller and simpler systems. If the existence of an entire system is dependent on the health of a few species, it may easily collapse when those

Figure 2: BEST Futures

Ecosystems require functional integrity to survive

Without wholeness and health ecosystems can be irreversibly damaged.

species are stressed. Systemic resilience is lost with the destruction of ecosystem biodiversity, increasing the likelihood of widespread biophysical collapse.

Human-induced stresses are threatening to degrade many major ecosystems beyond threshold points — the points at which additional degradation will trigger irreversible collapse. Because the long-term viability of human societies is utterly dependent on the long-term viability of the biophysical systems that support them, the long-term sustainability of human systems requires the maintenance and restoration of ecosystem integrity, biodiversity and resilience.

What are human needs?

Humans are the only species who use symbols to interpret their environments. We live in societal systems because we cannot survive without learning to use language and tools, and because we cannot physically or emotionally survive on our own.

Language, culture and social institutions co-evolved with the human brain. Our triune brains represent three stages of evolution: reptilian, mammalian and human.[9] Humans have three corresponding types of needs: physical needs, emotional needs

Figure 3: BEST Futures [8]

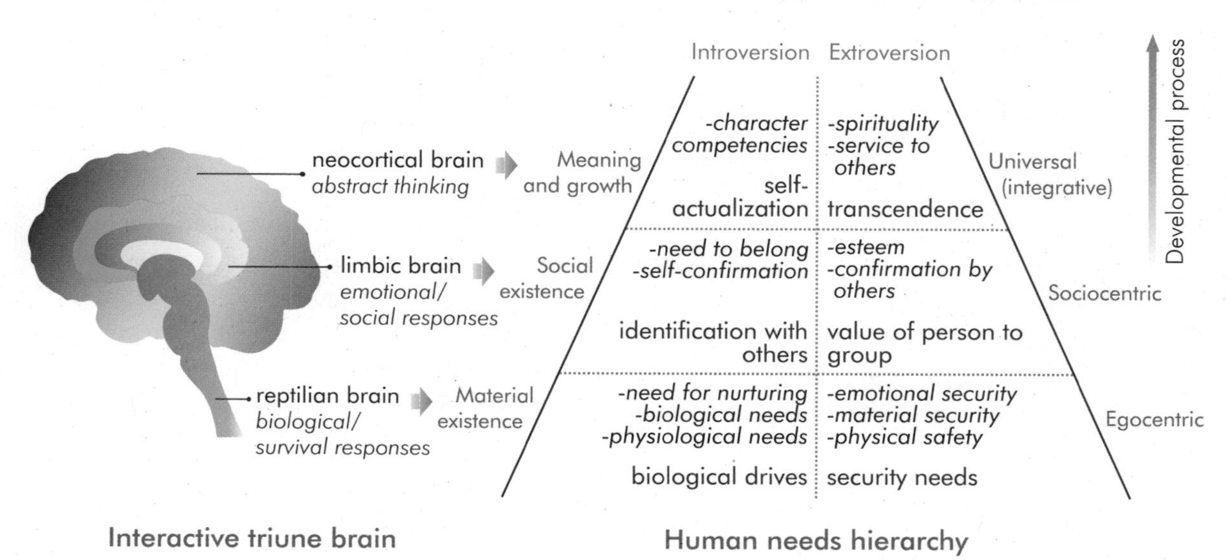

and needs for meaning and growth. These needs can also be described as individual existence needs, social existence needs and developmental needs.

In diagrams, material needs are typically placed at the bottom and spiritual needs at the top.[10] This is because people will normally take care of their needs for food, shelter and safety first, and only pursue intellectual and spiritual interests after other needs have been met.[11] However, this does not mean that people's needs for meaning or relating are less important than their physical or survival needs. Parents sacrifice themselves for their children, soldiers sacrifice themselves for their community or cause and people kill themselves when depressed. One of the harshest punishments someone can receive is being locked in solitary confinement. Humans only feel healthy and happy (fully functional), when all three levels of needs — material, social and creative — are being met.[12]

This is why international surveys indicate that rising incomes increase happiness the most in poor countries. Increasing income has much less influence on happiness once basic needs (for food, shelter, healthcare, education, economic security) are satisfied.[13] Passive consumption alone can not meet all our needs: if we wish to be healthy and happy we must be physically, emotionally, intellectually and spiritually active. And these needs can only be met through socializing, exercising and doing meaningful work.

The key to establishing a sustainable global system is realizing that it is not possible to meet all our needs through limitless consumption. Once real material needs are met, we will gain more happiness from improving the quality of our lives than from increasing the quantity of our possessions.

Genuine Progress Indicators

Nowadays, people know the price of everything and the value of nothing.

— Oscar Wilde,
playwright and poet
(1854-1900)[14]

The consumer society values individual material needs above all other needs. It is only interested in emotional, intellectual and spiritual matters to the extent that they can be owned, packaged and marketed as commodities or used to sell products. These values lead to forests being destroyed because it is more profitable to sell lumber than to preserve biodiversity, and wars being fought because it is more profitable to make weapons than to make peace.[15]

The consumer society is the product of the mechanistic and materialistic worldview of the industrial system, which believes that the universe is made up of discrete objects rather than interrelated systems. From this perspective, humans, animals, plants, mountains and rivers have no connection to each other and no

The real bottom line is survival

In practice orthodox economics treats social and environmental costs as *externalities*.
This means that values such as health and well-being are not included in economic modeling, planning or accounting.

Will Economists Realize that Fresh Air and Water Have Value?
Money is not the real bottom line.

Figure 4: BEST Futures

---

intrinsic value — they only exist to be manipulated and used for personal gain. This worldview creates an economic system in which natural capital is converted into manufactured and financial capital without taking into account environmental or social costs.

This incomplete accounting produces a global *race to the bottom*: because companies are not responsible for the health of their employees or the environment, they are under constant financial pressure to move their factories to locations where they face the fewest taxes, the lowest wages and the weakest environmental, health and safety regulations.[16] And since there are no common global standards, countries are under constant pressure to relax their standards in order to attract investment and remain economically competitive.

Of course, governments and businesses do not completely ignore problems like illness, crime or pollution. However, because political and economic decision-makers are focused on making short-term profits, they tend to neglect social and environmental issues unless they either represent commercial opportunities or have become unavoidable problems.[17] So resources are wasted and misallocated throughout the entire economy: for example, more money is spent on developing drugs than on disease prevention; more on marketing junk food than on developing healthy lifestyles; more on incarcerating criminals than on creating functional families and communities; more on waste disposal than on waste reduction; more on promoting commercial fertilizers and pesticides than on teaching organic farming; more on cleaning up toxic waste than on testing and regulating dangerous chemicals.[18]

National economies are further distorted by the national accounting system, which counts all financial transactions as economically productive and disregards all non-financial activities.[19] The result is that the costs of destructive or non-productive activities like war, crime, illness and speculation all add value to national economies, while unpaid productive activities such as parenting, housework or

Figure 5: BEST Futures

The difference between GDP and GPI accounting

With GDP accounting

= +

The economic activity created by accidents adds value.

= 0

Parenting and unpaid agricultural work have no value.

With GPI accounting

= +

All productive work adds value.

= –

Pollution and accidents reduce value.

volunteer work are not considered to be part of the economy and therefore have no value. This *Alice in Wonderland* approach to bookkeeping calculated that the Exxon Valdez oil spill added value to the US economy because it generated $5 billion dollars worth of work in clean-up costs. Even better (from the standpoint of traditional economists), the Iraq war added hundreds of billions of dollars to the US Gross Domestic Product. In the current global economy, the fact that both pollution and war destroy wealth rather than create it is irrelevant.

In order for a global economy to be sustainable, it must be able to accurately determine the real costs and benefits of human and biophysical activities. This means that accounting must be based on Genuine Progress Indicators (GPI) as well as Gross Domestic Product (GDP) or Gross National Product (GNP). While GDP and GNP only track monetary activities, GPI uses a triple bottom line (environmental, economic and social) that allows it to evaluate personal and public well-being. Mark Anielski explains how individuals, communities, companies and nations can calculate their real wealth in *The Economics of Happiness*.[20]

We will only have accurate economic accounting if we first determine essential human and biophysical needs and then assess the impacts of various activities on these needs. To do this we will need to value not only financial transactions, but also non-financial economic activities such as the costs of pollution, illness and crime as well as the contributions of the goods and services produced by ecosystems, unpaid work at home and community volunteers. We will then need to determine the extent to which each type of activity adds to or subtracts from environmental and social health and wholeness.

Much of this work has already been done by ecologists, biologists, medical researchers, sociologists and other scientists. We know (in general terms, at least) the carrying capacity of our planet. We know the various types of habitat that different species need in order to survive and the functional parameters of these habitats (climate, size, resources, predators, biological and chemical composition of the soil or water). We also know the essential needs of

---

Our gross national product ... counts air pollution and cigarette advertising, and ambulances to clear our highways of carnage. It counts special locks for our doors and the jails for the people who break them. It counts the destruction of the redwood and the loss of our natural wonder in chaotic sprawl. It counts napalm and counts nuclear warheads and armored cars for the police who fight the riots in our cities ....

Yet the gross national product does not allow for the health of our children, the quality of their education or the joy of their play. It does not include the beauty of our poetry or the strength of our marriages; the intelligence of our public debate or the integrity of our public officials. It measures neither our wit nor our courage, neither our wisdom nor our learning, neither our compassion nor our devotion to our country, it measures everything in short, except that which makes life worthwhile. And it can tell us everything about America except why we are proud that we are Americans.

— Robert F. Kennedy, US Senator (1925-1968)[21]

---

humans — what we need to be physically, psychologically and socially healthy and what makes us sick or dysfunctional as individuals or groups.[22] We know the causes of wars and how to prevent them.

We already have enough information to create a healthy, happy and sustainable global system: what we lack are caring and sustainable worldviews, values and institutions.

The requirements of a sustainable societal system

Living biological and social systems can only survive as long as they are able to interact with their external environments and acquire the energy and other inputs

Figure 6: BEST Futures

Individual needs and societal structures

Human needs

meaning and growth, social existence, material existence

relate to social structures

Institutions	Functions	
Religion/worldview	Meaning and direction	Meaning
Culture/aesthetics	Symbolic communication	
Government	Boundaries/regulation	Continuity
Education	Transmission of skills	
Family/community	Reproduction/integration	Basic structures & processes
Economy	Production	
Science/technology	Environmental control	

Universal Culture Pattern

Living Biological Systems
Human biological systems are composed of interdependent organic sub-systems such as the skeletal system, the digestive system, the nervous system and the reproductive system.

Living Social Systems
Living social systems are made up of interdependent social sub-systems. The basic structure of all societies is called the Universal Culture Pattern because societal systems require similar institutions in order to function.

they need to maintain and reproduce themselves. In order to carry out these tasks, all living systems require congruent, functional and environmentally relevant structures.

Because living social systems are human organizations, social structures serve individual needs as well as societal needs. While different types of societal systems have different institutions, every society requires the same basic set of institutions to carry out essential functions such as organizing economic production and transmitting skills from one generation to another. Although more complex societies have larger and more differentiated institutions, every society has some form of worldview, culture, government, kinship-based social organizations, educational system and economy.[23]

Human civilization will only be able to end scarcity and international competition over limited resources when it is able to satisfy the minimal physical, emotional, mental and spiritual needs of all the humans on the planet. This cannot be done by an industrial system designed for constant quantitative physical growth — a system that increases the quantity of things without taking into account environmental and social costs. Satisfying the full range of human needs will require replacing this dysfunctional system with a holistic system designed to improve the quality of people's lives. And because basic human material, social and spiritual needs cannot be achieved in a toxic and constantly degrading environment, a system that meets human needs must also be designed to meet the essential needs of the planet's biophysical systems and protect its millions of life forms.[24]

Since the real needs of both ecosystems and humans are for health and wholeness, the cultural requirements of a sustainable global societal system are similar to the biological requirements of a sustainable ecosystem: values and institutions must foster wholeness, interdependence, diversity and resilience. Global sustainability can be achieved if the world system is redesigned to meet these real needs. However, this will require a profound paradigm shift from viewing humans as outside of nature, to viewing our species as inside of nature. Instead of seeing reality as being composed of independent objects, we will have to realize that reality is composed of interdependent systems.

We will only be able to meet all our needs with a culture and institutions that support constructive behaviors. While current social structures facilitate competition, inequality, injustice and conflict, sustainable structures will need to facilitate cooperation, equality, justice and peaceful conflict resolution. The consumer society manipulates and markets our deepest feelings and values. It perpetuates fear and alienation and pits the individual good against the common good. We feel overwhelmed and grow numb to our own needs and to the sufferings of others. For these reasons we will only be able to heal our sick planet with healthier morals and a higher common purpose.

The model for this approach is natural ethics: relationships based on reciprocity. An example of this is the relationship between bees and flowering plants: the flowers feed the bees, and in return the bees pollinate the flowers. The different species help each other because they coevolved and need each other — without flowers there would be no pollinators, and without pollinators there would be no flowers. Once we realize that our human species is only one species among millions and that we need all the other species to maintain life on earth, then we can begin to develop ethical, mutually beneficial and sustainable relationships with the natural world and with each other.

The basic requirements of a sustainable global system are fairly simple. Instead of destroying our environment we have to restore it so that we leave our children, grandchildren and great-grandchildren a world that they can live in. This means that all economic activity must take place within the ecological carrying capacity of our planet. And in order to stop destroying the environment, the dominant global culture and economy must meet the needs of humans and ecosystems for health and wholeness. This can be done if we make a paradigm shift from the current mechanistic worldview to a systems worldview.

Sustainable development is not about maintaining quantitative growth, but about qualitative development. It is not about more of the same, because more of the same is unsustainable. It is about healing ourselves and our world. This will require the complete transformation of our society — from unethical to ethical; from paranoid to peaceful; from serving false greeds to meeting real needs.

7 Transformative Material Technologies

New scientific paradigms

THE COMPLICATED PROCESS OF SOCIETAL EVOLUTION usually starts with scientific or technological innovations. The domestication of plants and animals by our Stone Age ancestors triggered one of the biggest societal transformations in the history of our species. The new technologies were widely adopted because herding animals and cultivating crops were more reliable ways of producing food than hunting and gathering. However, it was not possible to incorporate herding and cultivating within the existing economy. In order to tend gardens nomadic bands had to settle down and establish permanent villages.

This enormous change in the way people lived marked the transition from the Old Stone Age to the New Stone Age. Herder-cultivator economies were able to support larger and denser populations. As communities grew in size, knowledge accumulated, new technologies appeared, economies became more diverse and productive and more complex cultural systems developed with new values, new forms of artistic expression and new social structures.[1] The domestication of plants and animals had far more than economic impacts: it began a process that transformed the entire societal system and our species' experience of reality.

Over the course of human history four major types of societal systems have evolved in response to changing environmental and social needs. We are now in the midst of another period of evolutionary transformation. Qualitatively different innovations are emerging that have the potential to completely change our relationships with our environments and with each other. Although technological changes alone cannot create a new type of societal system — congruent cultural and social structural changes are also needed — scientific and technological innovations are again driving the process of constructive change.

Technological changes can be either incremental or transformative. For example, developing a better radio is an incremental change that improves an existing system of communication, while inventing television is a transformative change that creates a new type of system. Many incremental changes often produce transformative change when either a tipping point is reached or an innovative breakthrough is achieved. Innovations that eventually overturn the dominant technology are called disruptive innovations or disruptive technologies. The radically different designs of disruptive innovations introduce new paradigms (patterns of organization).

The invention of the internal-combustion engine produced a paradigm change in transportation and production in the industrial world. Within a few decades the horse and buggy was history and carriage makers had either switched to making cars or gone out of business. The technology was so powerful that it created new types of technological systems (e.g. generators and planes) and new forms of social and economic organization (e.g. suburbs and freeways). However, this paradigm change was not in opposition to the dominant industrial worldview. The internal-combustion engine was introduced at a time when key sectors of the economy were still at a pre-industrial stage and was developed because it supported the expansion and consolidation of the industrial system. When it replaced draft animals with machines, it replaced Agrarian Age technology with more productive (and socially and economically congruent) Industrial Age technology.

A society will only begin the process of evolutionary transformation when disruptive social and material innovations start to challenge existing organizational patterns with more environmentally relevant and functional designs. These innovations build on each other and contribute to the emergence of a different, more powerful explanation of reality (of how the universe is organized). The development of a new worldview is central to societal evolution because a worldview is the dominant paradigm of a society — the primary pattern responsible for organizing a society's many institutions and economic processes into a single living system. Since new paradigms take time to develop and become widely adopted, a new worldview always begins to emerge long before it becomes the dominant organizing pattern of a society.[2]

The emerging worldview

The signs of evolutionary societal change are all around us. A new way of understanding reality is emerging along with many different types of paradigm-changing technologies. Just as the Agrarian Age's theocratic worldview was challenged in the 17th century by the emergence of rationalist science, the Industrial Age's mechanistic worldview is now being challenged by the emergence of systems-based science.

The idea that God had created a universe composed of independent objects was questioned in the early 19th century by Michael Faraday, James Clerk Maxwell and others, who began to replace the concept of mechanical force with the concept of electromagnetic force fields. The idea of a fixed reality was further undermined by the evolutionary theories of Jean Baptiste Lamarck and Charles Darwin. These theories created an intellectual and moral storm. Fritjof Capra describes their impact: "The discovery of evolution in biology forced scientists to abandon the Cartesian conception of the world as a machine that had emerged fully constructed from the hands of its Creator. Instead, the universe had to be pictured as an evolving and ever-changing system ..."[3]

The new scientific worldview was consolidated in 1905 by Albert Einstein's theories of relativity and atomic phenomena, and, a few decades later, by quantum theory. Modern physics replaced the Newtonian model of fixed objects and mechanical cause-and-effect with a model of relative perspectives and simultaneous interaction. Time, space, energy and matter are now understood to be completely interrelated. In this new paradigm, the universe is viewed as an indivisible whole — a metasystem composed of dynamically interacting subsystems.

Led by the physical and life sciences, other scientific disciplines are increasingly adopting a systems view of reality. For example, ecologists have been challenging the industrial worldview with research showing the interdependence of all life on Earth; mathematicians have been modelling the dynamics of complex and chaotic systems and computer scientists have been studying the evolution of networks. More and more scientific breakthroughs are the result of systems-based, interdisciplinary work. The new integral worldview represents a profound change from seeing and believing that things exist in isolation to seeing things as inseparable from their relationships and contexts. It is the shift from being able to see the trees but not the forest to being able to see both the trees and the forest.

The promise of new technologies

While humans have always been inventive, the pace of innovation is constantly accelerating as the size and density of populations increase, people become more educated and knowledge becomes more accessible. Tens of thousands of years ago new ideas spread very slowly, as people lived in small, isolated groups and had few ways to record and exchange information. Now the pace of change is so rapid that it is difficult for researchers to keep up with the latest theories and difficult for businesses to keep up with the latest technologies.

Innovations tend to come in waves because disruptive innovations open up new technological and commercial possibilities. For example, the practical application of

Figure 1: Natural Edge Project[4]

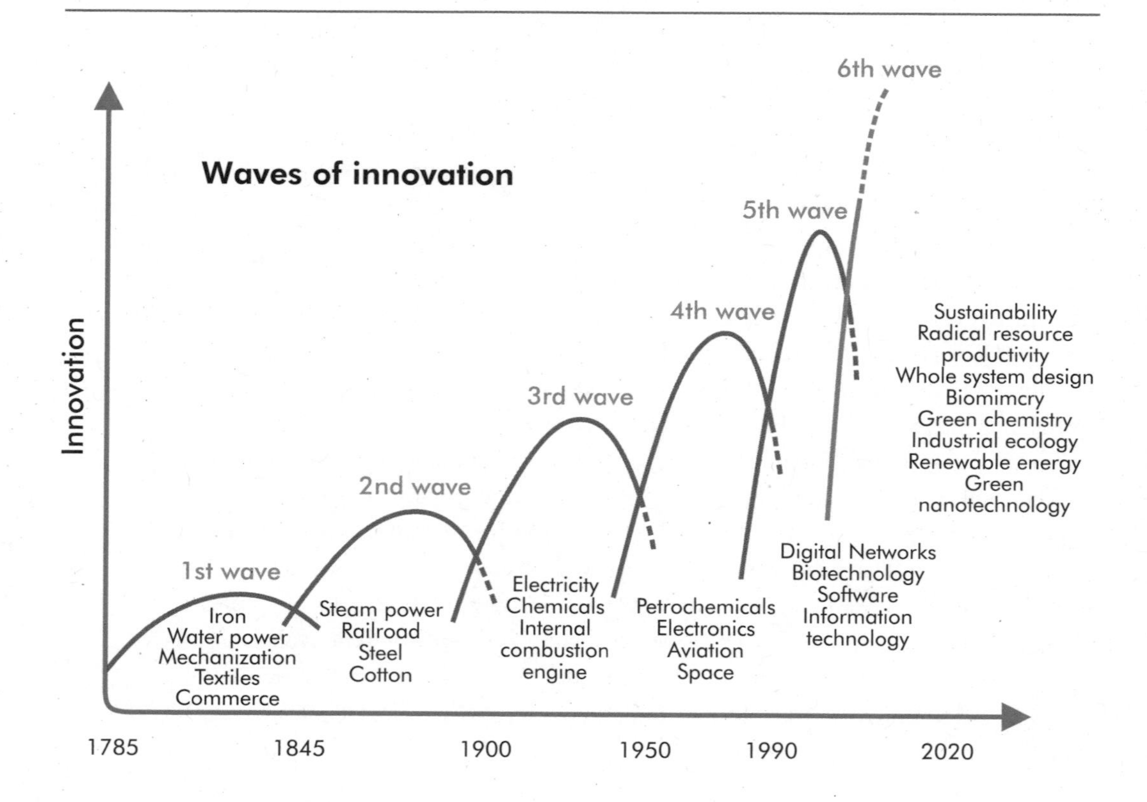

electrical theories did not just produce lights, but also electrical motors, batteries, telephones, movies and thousands of other new technologies. Innovative waves are also the result of a critical mass of technologies developing to the point where they can be combined to produce entirely new products. Telecommunication and computer technology had to exist before the Internet could be invented.

While we sometimes think of new technologies as the creations of scientific geniuses and mad inventors, before new ideas become commercial products they have to be researched, manufactured and marketed. This only happens if governments and businesses are convinced that commercial opportunities justify the large financial risks involved in developing a new technology. The Natural Edge Project has observed, "If the last wave of innovation, ICT (Information and Communications Technology), was driven by market needs such as reducing transaction costs, we believe that there is significant evidence that the next waves of innovation will be driven by the twin needs simultaneously to improve productivity whilst lightening our environmental load on the planet."[6]

There are only two possible ways to sustainably expand the global economy: by increasing the supply of renewable resources and by producing goods and services more efficiently (i.e. reducing the resources required to produce each unit of economic output). The challenges will be enormous — 10- to 20-fold eco-efficiency improvements will be needed by 2050.[7] Nevertheless, these challenges will have to be met because growing resource shortages and degrading ecosystems will leave us no choice: further expansion of resource consumption is impossible.

> The Stone Age did not end for lack of stone, and the Oil Age will end long before the world runs out of oil.
>
> — Sheik Ahmed Zaki Yamani, former Saudi Arabian oil minister[5]

Many business and government leaders are already aware of the economic advantages of developing environmentally friendly technologies. The *dot com boom* in Silicon Valley is being replaced with a *watt com* boom as venture capitalists invest in start up companies developing alternative energies.[8] On the other side of the world the Emirate of Abu Dhabi's Masdar Initiative plans to invest $200 billion in renewables in the next decade.[9] The question is not whether it will be technologically possible or economically profitable to create a sustainable economic system, but how much time it will take to completely redesign, retool and reorganize the global economy.

The obstacles to change are not technical, but social — the need to change the views, values and habits of governments, businesses and consumers. These values are reflected in the decisions being made — or not made — to develop and introduce new technologies. Although the cost of developing and introducing renewable energies will be enormous, they will not be any higher (relative to the size of the economy) than the costs of introducing steam engines, internal-combustion engines, gas and electricity. In all these cases governments accelerated technological modernization through subsidizing the development of infrastructure such as canals, railways, dams, highways and airports.

Let's have a look at some of the technological innovations that are supporting the evolution of a new type of society. By themselves they are not entirely incompatible with the views and structures of industrial civilization, but together they are beginning to build the foundations of an entirely different type of economic and social system.

Renewable energies

There is no shortage of energy in the world. The sun is pouring 80,000 billion kilowatts of energy onto the surface of the Earth every day, which is 10,000 times more energy than human beings currently use.[10] The solar energy heats the earth and the atmosphere, where it is converted into wind (and then wave) energy and powers biological photosynthesis. In addition, because the core of our planet is heated by radioactive decay, the Earth radiates geo-thermal energy. The technological challenge has been to develop technologies that can affordably capture this energy.

Many forms of renewable energy are being developed, including solar thermal power, solar photovoltaics, solar cooling, solar hot water and space heating, wind energy, wave energy, tidal energy, geothermal energy, bio-energy and small run-of-the-river hydro power. It is estimated that existing renewable technologies could supply many times the current global demand for power.[11] And hundreds of new technologies are under development.

Renewable technologies are still at an early stage of development and market acceptance. Nevertheless, wind and solar power are the fastest growing energy sources of Earth. In China, for example, by the end of 2007 40 million solar water heaters had already been installed, and wind power was growing at more than 80% per year.[12] Technological problems — such as the need for constant power output — are being progressively overcome.[13] Per unit costs will decline as the technologies mature and are mass produced.

Wind power is attracting major investments because it is already financially competitive with energy produced from fossil fuels. T. Boone Pickens, the legendary Texas oilman, decided to invest $10 billion into building the world's biggest wind farm because "I have the same feelings about wind as I had about the best oil field I ever found."[14] Other investors are gambling that the price of other renewable energies will rapidly drop. Search giant Google is investing hundreds of millions of dollars on an initiative (RE < C) designed to make renewable energy cheaper than coal.[15] Once this price is reached, consumers will have no reason to use polluting energy sources.

Most renewable energy sources would be competitive today if the environmental and health costs of fossil fuel pollution were factored into the price of energy. A project funded by the European Commission, ExternE, estimates that if electricity generators had to pay the full costs of pollution, the cost of producing gas would increase by 30%, and the cost of producing coal or oil would double. If direct and indirect subsidies to fossil fuels and nuclear power were also removed, there would be no need to support the development of renewable electricity.[16]

It is already possible to provide most of the world with dependable sources of renewable energy. For example, Stirling Energy Systems is building a solar thermal generating station in California's Mojave Desert that will provide 500 MW of power 24/7.[17] The Trans-Mediterranean Renewable Energy Cooperation (TREC) proposes building a series of similar solar power generating plants in North African and Middle Eastern deserts. These could provide those regions with clean energy and water as well as meeting all the power needs of Europe for 5-8 cents per kilowatt hour.[18]

Another European proposal is to build a wind-powered electricity supergrid stretching from Siberia to Iceland. Because a grid of this size would even out variations in wind speed, it would provide reliable low-cost power to 1.1 billion people in 50 countries while reducing Europe's carbon dioxide emissions by a quarter. Gregor Czisch, an energy systems expert at the University of Kassel, said, "We have the technical abilities to build such a supergrid in three to five years. We just need to commit to this big long-term strategy."[19] T. Boone Pickens has suggested that the United States could break free of its dependence on Middle Eastern oil by building wind farms along a north-south axis on the Great Plains, and by building solar generating stations on an east-west axis from Texas to California.[20]

The global economy will also need to replace the oil and natural gas used in boats, vehicles and aircraft with renewable energies. Since it makes neither environmental nor economic sense to produce ethanol from food grains such as corn, in the near future bio-fuels will be produced in ponds and on waste land from algae and other fast-growing plants.[21] Another major source of fuel is likely to be hydrogen gas, which produces no pollutants. Hydrogen can power either internal-combustion engines or more efficient fuel cells. However, whereas oil is a primary source of energy, hydrogen is similar to electricity in that it is an energy carrier that must first be produced from another energy source before it can be used. A cheap method for producing hydrogen — and for storing solar energy — is now being developed by Daniel Nocera, Matthew Kanan and other researchers at Massachusetts Institute of Technology.[22] The first country to develop a clean and renewable hydrogen-electric economy is likely to be Iceland, where geothermal energy is used to produce hydrogen.[23]

Conservation

Renewable energies are examples of technologies which increase the supply of renewable resources on the planet while simultaneously reducing both the consumption of non-renewable resources and pollution. But increasing the supply of renewable resources will not be enough to prevent environmental and economic collapse: the global economy also needs to reduce resource consumption. The easiest and cheapest way to lower consumption is through conservation. This can be done in three ways: through reducing waste, through reducing the demand for goods and through reducing the resources needed to produce each unit of goods or services.[24]

The best way to reduce resource consumption is to make fewer products. Demand can be reduced through individuals changing their values and adopting conserver rather than consumer lifestyles. This means purchasing only what is needed rather than everything that is desired. Demand can also be reduced through people sharing resources (e.g. using public transport instead of private cars), through rewarding companies for providing the best services rather than for selling the most products,[25] through making goods last longer and through planning to maximize the quality of life while minimizing wasted resources (e.g. locating employment and shopping close to housing to encourage walking and reduce the need for commuting).

Products can also be made much more efficiently. For example, while 19% of global electricity is being used for lighting, half of the lightbulbs being used in homes around the world are wasting 95% of the energy they use. Incandescent bulbs, which only turn 5% of the energy they use into light, can be replaced with compact fluorescent lamps which last longer and use much less electricity. Even better lights are now being developed. Light-emitting diodes (LEDs) will be more efficient because they are much smaller and focus light where it is being used.[26] Of course resource conservation is possible everywhere, not just in the use of electricity. For example, industries are increasingly reducing waste heat through combined heat and power generation (CHP), and architects are increasingly designing *green buildings* that use the sun to provide heat, hot water and cool air. Dana Christensen, associate director at the Oak Ridge National Laboratory in Tennessee, said, "I believe with technology pretty much available today or in the very near short term, if we could move those fully into the market, we could get a 30 to 40 percent reduction in greenhouse gases."[27] While using resources more efficiently reduces environmental damage, the main reason why businesses and consumers are buying more efficient products is to reduce costs. Green buildings not only decrease operating costs, but improve staff health and morale and increase productivity.[28] In *Green to Gold*, a practical guide for business leaders, Daniel Esty and Andrew Winston show how cutting edge companies are now using environmental strategies to innovate, create value, save money and build competitive advantage.[29]

More efficient and sustainable technologies do not have to be either high-tech or expensive. For example, traditional technologies can be used to restore damaged ecosystems and improve agricultural output.[30] Organic farming techniques can often be used to improve agricultural production with fewer inputs of water and oil-based fertilizers and pesticides. Envirofit International has developed cheap, durable, attractive and efficient stoves which cook food with far less wood and fewer emissions — the Shell Foundation is helping to send 10 million of these to India, Africa and Latin America.[31]

Nanotechnologies, biotechnologies and biomimicry

Over time, larger and more complex types of human societies have evolved. Each evolutionary leap has been made possible by the invention of new technologies (and supporting social institutions) capable of processing increasing quantities of energy and resources. The problem facing humanity today is that while Agrarian Age and Industrial Age civilizations developed through discovering new ways to increase resource consumption, this model of economic and social development is now unsustainable.

Human economies have expanded to the point where further economic growth is only possible if it is decoupled from the increased consumption of resources. This can be done through shifting from the increased use of raw materials and non-renewable energies to the more efficient use of resources; through shifting from the creation of larger and more powerful products to the creation of smaller and more effective products; and through shifting from centralized production to distributed production.

> ... [N]ature, imaginative by necessity, has already solved many of the problems we are grappling with. Animals, plants, and microbes are the consummate engineers. They have found what works, what is appropriate, and most important, what lasts here on Earth .... After 3.8 billion years of research and development, failures are fossils, and what surrounds us is the secret to survival.
>
> — Janine Benyus, author and innovation consultant[32]

The model for a sustainable economy is the natural world. Plant and animals efficiently process matter and energy from their environments at low temperatures. Every living biological system (including our bodies) is constantly disassembling and reassembling molecules into the thousands of complex chemicals needed to build and maintain life. In contrast, industrial processes require big facilities, high temperatures and large quantities of raw materials.

Stephan Gillett, an expert in nanotechnologies, said that human civilizations have not yet developed past a primitive *Paleotechnical Era*: "Conventional technology is depleting its resource base at an accelerating rate and creating an increasingly unliveable mess in doing so .... In fact, present technology is grossly inefficient and gratuitously dirty, and this results from an overwhelming reliance on heat — the 'Promethean paradigm'... Molecular nanotechnology [MNT] will allow us to supplant the Promethean paradigm and move beyond the Paleotechnical .... The five-millennium era of locating anomalous deposits to 'dig up and cook' for metals extraction is coming to its close, probably over the next few decades and certainly within the next century."[33]

Nanotechnology is a new field that enables scientists to control matter on a scale smaller than one micrometer and to construct materials at a molecular level. The combination of nanotechnology with biomimicry (copying natural processes), will allow better materials to be created with less energy, less materials and less waste.[34] For example, highly specific catalysts can be created that will copy the function of biological enzymes and generate only the desired product. Some of the products currently under development include extremely strong and lightweight carbon materials, paper batteries and windows that passively manage changing outdoor temperatures (and never need to be washed). Nanosolar, a California company, has invented a cheap way to produce solar power — a mixture of nanoparticles and metals that can be sprayed on to flexible foil materials and other surfaces such as roof tiles. The potential markets for these products are enormous

The information age is about smarter and smaller

As societal systems become more complex they require more energy

	kcal/person/day
Early hominids	2,000
Hunter-gatherer societies	4,000
Herder-cultivator societies	12,000
Agricultural societies	24,000
Early industrial England (ca. 1850)	70,000
Modern industrial US (ca. 1970)	230,000
Information societies (ca. 2050?)	?

In order to be sustainable, the information age must utilize energy more efficiently

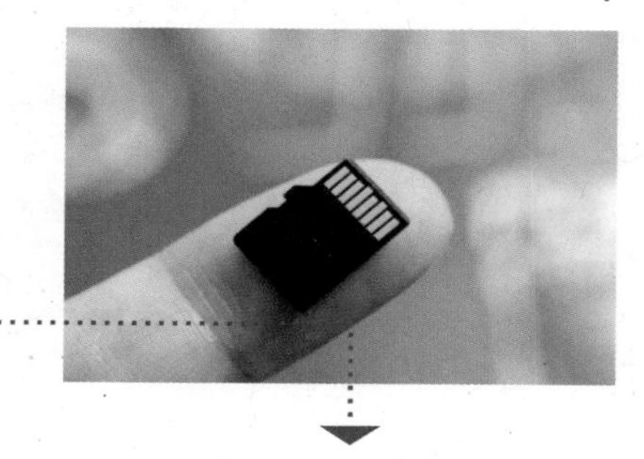

Industrial age = constant expansion / big and bureaucratic

Information age = increasing efficiency / small and smart

Figure 2: BEST Futures [35]

— before their first manufacturing facility was even completed, production was sold out a year in advance.[36]

Research in biotechnologies is also rapidly advancing. Scientists are not only modifying the genetic codes of existing plants and animals, they are also attempting to create new forms of life. Their goal is not simply to create crops that are more resistant to pests and droughts, but also to create new types of organisms that will act as living factories. These will be designed to make drugs and other complex chemicals, break down toxic chemicals, fabricate plastics and so on.[37]

The development of artificial organisms represents a complete paradigm shift in the way humans interact with the world. It raises important ethical and safety issues that must be addressed immediately — we know from our experience with computer viruses that it is only a matter of time before crackpots (and terrorists) acquire the technology and use it to create and release deadly new organic viruses. Nevertheless, there is no doubt that new biotechnologies are developing powerful tools that have the potential to help our species mimic organic processes and create extremely efficient, pollution-free, sustainable technologies.

Renewable energies, nanotechnologies and biotechnologies will allow us to increase economic growth while reducing material consumption to sustainable levels.[38] However, reducing our carbon emissions will not be enough to prevent runaway global warming because critical climate change tipping points were passed when average global temperatures rose above 0.9°F (0.5°C). If we are to prevent catastrophe we need to redesign the entire global economy so that it not only stops emitting more greenhouse gases, but begins to recapture carbon from the atmosphere. This is not as hard as it sounds — we just have to use and imitate natural processes. Cultivating fast-growing crops such as algae, planting billions of trees and developing agricultural

Figure 3: C. Blake [39]

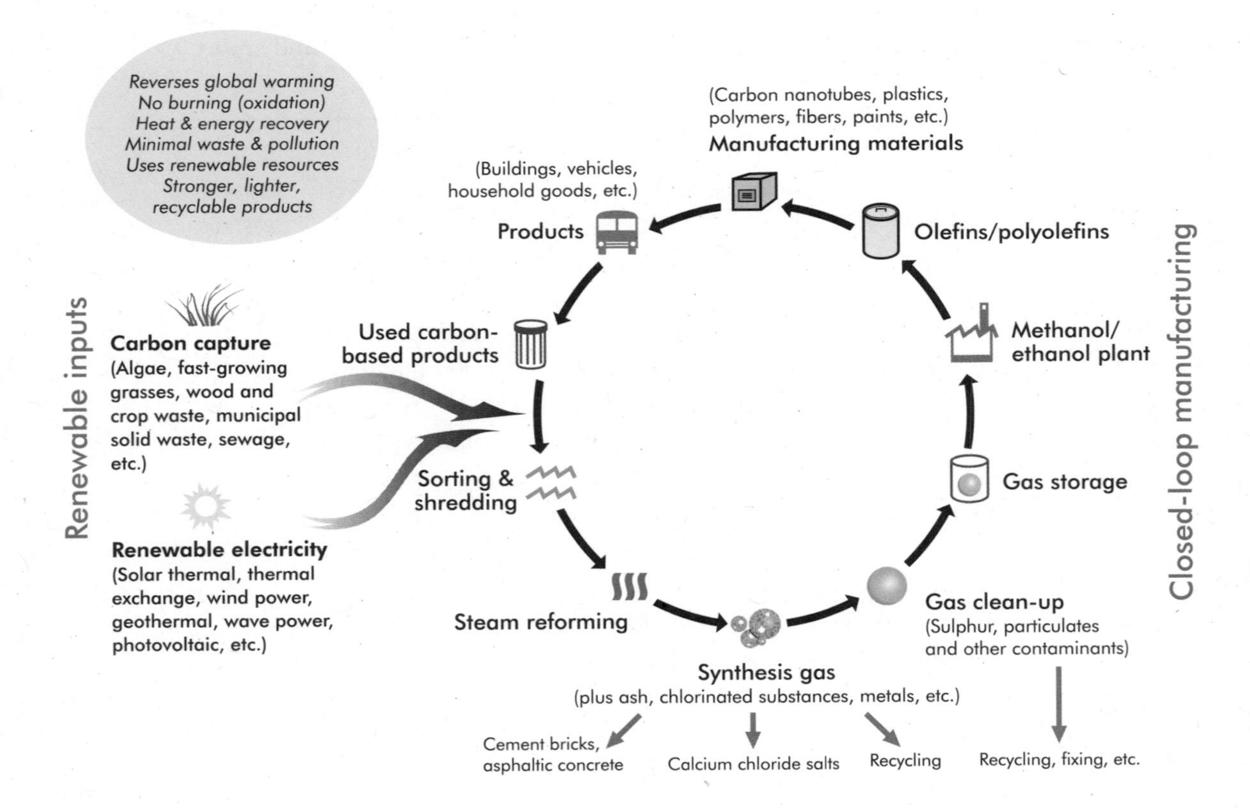

practices that increase soil carbon can take carbon dioxide out of the air. At the same time we can design manufacturing processes that do not waste carbon through burning it and releasing it into the environment as a pollutant, but instead use carbon stored in plants, coal, gas and oil to make fully recyclable products.

This technology already exists. Renewable energy can be used to break down organic wastes and coal into synthetic gases which can then be used either as fuels or as industrial feedstocks.[40] These feedstocks can in turn be used to manufacture plastics and other products (like carbon nanotubes) from which almost anything can be made from fabrics to structural beams, from cars to buildings. Our future cities will be made out of something that we have too much of — carbon — rather than increasingly scarce supplies of concrete, asphalt and steel.

Decentralized, distributed production

The global economy is not only extremely inefficient but enormously wasteful. For example, Britain imports and exports 15,000 tons of waffles each year, and exchanges 20 tons of bottled water with Australia. Argentinean lemons are shipped to Spanish supermarkets while local lemons rot on the ground.[41] Conservation involves not only avoiding the production of waste materials, but also minimizing the need to ship raw materials and finished goods long distances. This can be done through growing foods and producing goods as closely as possible to where they will be consumed. Local production can be encouraged through policies that add pollution taxes to the costs of transporting products.

Computer controlled manufacturing processes now allow small numbers of flexible machines to produce a wide range of different products. Because rapid and flexible manufacturing greatly reduces the cost of short manufacturing runs, it can decrease the need for large-scale, centralized production. In the long run, molecular nanotechnologies will be able to make almost anything from local materials, thus greatly reducing the energy and resources required for global transportation.

Distributed production can take many forms. Solar panels on houses can not only reduce reliance on the grid, but also sell power to it while homeowners are at work. Hydrogen-electric cars are also being developed that combine fuel cells with electric motors. In the near future the fuel cells may be able to be plugged in to the grid during off-peak hours and used to produce the hydrogen needed to power the cars. During peak hours the electrical flow could be reversed and the cars used to provide low-cost electricity for their owners' homes. This would reduce the need to build more large power plants. "The power grid of an entire country begins to look like the Internet rather than a mainframe. In fact, if all the cars on the road in the U.S. had fuel cells, you would have five times the electrical capacity of today's installed base."[42]

Computers and the Internet

The development of renewable and distributed technologies is made possible by the existence of computers and the Internet. Computers enable individuals to access and manage complex information, while the Internet allows knowledge to be rapidly distributed. These new capabilities do not simply improve the industrial system; they transform the way people think, communicate and interact. Until now, knowledge has been a scarce resource that was only available to powerful people and experts in governments, large businesses or universities. The introduction of information and communications technology is now enabling local producers and communities to also access the latest scientific developments and the latest market information. Distributed production is becoming possible because urban elites are beginning to lose their abilities to monopolize information and knowledge.

> Information is a verb, not a noun .... Information economics, in the absence of objects, will be based more on relationship than possession.
>
> — John Perry Barlow, Executive Chair of the Electronic Frontier Foundation[43]

Like biomimicry, the Internet is based on a post-industrial systems-based paradigm. The Internet adds value through connecting individual activities with an interactive network, which allows participants to develop social networks that share information and cooperatively develop ideas. In the new paradigm individuals benefit from sharing knowledge rather than hoarding information, and from encouraging critical feedback rather than discouraging criticism. Power, value and status increasingly come through generating distributed networks rather than through building centralized hierarchies.

Because the Internet is an interactive network, it encourages users to broaden their horizons and begin see the world as interrelated. In the process of exploring the web, communicating, learning and creating, people begin to see how reality is made up of interconnected processes, relationships and communities. Although most people have not yet made the shift to a systems perspective, the existence of the Internet lays the technological foundation for the emergence of both a systems-based economy and a systems-based culture.

Whole-systems design

If we wish to dramatically increase efficiencies, we will need to redesign our economy to encourage conservation instead of waste. The goal of whole-systems design (also called lean design) is to understand the entire life cycle of economic activities in order to create more efficient processes. A systems approach requires us to stop seeing things in isolation and instead start seeing them as parts of wider interactive systems.

For example, Josette Sheeran, executive director of the UN World Food Program, believes that global hunger can be overcome if we take a broader look at the problem: "We see up to half the food lost in developing countries simply because there's no way to get it from farm gates to markets. We see virtually nonexistent agricultural markets, so there's no place for buyer and seller to meet. These are things that can be solved."[44] Robert Watson, the director of an international study on agriculture, has also warned that agriculture must be viewed as more than a single issue: "We need to consider the environmental issues of biodiversity and water; the economic issues of marketing and trade, and the social concerns of gender and culture.... Agriculture is far more than just production of food ..."[45]

To take another example, we can view houses not as static objects, but as dynamic systems that are that are constantly exchanging energy and other resources with wider environments. Because traditional architects view houses as relatively isolated structures, they tend to focus on factors such as the desired size and layout, the cost of the land and building materials. This approach ignores the long-term costs of the resources used to operate and maintain the buildings, as well as social and environmental costs.[46] In contrast, architects using whole-systems design start by examining how houses will interact with their physical and social environments. Because they are designing houses to be supportive and healthy parts of wider communities, they take additional design issues into account, such as the resources used in the construction and maintenance of the houses, and how their design and location will contribute to social and personal health.

When we apply whole-systems design to housing, we see the need to change not only the way we build houses, but also the way we design cities and the global economy.[47] As resources become increasingly scarce in the future, cities will need to produce much of their own power, heat and food and recycle most of their water.[48] Designing *green* cities involves creating networks of small communities where people can work, shop and play within walking distance of their homes. This approach not only produces economic and environmental benefits, but also improves the quality of people's lives through improving physical and mental health, reducing crime and creating more lively and attractive neighborhoods.

Green buildings and green cities are not futuristic fantasies — they are being built now. However, they face the same problems as other developing, paradigm changing technologies: the initial costs of new products are high; designers and builders are unfamiliar with the concepts and lack skills; consumers are not aware of the new products and government regulations and subsidies are generally not supportive.

New technologies face multiple challenges: not only technical and financial, but also cultural. It's not enough to have a good idea — an idea cannot be developed without financing. But institutions are reluctant to finance a product for which there is no market. However, without a developed and proven product, it is difficult to develop market demand since most people — including most scientists, government officials, business executives and consumers — are familiar with the proven older technology and worry that the innovation either won't work or will negatively affect existing businesses. As a result, innovators are confronted with

Figure 4:

N. Davidson[49]

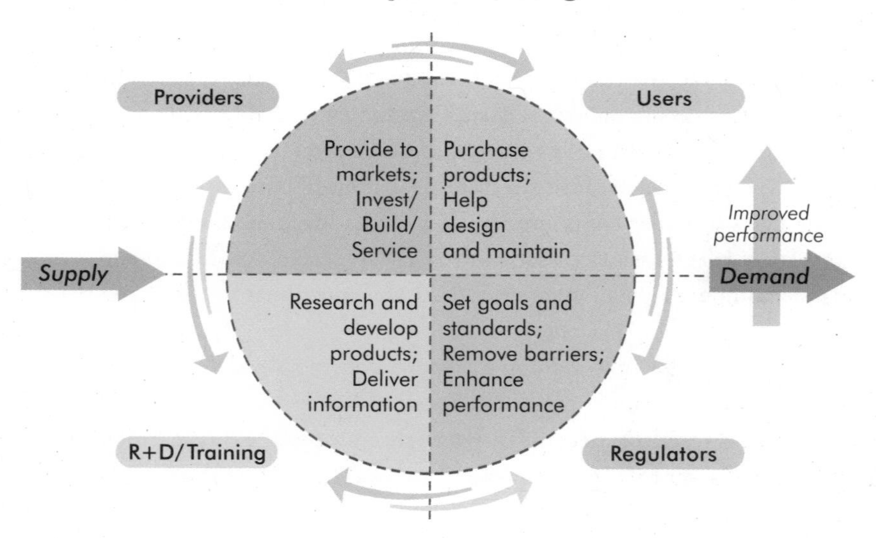

A whole systems approach to developing new technologies involves all stakeholders from the outset.

the chicken or egg dilemma: everyone is willing to support the new technology, but only after someone else introduces it.

Again the solution to this problem is to use a whole-systems approach. All stakeholders participate from the outset in the design, development, introduction, maintenance and upgrading of the technology. This helps to identify needs and problems, develop solutions, and mobilize support.

Whole-systems design can be applied to every area of society. The piecemeal approach of the industrial system means that many major problems cannot be solved because they are the product of many complex environmental, economic and social interactions. For example, although studies show that Australians could prevent nearly 70% of cardiovascular disease, 33% of cancers and 60% of diabetes by adopting healthier habits, the majority of Australians are not reducing their alcohol consumption, exercising more, eating better food or losing weight (although people are smoking less). If current trends continue, 75% of adult Australians may be overweight or obese by 2020.[50]

The *epidemic* of obesity is not confined to Australia; waistlines are steadily expanding in every industrialized and industrializing country. Two main reasons why governments cannot get to grips with this problem: obesity is a product of the consumer culture and the high-pressure, sedentary, urban lifestyle that it promotes;[51] and health care systems are designed to treat illness, not to maintain health.[52] Because many interacting factors contribute to increasingly unfit and overweight populations, health campaigns that only focus on one aspect of the problem (e.g. telling people to drink less, eat better and exercise more) are not very effective.[53] Complex social issues can only be solved with a whole-systems approach that simultaneously addresses all the factors causing problems.

Ecodesign

Ecodesign is a systems-based approach to economics that integrates human activities with natural processes. Fritjof Capra described it this way: "Ecodesign principles reflect the principles of organization that nature has evolved to sustain the web of life .... In nature, matter cycles continually, and thus ecosystems generate no overall waste. Human businesses, by contrast, take natural resources, transform them into products plus waste, and sell the products to consumers, who discard more waste when they have used the products. The principle 'waste equals food' means that all products and materials manufactured by industry, as well as the wastes generated in manufacturing processes, must eventually provide nourishment for something new. A sustainable business organization would be embedded in an 'ecology of organizations,' in which the waste of any one organization would be a resource for another."[54]

Ecodesign copies nature's methods. Nothing is wasted, new products are made with minimal resources and no permanent toxins are created. This process is described by William McDonough and Michael Braungart in *Cradle to Cradle: Remaking the Way We Make Things.*[55] They explain how products can be designed to be completely degradable and reusable either as biological nutrients that can re-enter the environment without depositing synthetic materials or toxins, or as technological nutrients which can continually circulate within closed-loop cycles. This is a very different concept from industrial recycling, which progressively degrades (*downcycles*) resources into lower quality materials and uses.

Ecodesign is still at an early stage of development. However, the principle of using the waste of one process as food for another is being increasingly applied in practice by organizations like the Zero Emissions Research and Initiatives (ZERI). One of their ecodesign projects is the Montfort Boys' Town in Fiji. Here waste barley mash is collected from a local brewery. The mash, mixed with straw or sawdust, grows mushrooms. Enzymes from the mushrooms transform the grain into nutritious food for pigs. The waste from the pigs is then sent through a biodigester to produce methane gas for lighting and cooking. After anaerobic treatment by the biodigester, the remaining waste is further treated by aerobic processes in pools. This produces algae, which become food for adjacent fish ponds which also grow crops. Other high-protein crops are grown on dykes surrounding the fish ponds. These biodiverse plants not only produce food but also help to biologically control pests.[56] The Boys' Town is an example of efficient, sustainable farming: a wide range of high-value agricultural products are produced with minimum organic inputs on a small area of land.

The limits of market forces

Books like *Natural Capitalism* argue that there is no fundamental clash between capitalism and environmentalism, and that it is in the interests of businesses to create products and provide services in clean, efficient and sustainable ways.[57] In fact many people believe that green technologies are the best business opportunities on the planet.

However, this does not mean that the current global economy can be made sustainable. Market forces are good at producing and distributing goods at the cheapest possible price (where there is open competition), but they are not designed to take into account social and environmental goals. For example half the food grown in Britain is thrown away by homes and supermarkets,[59] and pharmaceutical companies do not develop treatments for tropical diseases because most of the people suffering from these diseases are too poor to buy medicine. Bill Gates was shocked to realize that the global system doesn't work for much of the world's population: "We had just assumed that if millions of children were dying and they could be saved, the world would make it a priority to discover and deliver the medicines to save them. But it did not."[60] He is now donating his time and money to educational and health projects.

> You can't protect [the Amazon rainforest]. There's too much money to be made tearing it down. Out here on the frontier you really see the market at work.
>
> — John Cain Carter, founder of Alianca da Terra, an NGO that promotes sustainable ranching in Brazil[58]

Because companies will go out of business if they fail to make profits, they are under constant pressure to do whatever is the most financially rewarding. Unless it is going to improve the bottom line, it is unlikely that a company will invest in conserving energy and resources or in reducing pollution. They are even less likely to contribute to social goals such as sharing resources, preserving ecosystems or feeding the hungry. The problem with the marketplace is that Adam Smith's invisible hand is unprincipled: capital goes wherever it is the most profitable, and businesses tend to support government policies that will increase profits. This is why the oil companies and the vehicle manufacturers have consistently opposed regulations mandating improved fuel economy standards. A lot more money can be made selling gas-guzzling SUVs than small, fuel-efficient cars.

Although it is in the interests of businesses to produce products more efficiently, this doesn't always translate into environmental or social benefits. Garel Rhys of the Cardiff Business School was amazed to discover that cars have been getting heavier for the last ten years even though they are being built with lighter materials. While the industry blames the weight gain on regulations requiring more safety equipment, Dr. Rhys said that car designers have seen lighter vehicles as "a chance to load up with seven speakers."[61] Adding climate control systems, electric sunroofs and remote controls doesn't improve fuel efficiency standards, but it certainly increases profit margins.

As well, because profits can only be made where there are scarcities (since price reflects the relationship between supply and demand), corporations are constantly trying to control or monopolize productive resources and intellectual knowledge, and constantly centralizing wealth in heirarchical, undemocratic institutions. The big energy corporations have resisted the introduction of renewable energies for decades, not simply because they are competition, but also because they are decentralized technologies. If people can produce their own power, how can you make money from them?

However, it would be wrong to blame everything on industry. Consumers want more gadgets. In Britain, appliances like washing machines and fridges are becoming more efficient, but these power savings are more than wiped out by the rush to buy big flat-screen TVs and other high-tech electronics. By 2020 the equivalent of 14 power stations will be needed just to power consumer electronic devices.[62] This is a problem of consumer culture: advertising drives demand, and increasing demand drives increasing production. Nowhere in the economic or cultural system is there anything to keep resource consumption within environmentally sustainable limits. Markets encourage efficiency but not sufficiency; consumption but not conservation.

A further problem with leaving conservation solely up to market forces is what is known as *Jevons paradox*. When a good is produced more efficiently, its price tends to go down. As the price goes down, more people can afford to buy it, so sales and production go up. We can see this with electronics: as they become cheaper, more and more are sold. While market forces may encourage businesses to reduce the resources used to produce each unit of goods and services, they do not encourage either businesses or consumers to reduce their total resource consumption.

Richard Sanders, an Australian economist, commented that "economic efficiency is about people getting whatever they want (so long as they have the money) at the cheapest possible price. However, while it may be rational for individuals to want a particular thing, the impact of millions of people wanting that thing may be socially and/or ecologically irrational. For example, the private car is an individually rational form of transport while public transport is socially rational... [Because] the essence of the sustainability problem is that in the aggregate, humanity is able to demand natural capital via the market at volumes far beyond the carrying capacity of the planet ... there is no way within a market context to stay within the carrying capacity of the planet."[63]

Planning for sustainable development

Ecological bankruptcy is inevitable unless the process of continual environmental degradation is reversed. This means that the first law of a sustainable economy is that all human economic activities must take place within sustainable biophysical limits and support ecological health and wholeness.

Sustainability requires a completely different approach to economics. It will not be enough to use resources more efficiently: we will have to function within sustainable limits and stop producing toxic wastes. Instead of treating the environment as external to the economy — as a place where we extract raw resources and dump waste — we will have to recognize that the economy is a subsystem of the environment.[64] This involves changing our worldview: in reality the environment is not outside and separate from the economy; the economy is inside of and a part of the environment.

It is not difficult to figure out the requirements of a sustainable economy. The economist Herman Daly has proposed three simple rules to help define the sustainable limits of resource consumption.

- Renewable resources (e.g. soil, water, forests, fish) cannot be consumed at a rate greater than the rate of regeneration.
- Nonrenewable resources (e.g. fossil fuels, high-grade mineral ores, fossil groundwater) cannot be consumed faster than the rate at which a renewable resource can be substituted for it.
- Pollutants cannot be emitted at a rate greater than the rate at which they can be recycled, absorbed or rendered harmless.[65]

The key principles of sustainable development have also been defined by many governments and organizations. Although the following declarations define the principles in slightly different ways, they share similar values and a common perspective: the Earth Charter, the 1992 Rio Declaration on Environment and Development, the CERES Principles, the UN Global Compact, Natural Capitalism Principles, Natural Step Principles, Next Industrial Revolution Principles and the Hannover Principles. For example, the United Kingdom Sustainable Development Commission proposed:

> Sustainable development should be the organizing principle of all democratic societies, underpinning all other goals, policies and processes ....
>
> All economic activity must be constrained within [ecological] limits. We have an inescapable moral responsibility to pass on to future generations a healthy and diverse environment, and critical natural capital unimpaired by economic development ....

> Sustainable economic development means "fair shares for all", ensuring that people's basic needs are properly met across the world, while securing constant improvements in the quality of people's lives through efficient, inclusive economies .... Once basic needs are met, the goal is to achieve the highest quality of life for individuals and communities, within the Earth's carrying capacity, through transparent, properly regulated markets which promote both social equity and personal prosperity.
>
> Sustainable development requires that we make explicit the cost of pollution and inefficient resource use, and reflect those in the prices we pay for all products and services ....
>
> There is no one blueprint for delivering sustainable development. It requires different strategies in different societies. But all strategies will depend upon effective, participative systems of governance and institutions, engaging the interest, creativity and energy of all citizens ....
>
> Society needs to ensure that there is full evaluation of potentially damaging activities so as to avoid or minimize risks. Where there are threats of serious or irreversible damage to the environment or human health, the lack of full scientific certainty should not be used as a reason to delay taking cost-effective action to prevent or minimize such damage.[66]

Flipping the paradigm

Although our species knows what has to be done to avoid environmental disaster and although we already have the technological capacity to create a sustainable global economy, sustainable development is not yet the *organizing principle* of any country, much less the world system. This is because the global economy is organized around quantitative growth rather than qualitative development, and around competition rather than cooperation.

If we wish to get the most out of a limited sustainable supply of natural resources, it will not be enough to make increasingly efficient goods; we will have to also make extremely durable goods and share them. The market is not focused on cooperative economics (e.g. the development of public transportation rather than cars) or on the development of more durable products (e.g. appliances that last 50 years or more) because promoting conservation is less profitable than promoting consumption. The market is designed to promote conspicuous consumption, and conspicuous consumption is based on competition over the private ownership of material goods.

While competition is a natural aspect of being human — for example we love competitive sports and games — cooperation is also a natural part of our makeup. No human family, organization or society could exist without cooperation. Every house, street and machine we see is the product of a cooperative effort. Without the love and care of our parents and the support of our societies, none of us would be alive. In reality all of us combine private enterprise and socialism in our daily lives — almost all of our relationships with our families and friends are based on giving and sharing, while almost all of our relationships with strangers are based on selling or exchanging. No one charges their children for breakfast, lunch and dinner; the most die-hard capitalists save up their wealth so that they can give it away to their children when they pass on.

A mistake of socialism has been to assume (or pretend) that people are not motivated by personal interest and competition. *What's in it for me* is almost always a factor in people's decision making. Even when people are doing things for love, they usually want to receive love or at least appreciation in return. At the same time, a mistake of capitalism has been to assume that people are primarily motivated by materialism and narrow self-interest. Most people value their families and social relationships more than money, and most people are willing to make sacrifices for the people they love.

Developing a sustainable global economy is not about replacing capitalism with socialism. Traditional, industrial capitalist and socialist models are neither useful nor relevant. We do not need to choose between competition and cooperation, but we do need to determine their appropriate relationship. The problem isn't that the economy values competition, but that it values competition over cooperation. A family where competition is more important than cooperation is a dysfunctional, unhealthy family. A football game where competition and winning is more important than having fun and playing fair is no longer a game but a fight. The problem with the global system is not that competition exists, but that national and corporate competition has become more important than our collective survival.

In order to create a sustainable system we will have to flip the paradigm from cooperation within competition, to competition within cooperation. This will mean putting elected governments in charge of economic policies instead of allowing corporate interests to determine government policies. We can then create a conserver economy where the role of businesses is primarily to provide services rather than sell disposable products. For example, businesses could rent out large appliances and vehicles instead of selling them.[67] Then it would be in their interests to make durable machines that required few repairs.

Tax what we burn — not what we earn.

— Al Gore, environmental activist and former US Vice-President[70]

Sustainability requires individual, corporate and governmental accountability and responsibility. Part of redesigning the economy is shifting financial costs from individual and corporate activities that benefit society, to activities that do environmental, economic and social harm. In *Capitalism as if the World Matters* Jonathan Porritt proposed that governments not tax the value that people add and instead tax the value that they subtract.[68] This would mean eliminating personal income taxes, company taxes and value-added taxes and instead taxing pollution, speculative currency transactions and unearned wealth.

We will only be able to prevent global collapse if global resource consumption is kept within sustainable limits. Businesses and consumers are not able to determine these limits because markets are mechanisms for determining economic prices, not ecological sustainability. It will only be possible to reduce total resource consumption through both providing clear regulatory and financial guidance to markets, and changing consumer values and behaviors.

Regulating economic activities is neither a new nor a radical idea. Because our societies recognize that not everything can be efficiently managed by market forces, governments already build and operate roads, schools and emergency services and regulate every type of business from railways to restaurants. During the Second World War democratic governments took control of their economies in order to ration and mobilize human and material resources in support of war production. Every society considers that wars and natural disasters are too important to be left to private interests and markets. With the future of humanity at stake, we now need to ensure that markets serve the interests of society and not the other way around.

The evolutionary race

The renewable and systems-based technologies that are emerging have the potential to support a sustainable planetary civilization. But we are not there yet. The industrial system is still expanding, and each day ecosystems are further degraded or irreversibly destroyed. We are caught up in a race between the forces of destruction and transformation. If our unsustainable global system can evolve into a sustainable system quickly enough, human civilization and most of the species on the planet will survive. But if we change too slowly, the world as we know it may be doomed.[69]

Technological problems are easier to solve than social problems. The worldview, values and social institutions of the industrial system support competition, consumerism, inequality and exploitation. Because the rich and powerful elites that rule the world benefit from the existing system, they tend to resist change. In this they are supported by large sectors of the population who fear that rapid and radical changes could cost them their familiar jobs and lifestyles.

Their fears are justified. The consumer society cannot be transformed into a conserver society without massive change. To take only one example, molecular nanotechnologies promise to enable products to be locally manufactured with local materials. As these new processes are developed, much of the present transportation and fabrication infrastructure will become obsolete. While the majority of people and the biosphere will benefit from these changes, they will dislocate hundreds of millions of people who are currently employed in industrial manufacturing and transportation. People must be able to see that their lives will be improved by change or they will likely resist innovations and put pressure on politicians to prevent or slow down the introduction of disruptive technologies.

It will not be possible to prevent the catastrophic collapse of nature and society by technological innovation alone. Social innovation is also required to change people's views, values and social behaviors. Fortunately, the new systems-based technologies have not been developing in isolation — a holistic culture has also been evolving.

8 Transformative Ideas and Social Movements

History is now

ALTHOUGH IT MAY SEEM AS IF SOCIETIES CHANGE VERY SLOWLY, if we compare life today with life in the past we can see not only how much times have changed but also how the rate of change is accelerating. Only 10 generations ago (circa 1750 C.E.) the world was ruled by absolute monarchies, the opinions of most people didn't matter, women were controlled by their husbands, slavery was taken for granted, almost everyone was illiterate and the majority of people on the planet led poor, hard and short lives. Most people assumed that because this was the way it had been throughout recorded history, this was the natural state of affairs and the way it would always be.

However, within 100 years the rigid heirarchical world of the Agrarian Age was under siege. Democratic revolutions had taken place in a number of Western countries, and the movement to abolish slavery was spreading. Industrialization was rapidly increasing the wealth and power of an expanding merchant class. Within another 100 years (by 1950 C.E.) the world had fundamentally changed. Industrialized countries with pluralist forms of government were the leading powers in the world. Slavery was illegal almost everywhere, women had the right to vote in many countries, colonies were becoming independent and literacy and life expectancy were rapidly rising with the spread of public education and public health.

Over the next 50 years, the human world again experienced enormous changes in views, values, institutions and living conditions. By the year 2000, the industrial system had grown into an integrated global economy. Because this system now dominates and penetrates every economy and culture, you can buy soft drinks and listen to pop music in the most remote villages on Earth, and Mount Everest is littered with trash all the way to the summit. For the first time the majority of humanity is not only literate but also exposed to a global electronic culture. Powerful international movements have also emerged in support of peace, human rights and the environment.

Not all the changes have been beneficial. Spiritual values are being increasingly replaced by consumer values, and close-knit communities with transitory lifestyles. While industrialization is increasing living standards, longevity, knowledge and personal freedom, it is also increasing alienation, mental illness, drug addiction and crime.[1] But the extent and rapidity of the changes are undeniable. Even twenty years ago it would have been impossible to imagine today's world with its Indian call centers, Russian billionaires and Chinese fashion designers.

The fact that the world is changing is not as significant as the type of changes that are taking place. Not only has the industrial system been expanding into a global system, but within it a new type of civilization has begun to emerge. We have now reached a historical juncture: the coming decades will see either the successful evolution of a sustainable planetary system or the collapse of advanced civilizations. History is not just about something that happened then to other people — history is now. Each of us lives at the point where our personal history intersects with natural and social history.[2] We are the products of the past and the creators of the future. Just as the world we now live in was created by the actions of previous generations and our past actions, the choices we make now will shape the rest of our lives and determine the type of world that our children and their children will inherit.

The process of social transformation

Societal change takes place in two areas: in material technology and in social organization. Every human society has to be able to control and modify its environment in order to meet physical needs for safety, shelter, food, clothing and transport. Societies do this with material technologies — tools and related techniques. Every society also has to organize itself and meet needs for belonging and meaning. This is done with social technologies — views, values, symbols, social structures and related

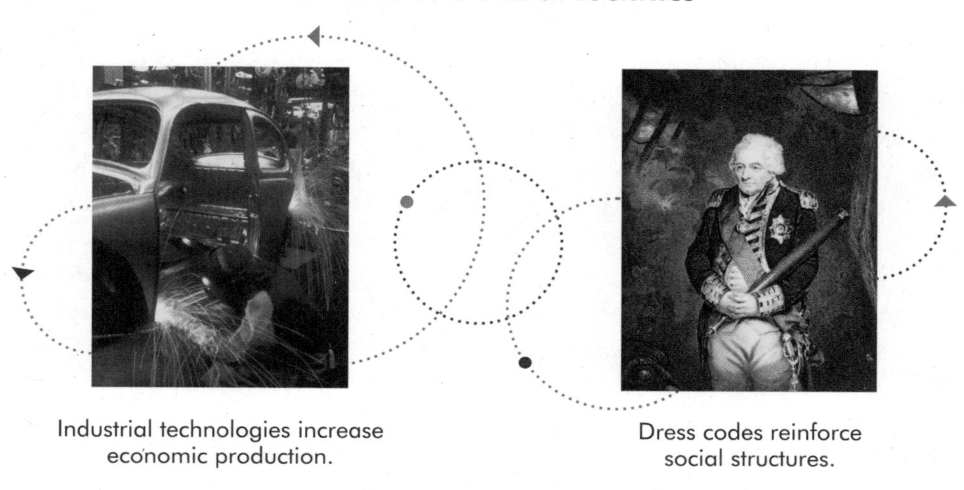

Figure 1: BEST Futures

techniques. We can call these two types of technologies material technics and societal technics.[3]

In order to be viable, societies must have not only functional and environmentally relevant material technics but also congruent social technics that are able to organize and support economic production. For example, if an agricultural civilization relies on irrigation to produce regular crops, it will not be able to survive unless it has a social system that is capable of mobilizing large numbers of people to build and maintain irrigation canals as well as plant and harvest crops. It was not possible for the hydraulic technologies of agricultural civilizations to develop first and for supporting social structures to develop later, since sophisticated social institutions are needed to build and operate complex irrigation systems. If we look at the historical record we can see that societal evolution involves the gradual emergence of various compatible paradigm-changing material technics and societal technics. Over time more and more congruent material and societal technics emerge, forming a critical mass of new technics with the ability to operate as a more functional type of societal system.

The process of societal evolution resembles the way in which a jigsaw puzzle with an unknown design is assembled. At the beginning when only a few pieces of the puzzle have been laid down, there is no coherent pattern. But as more pieces are connected, local patterns begin to emerge and it becomes easier to add new pieces. At some point the overall pattern of the puzzle becomes apparent. Now (barring a major accident) the completion of the puzzle is almost inevitable and the process of filling in the remaining gaps goes into high gear.

The industrial system evolved in this fashion. Thousands of different but compatible material and social innovations appeared and gradually coalesced into a new type of living social system. Over several hundred years agrarian economic systems were replaced with industrial economic systems (a process that is still not complete), while rigid social institutions based on heredity and caste were replaced with more flexible institutions based on merit and wealth.

The emergence of this new type of societal system was a lengthy and uneven process in which paradigm-changing material and social innovations appeared, interacted and then supported the development of further waves of innovations. With each wave the emerging elements of the industrial system became more complete, more consolidated and more functional. We can see from this process that societal evolution involves the simultaneous development of new material and societal technics — the economic base and the social superstructure develop in tandem.[4]

Societal change often starts with technological innovations that create new economic and political dynamics: for example, firearms took away the military advantages of aristocrats, and railways allowed the interiors of continents to be developed. But change isn't just driven by economic or technological factors. Societies change and evolve in response to human and social needs, which are not just needs for physical security and access to resources, but also needs for identity and meaning. Throughout history, many of the most significant social changes have been the result of humans' desire for more meaning, more community and more justice. Some examples are the spread of religions such as Hinduism, Buddhism, Christianity and Islam; democratic revolutions in the United States and France; communist revolutions in Russia and China; national independence movements around the world and peace, environmental and women's movements today. Movements for social change are often driven by many motives and factors: idealism and economic self-interest often go together, as was the case with many American revolutionaries, and as is the case today in India with Dalits, who are converting in large numbers to Buddhism and Christianity to escape their position as low-caste *untouchable* Hindus.[5]

Since social and political movements are frequently temporary alliances of diverse interests, social tensions do not end when the goal is achieved. Instead new dynamics develop that continue the process of social change. For example, while the American Revolution promised freedom and equality to *all men*, this was interpreted by different groups in different ways. Because the wealthy white merchants who led the revolution did not consider black slaves, women or Native Americans to be either men or equal, achieving independence set the stage for centuries of further struggle.[6]

Three major historical forces are shaping the world today. We can characterize them as dominant, reactive and emerging. The dominant force is the expansion and consolidation of the global industrial system. Most of the people in the developed world are deeply immersed in the consumer society. While some are very well off, the majority are working hard to pay their bills and maintain or acquire middle-class lifestyles. Most of the people who live in developing countries are also eager to get the benefits of industrialization, such as higher living standards, better health care, more educational opportunities and more personal freedom.

A second force is the reaction to modernization. Many people in developing countries are unhappy with the damage that industrialization is doing to their traditional cultures and communities. Some would like to either preserve their agrarian or pre-agrarian societies, or to create hybrid societies that combine traditional values and forms of social organization with Western living standards.[7] Traditional societies have been opposing Western attempts to control their resources and direct their affairs for hundreds of years. In general, reaction is much weaker today than it was 50 or 100 years ago, since by now most of the pre-industrial societies on the planet have been destroyed or severely weakened. However the opposition to integration is still significant. While some indigenous peoples are making creative efforts to create sustainable, non-industrial societies through blending traditional skills and values with emerging holistic values and structures, other groups are violently resisting the expansion of the industrial system with its consumer values.

At present, resistance is strongest in the Islamic world, where some traditionalists are violently opposing change. Although they appear to have broad support, much of it comes from nationalists who are not opposed to modernization, but rather to Western control over their resources and cultures.[8] Unless Islamic movements are able to create a viable, post-industrial Islamic alternative to the consumer society, their efforts will ultimately fail, since rigid social structures can only be maintained in isolated agrarian economies.

The third force is the emergence of material and societal technics that support the transformation of our unsustainable system into a sustainable system. These are developing not only in response to the invention of better scientific paradigms, but also in response to real human and environmental needs.

Of these three trends, the most significant are the dominant trend (because it is leading us towards inevitable environmental and social collapse) and the emerging trend (which holds the promise of societal transformation and survival). While the reactive trend frequently dominates the news because of the energy and prominence given to it by the misguided and misnamed *war on terror*, it has no long-term chance of success since the needs of most people cannot be satisfied by returning to pre-industrial lifestyles. Nevertheless, this trend is still dangerous, since appeals to return to the past flourish in the chaotic conditions of failed states, where people must rely on the support of extended families and traditional communities for their survival.[9] If the global system experiences catastrophic collapse, we can expect the rapid growth of reactive and reactionary social movements, some of them armed with weapons of mass destruction.

Societal evolution is an ongoing process. Since we live in the Industrial Age, we can see governments, corporations, schools and other institutions hard at work expanding and consolidating the industrial system. Not only is the global economy constantly increasing in size, but it is constantly being rationalized and integrated. The job of business schools around the world is to teach managers how to increase labor productivity, market new products, increase profit margins and engineer corporate mergers and takeovers.

The process of global expansion involves absorbing and integrating the remaining pre-industrial economies. Since industrial economies cannot function efficiently with rigid, bureaucratic institutions (this is ultimately why communism failed), developing countries are under constant pressure from the developed countries, the World Bank and other international organizations to create more open and flexible social structures.[10] In order to encourage capitalist markets, governments and communities are also under pressure to privatize public services and communal property.

However, at the same time as the global industrial system is becoming stronger, more uniform and more tightly interconnected, it is also creating new paradigm-changing material and societal technics. As we saw with the internal combustion engine and the Internet, technological innovations can either reinforce a societal system or introduce new directions and stresses. Similarly social innovations can also either strengthen an existing system or encourage the emergence of a completely new type of system. Many different societal innovations are now supporting the transformation to a sustainable global system. These include holistic and integrative values and views, inclusive and multirelational art forms, new forms of cooperative and participatory social organization.

Growing awareness and rising expectations

The scientific and technological advances of the Industrial Revolution have allowed populations to grow and living standards to rise. Knowledge has also spread with better education, increasing longevity, improving communications and urbanization. Industrialism and the consumer society promote the concept that progress is beneficial, that learning and scientific curiosity is good, and that change is possible. Schools and advertising encourage people to do more and to go further: to learn, work, achieve and acquire. As a result, horizons are constantly expanding and expectations are constantly rising.

Information technologies

Asian girl working with a computer

Computers and the Internet are changing global values and raising expectations.

Figure 2: BEST Futures

Globalization has greatly accelerated this process. The development of an integrated global economy not only increases trade, but also increases the exchange of information and culture. It not only breaks down commercial barriers and weakens traditional economies, but also breaks down social barriers and weakens traditional social and political institutions. As a result people in developing countries are increasingly adopting the values of the developed world and becoming dissatisfied with their own societies. But the cultural exchanges go both ways. People in the developed world are also becoming more aware of the existence of other views and values and more aware of the failings of the global system.

National borders are becoming more permeable with consumer products, television, the Internet and the rapidly growing number of people who can afford to travel. The pace of change is astonishing: around the world 1000 mobile phones are being hooked up every minute,[11] and over 220 million Chinese now use the Internet.[12] The effect of increasing connectivity is to shrink our sense of time and space and, for the first time in human history, to create a *global village* of billions of people who share common values and experiences. An example of this is international sporting events — the 2006 World Cup was watched on television by an estimated 17% of the world's population.[13]

Increasing trade and travel has led to the development of a de facto international language. Although more people speak Mandarin, Hindi and Spanish as their first languages, all over the world people now study English as a second language.[14] Not only are people communicating more with people from other countries, but international migration is also at historically high levels. On average, 8% of the population of developed countries are now immigrants.[15] The significance of this is that many countries are becoming increasingly multicultural. While this often increases social tensions, it also increases people's appreciation of the value of diversity — a process that often starts with acquiring a taste for different types of food and music.

Globalization is changing people's sense of identity. In nomadic hunter-gatherer clans, the world and one's identity revolved around the extended family or clan. In herder-cultivator villages, one's identity came from the tribe. In agrarian civilizations, life and loyalty was defined by the supra-tribal authority of the king. In industrial civilizations, we identify largely with our nations — which is why there is so much interest in international sports competitions. However, the *modern* national identity of the Industrial Age is now beginning to be replaced with a *post-modern* transnational identity. More and more we see ourselves as being not only citizens of a country, but also citizens of the world. This shift is reflected in changing global values.

Changing global values

World values surveys study the values people have in different countries and how their values change over time. These surveys indicate that social values are closely linked to a society's level of economic development. Agrarian societies are usually organized around strong religious values. They have collectivist social structures in which the interests of the group are more important than the interests of individuals. Industrial societies are organized around rationalist and materialist values. They have individualist social structures in which a person's rights are more important than the rights of the collective.

Since societies have interdependent economic, political and cultural structures, it is not surprising that the surveys indicate that values change as societies industrialize. Because the process of industrialization follows predictable patterns (e.g. populations become more educated and urbanized), values also change in coherent patterns. Many social scientists refer to this as the shift from *pre-modern* (pre-industrial) to *modern* (industrial) values. In advanced industrial societies a further shift is occurring to values that support individual autonomy, diversity and self-expression. Some social scientists refer to these values as *post-modern* or *post-industrial.*[16]

Figure 3 illustrates how industrialization is changing global values in two phases.

1. Agrarian and pre-agrarian societies are increasingly industrializing both because they are under economic pressure to modernize and because they are attracted to the benefits of the consumer society. With exposure to the global industrial economy, their traditional/religious values increasingly change to rational/legal values.
2. Values are also changing in industrial societies. Needs for individual growth become more important as incomes rise and economic survival becomes more assured. The members of advanced industrial societies increasingly shift from survival/materialist values to self-expression/post-materialist values.

Ronald Inglehart explained why values are changing.

> The unprecedented wealth that has accumulated in advanced societies during the past generation means that an increasing share of the population has grown up taking survival for granted. Thus, priorities have shifted from an overwhelming emphasis on economic and physical security toward an increasing emphasis on subjective well-being, self-expression and quality of life ....
>
> Self-expression values give high priority to environmental protection, tolerance of diversity and rising demands for participation in decision making in economic and political life. These values also

reflect mass polarization over tolerance of outgroups, including foreigners, gays and lesbians and gender equality. The shift from survival values to self-expression values also includes a shift in child-rearing values, from emphasis on hard work toward emphasis on imagination and tolerance as important values to teach a child. And it goes with a rising sense of subjective well-being that is conducive to an atmosphere of tolerance, trust and political moderation .... This

Figure 3: The Inglehart-Welzel Cultural Map of the World [17]

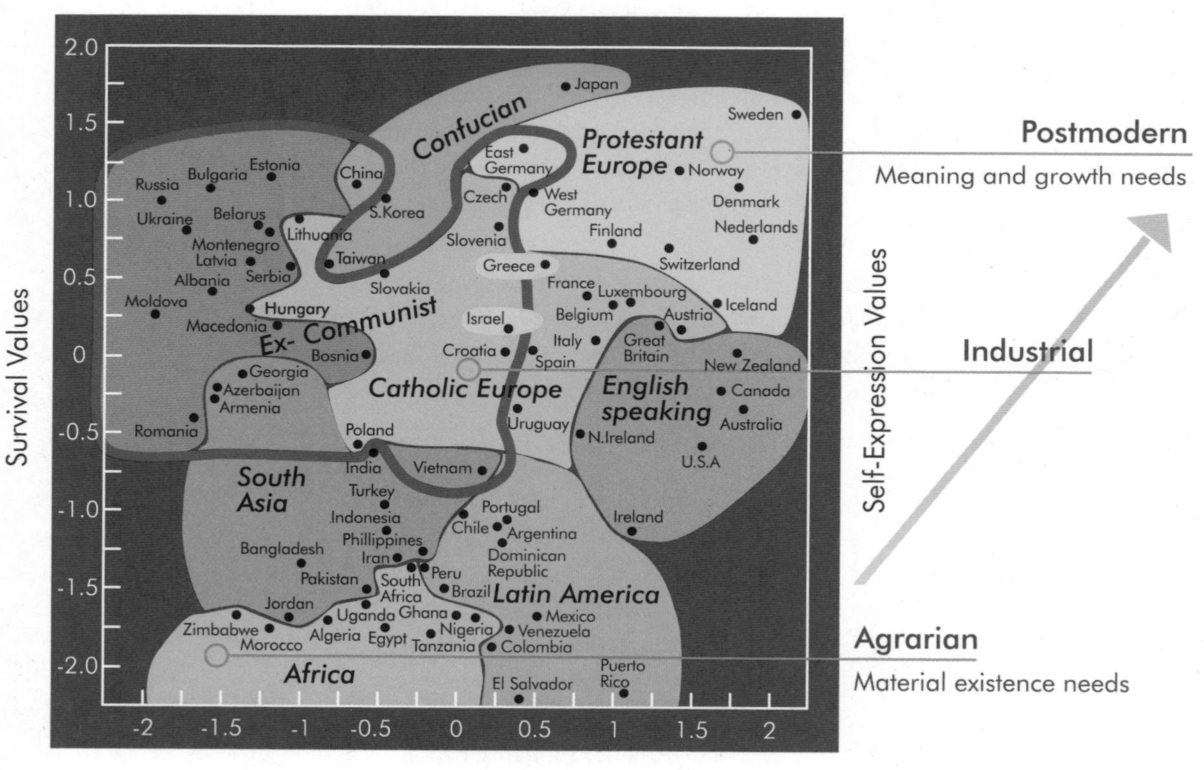

> produces a culture of trust and tolerance, in which people place a relatively high value on individual freedom and self-expression, and have activist political orientations. These are precisely the attributes that the political culture literature defines as crucial to democracy.[18]

The Cultural Map of the World shows how values change as countries industrialize. These are averages since countries are composed of different groups (and different individuals), whose values are changing at different rates and in different ways.

Differences in views and values create social tensions and conflicts: within relationships, between generations, within countries and in the world. Depending on how they are managed, the differences can spur creativity and produce constructive outcomes, or result in violence and destruction. Changing values also mean that individuals and nations approach conflict in different ways. In agrarian (pre-modern) societies, people have a duty to follow their leaders and protect their community. If a community feels wronged, tradition usually demands that all members help their group get either compensation or revenge. In advanced industrial (post-modern) societies, people are less closely identified with communities and more autonomous. As a consequence they are more likely to examine the different aspects of an issue before they take sides and more likely to seek peaceful solutions to conflicts.

European history demonstrates how values have changed over the last one hundred years. Before the First World War it was taken for granted that differences between European powers would often have to be settled by war, and millions of men were willing to fight and die for *God, King and Country*. But by the 1960s values had changed.[19] Not only were Europeans tired of the destruction of war, but their economies and cultures were becoming increasingly interconnected. As a result, more and more people were coming to believe that they were not just citizens of their own country, but also of a wider Europe. In this climate it became possible for politicians from both the left and the right to work for the economic and political integration of Europe. It is now almost impossible to imagine millions of young French and Germans volunteering to fight and kill each other for nationalist causes.

Jeremy Rifkin describes the differences between the *American Dream* and the *European Dream*. The *American Dream* emphasizes material freedom and self-reliant autonomy: with wealth comes exclusivity and security. The United States' focus on using national power to promote constant economic expansion epitomizes the values of modern industrial civilizations. The current *European Dream* is more about community and interdependence: inclusivity brings security. The European Union (EU) strives to function as a transnational co-ordinating body rather than as a national government. Its constitution refers to universal rather than national rights and gives more emphasis to promoting peace, social security and sustainable development than to private property.

Economic and political power within the EU still rests with its industrial nation states and European countries still have consumer cultures. Nevertheless, the European experiment represents a significant shift towards post-modern values and structures that support transnational and cooperative solutions. For these reasons the European Union may be in a better position to manage the transition to a sustainable societal system than the United States.

Not all post-industrial values support the emergence of a flourishing and sustainable global system. While the increasing emphasis on individuality and freedom from traditional restraints means that many people value diversity, it also means that many are more selfish. Although the number of idealists is increasing,[21] the number of self-centered hedonists is also on the rise.

Figure 4: BEST Futures [20]

Different policies

The US has:

- more individual freedoms
- a stronger military
- more individual material wealth
- more poverty
- more violent crime
- more prisoners per capita (8 x more)

Europe has:

- more social security
- more doctors per capita
- longer life expectencies
- more economic equality
- longer paid vacations (3 x more)
- shorter commutes

Every age has its good qualities as well as its limitations and faults. Previous generations may not have had as much personal freedom as people do today, but they were more connected to their communities and had a stronger sense of duty to family and society. Now many people in the rich world not only have trouble committing to their partners, but also have little sense of responsibility to their parents, their children, their community and the wider world. Although many young people are joining social movements and working for a better world, others are just dropping out — not watching the news, not voting, and not caring.

Nevertheless, values such as tolerance, diversity, inclusivity, respect, environmental protection and participatory democracy are steadily increasing. Values around the world are converging on many issues such as the need for gender equality, better governance, improved human rights and the elimination of extreme poverty and international conflict. The publics in countries as diverse as Canada, Sweden, Spain, Peru, Ukraine, China and India now consider pollution and environmental issues to be the world's number one problem.[22] In 2007, four out of five people interviewed in 21 countries said that they were prepared to make personal sacrifices in order to prevent global warming.[23]

The power of social movements

Social movements are not new phenomena — for example, religious movements have existed for thousands of years. In the 19th and 20th centuries powerful movements in many countries advocated political rights, worker's rights, children's rights, women's rights, minority rights and many other causes.[24] Without these long struggles the majority of people in the developed world would not now enjoy the right to vote, public education, public health care, pensions, holidays, decent wages and a high standard of living. While it is easy for people in the rich world to take these benefits for granted, they still do not exist in much of the poor world, where child labor is common and health care, holidays and pensions often non-existent.

Strong political and social movements have also existed in the developing world: in particular, movements for national liberation, socialism, democracy, women's rights and human rights. They have not been able to achieve as much as movements in the developed world for many reasons, including the support of Western governments for authoritarian neo-colonial regimes, the failure of communism and the unequal global economy. Nevertheless, the influence of social movements is steadily growing.

In his book *Blessed Unrest*, Paul Hawken argues that the hundreds of thousands of non-governmental organizations that have sprung into existence over the last few decades constitute the main force for change on the planet today. The spontaneous emergence of so many autonomous, grassroots organizations indicates that global values are converging and supporting constructive transformation. These diverse groups are progressively networking and developing "solutions to disentangle what appear to be insoluble dilemmas: poverty, global climate change, terrorism, ecological degradation, polarization of income, loss of culture, and many more."[25]

To support this process the Natural Capital Institute has established Wiser Earth, a community directory and networking forum of over 100,000 organizations that connects non-governmental organizations and individuals addressing the central issues of our day."[26] Let's look briefly at the impact of social movements in four areas.

Women's Rights

Although women do not yet have political or economic equality with men in any country, women's educational levels, rights and power are steadily increasing. This is changing the world in many ways: better educated women are having fewer and healthier children, families where women work outside the home have higher standards of living, and women are playing important roles in ending conflicts.[27] For example, the majority of university students in both Iran and the United States are now women.[28] As managers and other professionals in these two countries retire from the workforce, more and more senior positions will be filled by skilled women. This change will inevitably have an influence on American and Iranian political, social and economic policy-making — and quite possibly improve the relations between the two nations.

Peace

The influence of peace movements has also been growing. Organizations like the Campaign for Nuclear Disarmament and Greenpeace have been instrumental in raising awareness of the dangers of nuclear war and have had significant success in pressuring countries to ban nuclear testing, landmines and cluster bombs.[29] Although they weren't able to prevent the invasion of Iraq, millions of people in 600 cities around the globe took part in demonstrations against the war on February 15, 2003. These were the largest mass protests in history.

We can see the extent to which views and values are changing — in part in response to the work of peace activists — by looking at the difference between President Roosevelt in the Second World War, President Johnson in Vietnam and President George W. Bush in Iraq. Americans have been progressively less willing to tolerate heavy casualties in war or to volunteer to fight wars.[30] After Vietnam the national draft became politically unpopular and military officers lost their trust in draft troops. It has become increasingly difficult for the US to maintain a large army at war for an extended period of time. Whereas President Roosevelt could send millions of troops overseas during the Second World War, President Bush had trouble keeping 150,000 soldiers and marines in the field. After four years in Iraq the American public was sick of the war and the American military was exhausted.[31]

Social Justice

Social justice movements have also grown in size and influence. Organizations like Oxfam (anti-poverty), Amnesty International (human rights) and Transparency International (anti-corruption) are taken seriously enough by governments to have observer status at many international political conferences. While abuses such as extreme poverty, child slavery and torture still continue, governments around the world are increasingly under pressure to eliminate the worst abuses. An example of the success of social justice movements has been the establishment of the International Criminal Court in 2002 to prosecute individuals for crimes of genocide, crimes against humanity, war crimes and the crime of aggression.[32] This represents a significant advance in efforts to establish accountable global governance and the global rule of law.

Environment

The fastest growing movement in the world is the environmental movement. Its growth is being spurred by the increasing awareness of how environmental degradation is threatening our survival. For example, Wal-Mart's chief executive, Lee Scott, realized that climate change is a real danger after he saw the damage done by Hurricane Katrina. As a result Wal-Mart has changed some of its practices. It is now committed to using 100% renewable energy and stocking sustainable fish.[33]

Like other social movements, the environmental movement is introducing paradigm-changing values that transcend the narrow selfishness and mechanistic world view of the industrial system. Environmentalism asks people to think of the common good and their relationships with other living beings: to see themselves as part of interacting human and biophysical systems.

The emerging integral worldview

In order to be sustainable, the global economy will have to be organized around a worldview that understands that human economies are subsystems of the environment. The emerging worldview is often termed holistic (because it sees the world in terms of whole systems instead of just parts) or integral (comprehensive, interconnected and integrative). Chapter 7 outlined how a systems approach started to appear in the physical sciences in the 19th century and how it is now spurring the development of new material technologies and economic processes. Because economic and social processes are interdependent and interactive, similar developments have also been taking place in the social sciences and in culture, politics and social organizations.

Since humans use symbols to interpret reality and communicate with each other, every society expresses itself through language and art. It is not surprising then that elements of a holistic culture also began to emerge in the 19th century. Impressionist painters in the 1860s broke away from the tradition of trying to objectively reproduce reality, and instead created subjective impressions of the outside world. Cubism in the early 20th century went further by proposing that reality could be better understood by simultaneously observing it from multiple perspectives. Over time mechanistic, linear views were also challenged in other art forms, such as jazz and modern dance.

The real paradigm shift came with the development of movies, which can be described as a post-industrial art form, even though photographs and films were industrial technologies. Movies permit viewers to see distant and imaginary scenes, to see them from multiple perspectives and to move backwards and forwards in time. They challenge the idea that reality is objective and fixed — by bridging space and time, by allowing people to experience other lives and cultures and by showing not only what exists but also what is possible. This process has continued with the spread of the Internet and the development of interactive virtual worlds.[34]

In the political area, the isolation and absolute independence of nation states began to erode with the creation of international organizations. The International Telecommunication Union was first established in 1865 to standardize international wireless signals, followed by the creation of the Universal Postal Union in

1874 to facilitate the international exchange of mail. The process of global economic integration has been steadily accompanied by a rapid rise in international communication and a rapid increase in the number of international governmental and non-governmental organizations.

The environmental movement has its roots in the writings of American Transcendentalists such as Ralph Waldo Emerson and Henry David Thoreau in the first half of the 19th century. Their ideas inspired John Muir and others to organize the Sierra Club in the 1890's — the world's first politically active environmental organization.

As the world economy becomes more interconnected and interdependent, values and social structures also become more transnational. Barring the catastrophic collapse of the global economy, the trend towards increasing international integration is irreversible. Although the power of the United Nations is still extremely limited, more and more people around the world are realizing that it will be impossible to solve global problems without better global governance. This is part of the shift from the strictly national awareness and identity of industrial societies to a more inclusive global awareness and identity of the emerging holistic civilization — the awareness that we are not just citizens of particular countries but also that we share a common destiny on a finite planet.

The evolution of the industrial societal system was supported and accelerated by the development of new theoretical paradigms. While scientists such as Isaac Newton developed theories that challenged traditional views of the nature of physical reality, philosophers like John Locke and Adam Smith developed theories that challenged the social and economic foundations of the agrarian system.[35] These powerful new paradigms interacted and coalesced into the general worldview of the industrial system. Now new systems-based theories are developing in both the natural and social sciences that are in turn challenging the mechanistic views of industrialism and supporting the emergence of a holistic societal system.

While mechanistic science focuses on individual components, a systems approach views the world in terms of relationships and integration. Reality is seen to be composed of nested levels of systems (holarchies), each of which is an integrated whole whose properties (characteristics and abilities) do not exist at the level of its parts or at any other lower level. Evolution involves the emergence of new, previously unknown properties through the reorganization of existing structures into new systems. While systems theory is most developed in the physical and life sciences, it is being increasingly applied to understanding societal change and evolution.[36]

The concept of systems isn't new. Elements of systems theory are found in Aristotle's *Great Chain of Being* — the idea that reality is a hierarchy of integrative levels beginning with God. However the concept of evolving, self-organizing biological and social systems first developed in the 19th century, where Karl Marx and Friedrich Engels famously applied it to historical processes.[37] In the last century thousands of scientists and philosophers have contributed to our understanding of social systems, including Sri Aurobindo, Don Beck, Ludwig von Bertalanffy, Kenneth Boulding, Nikolai Bukharin, Joseph Campbell, Fritjof Capra, Jean Gebser, Steven Jay Gould, Jürgen Habermas, Marvin Harris, Paul Hawken, Hazel Henderson, Chris Holling, Erich Jantsch, Carl Jung, Ervin Laszlo, Gerhard Lenski, James Lovelock, Niklas Luhmann, Joanna Macy, Henry Margenau, Lynn Margulis, Humberto Maturana, George Herbert Mead, Donella Meadows, Arne Næss, Joseph Needham, Talcott Parsons, Ilya Prigogine, Elisabet Sahtouris, Jan Smuts, Joseph Tainter, Richard Tarnas, Alastair Taylor, Valentin Turchin, Alfred North Whitehead and Ken Wilber.[38]

New theories are not just interesting ideas; they are practical and powerful tools for changing the world. For example, Newton's ideas helped fill the world with machines; Einstein's ideas helped build atomic bombs; Locke's ideas helped create democratic movements and Marx's ideas helped create socialist movements. A new theoretical paradigm is adopted and developed because it provides more answers and better solutions than the previous paradigm, and a new worldview evolves because it is able to organize a more functional and conscious societal system than the previous worldview. The reason why the holistic worldview is now emerging and spreading is because it can provide more answers and solve more problems than the industrial worldview. By integrating the natural and social sciences — as well as science and faith — it gives people powerful tools for understanding their realities, developing sustainable solutions and taking control of their destinies.

It is important to note that worldviews are not monolithic. Although there are very different political tendencies within industrialism — fascist, conservative, liberal, socialist, communist and anarchist — they all reject the caste-based social structures of agrarian societies and all support industrialization with its model of constant economic expansion. In the same way, while different models of a holistic societal system are likely to emerge (e.g. capitalist, socialist, Islamic, indigenous), they will all be characterized by their rejection of social structures that are based on the domination and exploitation of humans and nature and will all support environmentally sustainable economies. And just as all the industrial models are based on a mechanistic view of reality, all holistic models will be based on a systems perspective.

The spread of deep democracy

The mechanistic worldview of the Industrial Age sees reality as being primarily composed of inert objects that move or change in response to external pressure. In the integral worldview, reality is made up of dynamic, self-organizing systems. The promise of the emerging Information Age is that it will increasingly replace force with knowledge, and mechanistic external pressure with organic internal self-organization. In technology this means replacing methods that brutally exploit nature with methods that mimic and cooperate with nature, and in social organization it means replacing bureaucratic institutions with self-organizing and self-regulating communities.[39]

Throughout human history only powerful people and specialists (e.g. priests and bureaucrats) have had the resources and expertise to access written information. Information has been centralized not only because it has been expensive to acquire and record, but also because knowledge is power, which means that whoever controls the information and the records of a society has control of its political, economic and cultural institutions. In recent times the monopoly of knowledge and power by elites has begun to erode, in particular because the Internet is allowing more and more people to access, share and create knowledge.[40] Once people understand where they fit in the natural and social worlds and have the means to solve problems, they can fully participate in social life. This is the beginning of deep democracy — the development of an inclusive, participatory and qualitatively more aware political process.

Because the Internet functions as an interactive system, it encourages people to think and act in terms of cooperative networks.[41] Examples of this include peer-to-peer (P2P) collaborative projects like Wikipedia, which grows in an organic fashion due to the contributions of thousands of self-organized volunteers, and social networks like MySpace, Orkut or Facebook, where people create and share information. In the industrial economy, where businesses manufacture and sell finished products, the consumer is a passive participant in the process. With the

Internet, a profound shift is taking place. Businesses are increasingly providing the infrastructure for people to create their own products. However, most of the products that are being created are neither material nor commercial. Cyberspace is a world of relationships where people mostly share information, develop communities and generate new ideas.

This is the beginning of a cultural transformation from people being consumers and observers to people being creators and participants. Of course this process has only begun, and exchanging photos on a social networking program is a long way from changing the world. However, three interacting processes are underway.

- As networking spreads, the exchange and creation of knowledge is accelerating quantitatively and developing qualitatively.
- As social and environmental activists spread knowledge about global issues, awareness of the urgent need for collective action is rising.
- As the integral worldview develops and spreads, people increasingly understand how they can constructively change their own lives and the world.

The interests and inertia opposing change are enormous, and the time we have left to avoid disaster is fast disappearing. Yet the task of transformation is possible. All over the world pieces of the puzzle are appearing and being connected together. The solution to global problems is becoming clear. A sustainable planetary system is rapidly evolving.

9 Constructive and Destructive Responses to Crises

Forecasting the future

We all need to know what the future will be like and how different decisions are likely to influence events. Predicting global futures is not easy because societal systems, like weather systems, have complex and chaotic dynamics. However although precise forecasts are not possible, the probability of various outcomes can be estimated. While meteorologists cannot predict with certainty what the weather will be like in a particular place on a particular day, they can tell you the likelihood of the day being sunny or rainy, windy or calm and hot or cold. They can also predict trends — how weather patterns are likely to change over time.

In the same way, doctors and psychiatrists make decisions based on estimates of the probable future health of patients, farmers predict harvests and economists predict interest rates. In fact all of us, all of the time, are making predictions about what will happen based on past experiences and our assumptions of how the world works. We predict that we will be able to safely cross a road based on assumptions about how the drivers of nearby vehicles will act, and we go to work, get married, have children and buy houses based on assumptions about what our lives will be like in the future.

Change can be incremental or non-linear. Examples of incremental change are accelerating cars and population growth, while examples of non-linear change are car accidents and evolution. The non-linear dynamics of complex and chaotic systems can result in small events having large effects. One example is getting pregnant. Another example is the assassination of the Austrian Archduke that set off the First World War. Because two rival alliances were already positioned to attack each other, this otherwise minor political event triggered a war that caused the deaths of nine million people.[1]

In addition, non-linear dynamics can produce system-state changes. A common example is boiling water: adding heat gradually raises the temperature of the water until a tipping point is reached at 212°F (100°C). At this point there is too much energy for the water molecules to maintain their structures, and the liquid turns into gas. Developmental and evolutionary stages in living systems (e.g. emergence and extinction) also involve system-state changes.

Incremental trends such as growing populations, growing consumption, and degrading environments are easily forecast. It is not as easy to predict non-linear developments such as the development of new technologies or values or the sudden collapse of ecosystems. Nevertheless, it is possible to see that we are approaching tipping points, even though we may not know precisely when they will occur or the exact way in which many complex and interacting events will develop.

The difficulty of forecasting is shown by a conference held by the US National Intelligence Council (NIC) in 2002 to predict global energy needs in 2015. The conference decided that four future scenarios are likely. In the first, a "headline-grabbing environmental disaster" mobilizes public opinion in the developed world, causing politicians to aggressively develop alternative energy sources with the result that the demand for oil drops sharply. In the second scenario, a series of technological breakthroughs in alternative energy sources and energy efficiency also lead to substantial declines in the need for oil. In the third, global oil production peaks sometime between 2010 and 2015, pushing oil prices up to $40 a barrel and pushing the global economy toward recession. And in the fourth scenario, a US invasion of Iraq backfires and causes the overthrow of "relatively friendly Arab governments by nationalist Islamic regimes." Global oil production is cut, and the price of crude oil reaches US$50 dollars a barrel, bringing and end to a global economy based on cheap oil.[2]

So how accurate have these predictions been? We'll have to wait until 2015 before making a definitive judgement, but by the time this book was completed in July 2008 the price of oil was over $140 per barrel. This was due to a number of interacting factors, among them increasing global demand, a falling US dollar, declining oil production in most of the world, reduced production in Iraq and Nigeria due to conflicts, and the reluctance (and/or inability) of other OPEC nations to substantially increase their production. The era of cheap oil is already over.

Such predictions show both the potential and the limitations of forecasting. On the one hand it is possible to predict trends, the range of likely outcomes and general time frames. On the other hand, there are too many variables to make precise predictions. In particular, it is impossible to know the role of wild cards such as natural disasters, pandemics, innovative breakthroughs, political or economic crises and inspired (or terrible) political leadership.

Moreover, the accuracy of the predictions depends on the accuracy of the theories used by the forecasters, since the quality of the final results is determined both by the quality of the input data and the quality of the methods used to process the data.[3] Although the 2002 National Intelligence Council energy forecasts made many reasonable assumptions, in my view their analysis was fundamentally flawed because it failed to take into account essential factors such as environmental limits to growth and the non-linear dynamics of biophysical and societal systems. Because the industrial worldview emphasizes parts over wholes, it tends to view issues in isolation. For example, for years both the UN and the US National Intelligence Council have recognized that water scarcity is a growing problem in much of the world, and yet both continue to predict that farmers will be able to rely on water-intensive agricultural technologies to constantly increase harvests and meet the needs of an expanding global population.

If we wish to make relatively accurate forecasts, we need to take a systems approach and examine how different factors interact with each other. We also need to recognize that globalization is creating an increasingly interconnected and interactive global economy. As the world system becomes more and more integrated, it also becomes increasingly unstable; now crises can rapidly spread from one country to another.

The large number of interacting problems facing humanity in the coming decades multiplies the probability of major crises. In addition, interacting problems can produce cascading crises (e.g. ecological crises can trigger economic crises which can trigger political crises). The danger is that the number, size and complexity of local and global crises will continue to grow and converge in the coming years, eventually producing unmanageable fluctuations and potential system failure.[4] For these reasons proactive measures will have to be taken to prevent crises developing to the stage where they are uncontrollable.

Probable responses to growing energy shortages

The global supply and distribution of energy is already inadequate. Forty percent of the world's population currently lack modern fuels for cooking and heating. Of these, 1.6 billion have no access to electricity at all.[5] Adding to existing problems are rapidly increasing global energy prices which are rising in response to the narrowing gap between supply and demand. The National Petroleum Council, which includes the heads of the world's biggest private oil companies, has now issued a report stating that "the global supply of oil and natural gas from the conventional sources ... is unlikely to meet growth in demand over the next 25 years." The International Energy Agency agrees that oil prices will continue to rise with supplies becoming "extremely tight" by 2012.[6] In April, 2008 the president of the Organization of the Petroleum Exporting Countries, Chakid Khelil, warned that the price of crude oil could reach $200 a barrel.[7]

Prices are rising as the planet gets closer to the point of peak oil and gas production. As global demand increasingly outstrips supplies, energy prices will continue rising unless alternative energy sources have been developed. This will further restrict access to essential products such as electricity, heat, transportation, clean water, sanitation and fertilizers, and will raise prices for food and other products. Because the richer people in the world can afford to pay rising prices, the worst effects of increasing energy shortages will be felt by the poor and the middle classes in developing countries.

Figure 1: BEST Futures

Megacities need affordable energy

◁◁◁ Tehran

Hong Kong ▷

Since rising prices will result in growing unemployment and declining services, continuing increases in the cost of food, energy and other essential commodities will cause hunger, disease, crime and social unrest to increase sharply in the megacities of the developing world.

Increasing energy prices will slow economic growth in the short-term. However, increasing prices will also encourage energy investments and mobilize political interventions to avoid future problems. These responses can be categorized as constructive or destructive in terms of the long-term affects they will have on the global environment and economy.

Constructive energy responses involve the shift to renewable energies, increasing energy efficiencies and conservation. Rising prices are encouraging all these strategies, as are the fears of energy-importing countries that their dependence on foreign resources makes them increasingly vulnerable to economic blackmail and crippling supply disruptions. These economic and strategic concerns are driving governments to support policies that reduce oil and gas consumption.

Destructive energy responses involve increasing pollution through building more coal-fired energy plants,[8] increasing hunger through using food crops to produce biofuels and increasing conflicts through attempting to control foreign oil and gas producing regions. Paul Rogers, a professor of peace and conflict studies at Bradford University, points out that if the US hopes to control the Middle East and its resources, it will have to stay and fight wars for decades.[9] Although it would cost less (in lives, wealth and reputation) to invest the money spent fighting wars on achieving energy independence, entrenched interests and beliefs in the United States' right to dominate world affairs make it difficult for American politicians to advocate complete military withdrawal from the region.[10]

At some point sharply rising energy costs are also likely to trigger a global depression. The problem is that at this time no international bodies, governments or businesses have concrete plans for making the massive investments required to meet global energy shortfalls, let alone for making global energy production environmentally sustainable.

Probable responses to growing water and food shortages

The global food and water situation is similar to the energy situation — shortages are growing and prices are rising. In 2006, 850 million people were malnourished and one-third of the world's population lived in water-stressed regions.[11] Because of climate

Food shortages

Food shortages are ultimately not technical problems, but political and social problems. Hunger is always linked to poverty. Enough food could be sustainably produced for everyone if the environment was protected and people throughout the world had access to education and their fair share of resources.

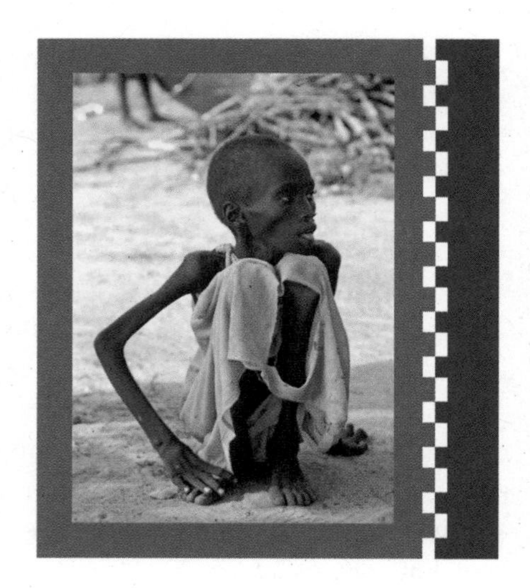

Figure 2: BEST Futures [12]

change, overuse of aquifers, environmental degradation, and growing demand, current projections are that water scarcity will affect two-thirds of the world's population by 2025. The same factors mean that food will also become increasingly scarce.

The 2007/8 harvest was dismal, with droughts and floods damaging crops around the world. It marked seven out of eight years in a row that global grain production had fallen short of demand. In 2007 reserves of the world's three main grains — rice, wheat and corn — fell to their lowest levels in more than two decades.[13] Michael Lewis, the head of commodities research at Deutsche Bank, believes that grain prices still have much further to rise, since India recently became a net importer of wheat and China will soon become a net importer of corn.[14] In early 2008 the World Bank said that average global food prices had risen 83% in three years.[15] The United Nations warned that nearly 40 countries were facing critical food shortages and blamed dwindling food supplies on the combination of crop failures caused by global warming, natural disasters, wars, and farmers increasingly growing grains for animal feed and biofuels instead of food crops.[16]

The rising cost of agricultural products hurts the poorest people in the world the most and increases the numbers who are chronically malnourished. It also has an inflationary impact on the global economy as a whole — for example, the rapidly rising price of cotton is driving up the cost of clothing.[17] Although the rising costs of fertilizer and other inputs has meant that many poor farmers have reduced planting,[19] rising prices are also raising farm incomes and stimulating investments in agriculture. While this will improve productivity, at the same time it will encourage farmers to grow more crops and raise more livestock on marginal soils and put more pressure on scarce water resources. Unless measures are taken to prevent and reverse environmental degradation, short-term improvements in production will inevitably be followed by long-term declines as soils degrade, water tables fall and the climate becomes warmer and more erratic.

Living on a dollar a day is defined as extreme poverty. A typical budget for a five-person family in Bangladesh earning $1 per day per person has been: $5 = $3 spent on food + $0.50 spent on energy +$1.50 spent on non-foods. A 50% increase in food and energy costs forces them to cut $1.75 from their daily expenditures.[18]

Experts are warning that if global warming isn't stopped, the terrible harvests of 2007 may just be a beginning.[20] In the view of Darrin Qualman, Research Director for the National Farmers Union of Canada, "the converging problems of natural gas and fertilizer constraints, intensifying water shortages, climate change, farmland loss and degradation, population increases, the proliferation of livestock feeding, and an increasing push to divert food supplies into biofuels means that we are in the opening phase of an intensifying food shortage ... If we try to do more of the same, if we try to produce, consume, and export more food while using more fertilizer, water, and chemicals, we will only intensify our problems ....We need to create a system focused on feeding people and creating health .... Diversity, resilience, and sustainability are key."[21]

There have been a wide range of responses to growing food shortages: for example the UN has set up a food crisis task force to raise donations to feed the increasing numbers of hungry people and to provide assistance to poor farmers who can no longer afford the rising costs of agricultural inputs, while the government of China is considering leasing fields in Latin America, Australia and the former Soviet Union to replace farmland lost to urban and industrial development.[22] However, these measures will only provide short-term relief as they do not address the underlying causes of the food crisis.

Long-term solutions to water and food shortages will require conserving water and soils, stopping destructive farming practices, restoring degraded lands, banning toxic chemicals and taxing pollution, using aquifers sustainably, preventing the spread of deserts, increasing forest cover, promoting healthier low-meat diets, ending subsidies on meat production and prohibiting factory farming. Many of the solutions are relatively simple and inexpensive: teaching poor farmers techniques such as crop rotation and organic farming has been shown to increase harvests by almost 80% while reducing the need for water, pesticides and chemical inputs.[23]

Vandana Shiva, a physicist and ecologist, said, "A green revolution based on nitrogen fertilizers in 2008 is a recipe for emissions of nitrogen oxides, further instability of the climate and further hunger and starvation .... There is a very short term solution [to the food crisis] — give up the industrial agriculture using fossil fuels, high cost imports. Give up the forced linking with an international commodity market. Allow farmers to grow and give them a just price. We can solve the problem tomorrow. I work with 400,000 farmers in India growing organic food. We have doubled yields and doubled output on farms. Nobody is dying in villages where there is organic farming."[24]

Unfortunately, most governments (with the encouragement of multinational agro-businesses) are promoting more of the same destructive practices. Policies and subsidies encourage inefficient, water- and energy-intensive agriculture. Tropical forests are being cut down to grow crops for biofuels. Farmers are being encouraged to produce more meat using industrial farming techniques that consume scarce supplies of grain and energy, pollute the air and water and breed dangerous diseases.[25] Subsidized fishing fleets are systematically destroying fish stocks. The longer these unsustainable practices continue, the harder it will be to prevent massive declines in food production. For example, India's 21 million wells are lowering water tables in most of the country and increasingly these wells are running dry. Tushaar Shah, head of the International Water Management Institute's groundwater station in Gujarat, said of India's water situation, "When the balloon bursts, untold anarchy will be the lot of rural India."[26]

Governments have been as slow to respond to growing water and food shortages as they have been to respond to growing energy shortages and climate change — and for similar reasons. Conventional economic wisdom has argued that markets will respond to oil, gas, water or food shortages by raising prices, which will encourage more investment, which will in turn produce more resources. In the short term, rising prices will raise production by providing the majority of the world's farmers with more capital and encouraging farming on marginal soils. A few years of good weather may also improve harvests.

However, these will only be temporary solutions because the underlying cause of growing food shortages are biophysical limits to growth. In the past, there have always been more oil fields to be drilled, more rivers to be dammed, and more fertile soils to be cultivated. The global climate was in equilibrium, which meant that years of poor harvests were always balanced by years of good harvests. Now we have destabilized the climate and are running out of cheap energy, fresh water and arable land. These problems cannot be solved by market forces.

As food prices rise and food and water shortages create growing regional political and economic problems, decision makers will become increasingly concerned and start looking for solutions. However, their responses may be too late to prevent major crises. And unless they recognize the need for environmentally sustainable policies, the actions they take will only make problems worse.

Probable responses to growing economic crises

A global depression could be triggered by a sharp rise in energy costs and food prices. Another potential cause of future economic crises is the combination of an unbalanced global economy and competition over increasingly scarce resources.

The United States is a superpower for three main reasons: it has the largest economy, it has the strongest military and most international banking and trade is conducted in US dollars. Its wealth and power give it cheap access to raw materials: although Americans are 5% of the world's population, they consume approximately 25% of the world's resources. However, the US is not just the world's only superpower; it is also the world's only super-ower. Americans have an unsustainable economy: in 2007 the United States imported $712 billion more than it exported.[28] In effect the entire US military is financed on credit. This deficit is only possible because Asian central banks prop up the dollar through buying US treasury bills. For years China, Japan and other countries have followed policies that undervalue their own currencies in order to help their manufacturers capture increasing shares of the US market.

> We continue to run deficits, and a larger share of our income goes to support this. Our attitude seems to be, "Lord, give us the strength to resist temptation, but not quite yet."
>
> — Brad Setser, former US Treasury official[27]

The huge American trade deficit is not new. The only reason why a massive global readjustment has not occurred is because it is not in the interests of any country to put the US economy into a major recession.[29] A major contraction of the American economy will affect every other economy, and countries like China and India do not want to risk rising unemployment and the social unrest that will follow. This is why central banks quickly intervened in August, 2007 to prevent problems in the US mortgage markets from spreading around the world. Their infusion of hundreds of billions of dollars worth of new credit into the global financial system did nothing to solve underlying structural problems, but it did prevent panic.

Market economies always go through cycles of expansion and contraction (boom and bust). The contractive (recessionary) phase is sometimes referred to as a period of *creative destruction* because it allows markets to adjust prices and weed out less efficient (or less well financed) investors and companies. This occurred when the dot-com bubble burst in 2000, destroying the paper assets of many high-tech start-ups whose dreams were bigger than their cash flows. However, because this crash had little impact on the structure of the US economy, it was able to recover relatively quickly. The latest recession has been triggered by the simultaneous bursting of property and credit bubbles.[30] What makes this recession different from earlier ones is not only its size,[31] but more significantly, that it marks a turning point: the end of the era of cheap resources.

The Bush administration was able to simultaneously fight foreign wars, lower taxes and maintain a booming economy through the magic of credit. They lowered interest rates and allowed US families to borrow against the rising value of their homes, and they expanded the nation's borrowing power through printing money and selling treasury bills to foreign banks. The problem facing the next President (and all other world leaders) is that it is no longer possible to paper over growing cracks in the global economy.

The fundamental problem is that the US economy and the country's high living standards depend on the ability of an artificially high dollar to import cheap goods and raw resources. Efforts to reduce the deficit by lowering the value of the US dollar not only increase the cost of imports but also risk triggering a stampede into Euros and other currencies. This vicious cycle may have already begun, and it may be uncontrollable. Jeffrey Garten, a professor of international trade and finance at Yale, explained why current economic strategies aren't working: "Almost no conceivable dollar depreciation would be enough to significantly narrow the trade gap, simply because the value of imports is twice that of exports .... Because the United States needs to borrow more than $2 billion each day from abroad to finance its imports and service its $3 trillion foreign debt, a dollar that plummets too fast would necessitate a rise in U.S. interest rates to attract foreign funds. We could have the worst of all worlds — inflation, high interest rates and recession."[32]

The global economy is immensely rich and powerful, and it is increasingly developing safeguards to manage economic crises. For these reasons it may be able to prevent (for a while) growing financial inequities from suddenly creating an economic earthquake that could level large sections of the global economy. However, the stresses are constantly increasing, and with them the dangers of uncontrollable fluctuations in currency prices and stock markets. For years many analysts have worried that US economic problems could trigger a global depression. For example, in 2005 Lau Nai-keung warned in the *China Daily* that "China should make extra efforts to rekindle internal consumption, and diversify its market really fast before the great U.S. bubble bursts."[33]

In 2007 the US dollar weakened, the price of oil and other key commodities (such as iron ore and cobalt)[34] began to rapidly rise and the domestic credit bubble burst. It may already be too late to avoid a global recession and the beginning of escalating economic and political crises. The global economy is running out of cheap resources, and these problems cannot be solved though financial adjustments.

Probable responses to conflicts over scarce resources

The financial and social inequality of the global economy is destabilizing and dangerous.[35] Growing income gaps between rich nations and poor nations and within countries like the United States, China and India can be managed as long as average incomes keep rising. But if shortages of essential goods and rising prices lower the standards of living of hundreds of millions of people — if people who are middle class today become poor tomorrow, and people who are poor today become hungry tomorrow — then there will be massive social unrest. In early 2008 rising food prices triggered protests and riots around the world.[36] Most people will tolerate bad government if they have economic security and hope for a better future. But if they lose that hope, then anger and despair can easily be channelled into intergroup violence and/or demands for radical political change.[37]

While governments can use economic measures, laws and force to stabilize financial and political disturbances, there is little that they can do to solve problems caused by biophysical limits to growth and environmental degradation. Resource shortages restrict the supply of goods with the consequence that they must be rationed either with higher prices that make them unaffordable for poorer consumers, or by limiting availability — e.g. by turning off the supply of water or electricity for part of each day. Because resource shortages and other environmental problems cannot be resolved by the current global system, they are likely to be the root causes of increasing global economic crises.

Countries are becoming increasingly concerned about their access to water, energy and mineral resources. For example, both China and India are making major investments in African resources: by 2010 China will probably be the continent's major trading partner.[38] Fears of being excluded from critical supplies are leading to new strategic alliances and a new arms race — in 2007 the US, Japan, India, Australia and Singapore held joint naval exercises, while Russia and China held joint military exercises that were observed by the leaders of the Shanghai Cooperative Organization (China, Russia, Kazakhstan, Uzbekistan, Tajikistan and Kyrgyzstan).[39] The focus of this competition is control over the oil and gas resources of the Middle East and Central Asia — objectives clearly articulated by the former US Secretary of State Zbigniew Brzezinski in his book *The Grand Chessboard*.[40]

We should not forget that competition over resources helped start two world wars. We are now in a situation that is similar to the years leading up to the First World War — while no country desires war, no major power believes that it can afford to be denied access to critical resources. This is the American dilemma in Iraq — political leaders do not want to stay and be bogged down in an unending war, and yet they are afraid that if they withdraw from the region they could lose access to vital energy supplies.[41] If global resource consumption continues to expand, competition over increasingly scarce resources will grow. The competition will initially be primarily political and economic in nature, but as prices rise and economies are destabilized, there will be more and more willingness to use military means to guarantee access to strategic resources. The danger is that at some point competition and confrontation will escalate into a war involving major powers armed with weapons of mass destruction.

The alternative to future resource conflicts is to eliminate the potential causes of war. This strategy (called developing sustainable security) will require the major powers to reduce their military budgets and use the savings to rapidly convert their unsustainable industrial economies into sustainable economies.[42] This constructive approach could reduce their reliance on imported resources, avert economic and political crises and prevent future wars.

Probable responses to climate change and the loss of biodiversity

Although growing resource shortages will provoke massive economic and social problems, the real dangers to humanity are climate change, failing ecosystems and the increasing loss of biodiversity. While water, food and energy scarcities will bring misery to billions of people, runaway climate change, unless stopped, is likely to drive most species to extinction, cripple agricultural production, provoke wars,[43] cause mass starvation and end advanced civilizations. The British International Institute for Strategic Studies (IISS) warns that that if climate change is allowed to continue unchecked, it will have catastrophic affects "on the level of nuclear war."[44]

On the positive side, awareness of the dangers of climate change is rapidly growing. While most politicians and business people either ignored or denied the reality of global warming a few years ago, it is now on the agenda of almost every international conference. Hundreds of billions of dollars are being invested every year in developing cleaner and more efficient technologies. Immense political and technological momentum is building to stop global warming.[45]

On the negative side, the present pace of change may be too slow to avert catastrophe. Politicians are paying little attention to the accelerating rate of species extinction,[46] and powerful vested interests are lobbying against constructive action. Market forces mean that the value of species rise as they become rarer, which increases the incentive to destroy the remaining forests and harvest the last fish.

Worse, not enough is being done to prevent runaway global warming. Leading scientists now warn that we have 10 years at a maximum to stop climate change. Even this short time frame may be optimistic given that almost every study shows that global temperatures are rising much more quickly than predicted. Some scientists fear that the Arctic sea ice is melting so quickly that the Earth's climate may have already crossed a critical threshold beyond which stabilization is impossible.[48] James Lovelock commented, "Deadly it may be, but when we pass the threshold of climate change there may be nothing perceptible to mark this crucial step, nothing to warn that there is no returning."[49]

A diagram by David Wasdell, the Director of the Meridian Programme, explains the stark choices now facing humanity.[50] Figure 3 shows the tipping point, or watershed, that we are rapidly approaching (represented by the ridge running from left to right). In front is the green area of historically stable global temperatures, where negative feedback processes (cooling factors) cancel out positive feedback processes (warming factors). The system becomes increasingly unstable as temperatures rise, until a major tipping point is reached (at the summit of the ridge). Here the permafrost begins to rapidly melt and release hundreds of billions of tons of methane gases. If we pass this summit, we will enter a zone where global warming, represented by the red slope, will be irreversible. At this stage it will be impossible to prevent global temperatures from rising year after year, and the catastrophic collapse of the environment and human societies will become inevitable.[51]

Figure 3 illustrates three future climate scenarios. If we continue with business as usual we will quickly reach the point of no return (red zone). Following through with the planned Kyoto agreements (50% global reductions in emissions by 2050) will still fail to stabilize the global climate — all we will do is delay disaster by a few years. Our only chance for survival is to quickly act to reverse global warming and restore global temperatures to a stable level (green zone).

These three scenarios — business as usual leading to rapid collapse, gradual reforms which only delay collapse and transformative changes that enable survival — hold true not only for climate change but also for other major global issues such as growing resource scarcities and biodiversity loss. How these scenarios develop will determine our futures, our children's futures and the futures of all the myriad creatures that share our planet.

Figure 3:

D. Wasdell[47]

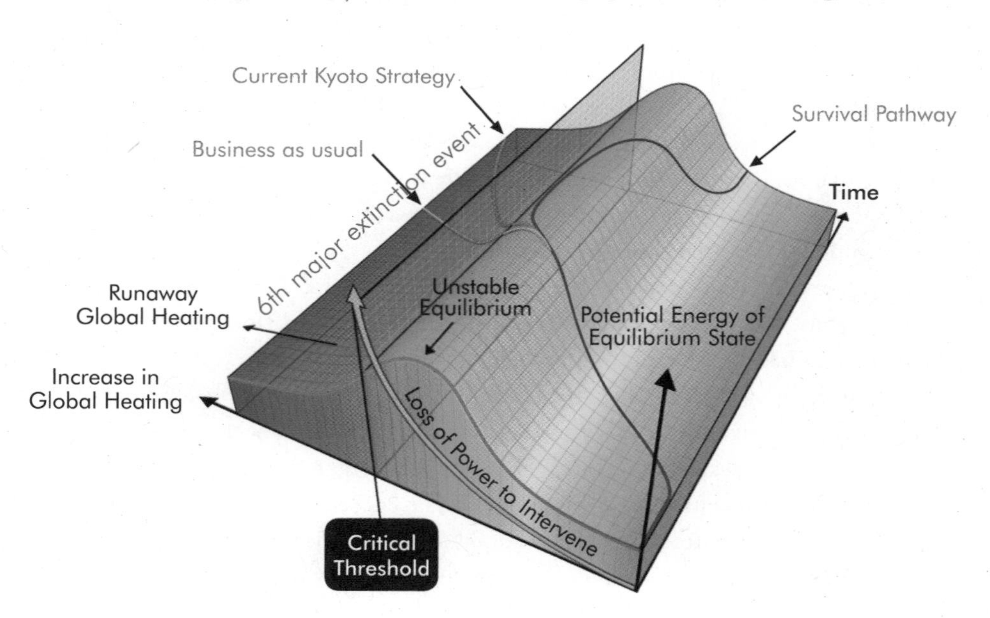

10 Future Scenarios

> Obviously, the first thing we need to do is act, and act fast. Every day we wait, another 30,000 children needlessly die; between 100-150 plant and animal species become extinct; 70,000 hectares of rainforest is destroyed and another 150m tonnes of CO_2 is released into the atmosphere. Meanwhile, another $3.0bn (£1.5bn) is spent on arms and weapons of mass destruction.
>
> — Eamon O'Hara, policy advisor[1]

The two main trends — collapse and transformation

AT PRESENT WORLD AFFAIRS ARE BEING SHAPED BY THREE MAJOR FORCES: industrialization, the reaction to industrialization and emerging post-industrial views, values, structures and technologies. Although stories about the rapid expansion of the consumer society and its battles with Islamic traditionalists have dominated the news for decades, environmental problems will increasingly shape national and global events. This is because resource shortages and failing ecosystems will not only limit the ability of the global economy to expand, but cause it to contract. The result will be growing economic and social crises, which will in turn provoke more local, regional and international conflicts. As a consequence, the industrial system will weaken while both pre- and post-industrial forces will strengthen.

These three forces will increasingly converge into a single dynamic because there are only two possible outcomes for humanity — either we will continue to destroy our environments until our economies and societies completely collapse, or we will create viable societies and survive. The third alternative, a return to a pre-industrial past, is not realistic for most of the people on Earth. The population of the world is now too large and the environment too damaged to permit most people to live in hunter-gatherer, herder-cultivator or even agrarian societies (i.e. societies dependent on traditional farming technologies). However, it is possible now and will be possible in the future for a minority of the world's population to live in sustainable pre-industrial communities where they can maintain, restore and develop their traditional lifestyles.

This means that there are only two options available for traditional groups opposed to modernization: they can either develop sustainable solutions that are appropriate for their cultures and physical environments, or they can make futile attempts to restore environmentally and socially obsolete social systems. The latter efforts will only increase chaos and accelerate the collapse of the global system into failed states and warring tribes.

Figure 1: BEST Futures

The choice: collapse or transformation

Growing social unrest

If we continue with business as usual our unsustainable global system will collapse.

Solar panels

In order to survive, we must create an environmentally sustainable system.

As a result we can say that all events occurring on the planet today are part of two fundamental, major trends: they either support continuing environmental degradation and unsustainable outcomes, or they support the transformation of the global system and sustainable outcomes. At present the destructive trend towards collapse is the dominant trend and the constructive trend towards transformation is the emerging trend. Our futures will be determined by how these two interrelated trends develop.

The three possible future scenarios

Over the coming decades, the two major trends can produce three possible scenarios: rapid collapse, delayed collapse and (if major irreversible ecological damage has not yet occurred) transformation.

Rapid Collapse

The majority of the world's political and business leaders resist making major changes and continue with business as usual. As a consequence, the pace of environmental destruction will increase and resource shortages will rapidly worsen. The response to shortages will be to increase the rate of exploitation of the planet's remaining natural capital, a process that will accelerate the destruction of major ecosystems. At some point in the near future cascading environmental, economic and political crises will become uncontrollable. This will cause irreversible damage to social and biophysical systems and bring about the catastrophic collapse of industrial civilization.

Delayed Collapse

The majority of political and business leaders proactively introduce environmentally friendly technologies and provide emergency economic support to prevent unrest and conflict. These efforts will temporarily stabilize the industrial system and slow the pace of global warming and environmental destruction. However, attempts to improve the system without making fundamental changes to its unsustainable culture and economy will fail. The environment will continue to degrade, and efforts to manage crises will consume more and more scarce resources. Although system failure will be delayed, the eventual result will be the same as in the first scenario: the inevitable collapse of major ecosystems and human societies.

Transformation

As regional and global crises grow and the world economy begins to fail, it becomes increasingly clear to people all over the world that the current global system is unsustainable and heading for catastrophic collapse. More and more people will then question the destructive values and institutions of the industrial system and begin to look for constructive alternatives — pathways to survival. Large numbers of people will be attracted to the developing systems-based vision of a sustainable future. The emergence of this new paradigm will enable the rapid constructive transformation of global views, values, technologies and social structures.

The strengths and weaknesses of the global system

Industrial civilization is in a period of accelerating change. The major forces driving constructive change are rising educational levels, the growing exchange of information and the development of more sustainable and distributed information, energy and productive technologies. The US National Intelligence Council notes that "future technology trends will be marked not only by accelerating advancements in individual technologies but also by a force-multiplying convergence of the technologies — information, biological, materials, and nanotechnologies — that have the potential to revolutionize all dimensions of life."[2] Although television and the Internet are spreading unsustainable consumer values, they are also developing a global awareness of democracy, the environment, peace and human rights.

The major force driving destructive change is the continual expansion of the industrial system with its environmentally unsustainable patterns of production and consumption. We can expect climate change and increasing resource shortages to intensify international competition and conflict. Major social conflicts are also likely to occur within countries as the result of global economic crises, growing inequality and an increasing gap between rising expectations and declining standards of living.[3] Dominique Strauss-Kahn, managing director of the International Monetary Fund, has warned that the food crisis poses questions about political stability and the ability of democratic governments to survive. He said, "As we know in the past, sometimes those questions lead to war. We now need to devote 100% of our time to these questions."[4]

The enormous size, strength and diversity of the industrial system give it resilience and a tremendous ability to manage crises and resist structural change.

Because billions of people depend on the global economy for their security, their first reaction to crises will be to rally behind the system. In addition, political and economic leaders will react to crises with actions designed to strengthen social institutions and prevent global economic collapse. These responses will include macroeconomic interventions, appeals to cultural values and the control of information and dissent.[5] They know that social unrest (and ultimately, social breakdown) will develop if governments fail to manage economic crises and provide people with the necessities of life. Leaders will also proactively intervene to prevent crises through introducing new policies and supporting the development of new technologies. But unless these initiatives create an environmentally sustainable system, economic collapse and social breakdown is inevitable.

Industrial societies are incomparably richer, stronger, more productive and adaptable than Stone Age or agrarian societies because they have pluralistic institutions. These permit more critical feedback and innovation and allow individuals and groups to meet a wider range of needs and to respond more quickly to changing circumstances. However, industrial societies also have competitive, heirarchical social structures that encourage inequality and exploitation. This means that the benefits of industrialization are unevenly distributed: while some countries and groups are wealthy and have the skills, infrastructure and physical resources to manage severe problems, other countries and groups are very poor. The extreme inequality within the global system makes it vulnerable to crises.

As global crises increase, we can expect the global system to break down first at its weakest links — regions containing large populations of the people most vulnerable to climate change and resource shortages, such as subsistence farmers in Africa who are facing growing droughts or urban slum dwellers in India who are facing rising energy and food prices. For example, in early 2008 the World Bank warned that rising food and energy costs had put the economy of Pakistan at risk of collapse.[6] Pakistan's many economic and political problems would be further exacerbated by a global recession. If Pakistan does suffer economic collapse, it may not only break down but also break apart, becoming another failed state like Somalia or Afghanistan. Complete collapse would not only bring more misery to its people, but further destabilize the region, especially if its nuclear weapons were to fall into the hands of terrorist groups. Although the increasing suffering of the world's poor will have little initial impact on the economies of more developed countries, as the crises deepen (e.g., as a recession turns into a global depression), no country will be able to avoid major shocks. Every country in the world will have to deal with growing environmental, economic and social crises, although each will be affected differently.

The wealth of the United States and its relatively open political system gives it more ability to manage economic and environmental shocks than developing countries. Nevertheless, global inequality is not just a problem for poor countries. The economies of rich countries depend on developing nations providing them with a constant and ever-expanding stream of cheap resources and consumer goods. Because the industrial system is not designed for a world with finite resources, it cannot adjust to the rapidly rising price of commodities. Increasing resource shortages will not only prevent development and increase hunger and poverty in the poor world; it will also slow down and then stop economic growth in the rich world.

The United States needs to maintain global political and economic inequality in order to continue importing enormous quantities of goods and resources at artificially low prices. For this reason, encouraging the democratic self-rule and economic independence of other countries is often at odds with American economic and strategic interests. Since national security and economic growth are the top priorities of every American administration, if there are conflicts between foreign policy goals, the lesser objectives of promoting democracy abroad are usually the first to be sacrificed.[7] The only way out of this dilemma, and the only way out of an unending Middle East war, is for the United States to eliminate its dependency on cheap imports and its need to control the politics and economies of other countries.

US politicians, diplomats and generals usually interpret *sustainable security* as the creation of a militarily secure situation that protects US interests,[8] rather than as the creation of a sustainable global community that is free of violent conflicts.[9] Chris Abbott of the Oxford Research Group stated that the main mistake of the Western *war on terror* has been to assume that insecurity can be controlled through military force. Instead of focusing on "fighting the symptoms", it is necessary to "cure the disease" through removing the causes of conflict.[10] Security does not come from having many weapons but from having few enemies. The US will only be able to develop peaceful and mutually beneficial relationships with the rest of the world when it creates a sustainable and efficient economy that requires only a small fraction of the resources that it now imports, and when it practices foreign policies that help people to meet their real material and identity needs — including the need to control their own affairs.

Growing populations, developing economies and degrading ecosystems will ensure that resources are increasingly scarce and increasingly expensive. We do not have decades to decide how we will prevent global environmental and economic crises: they have already begun. Actions need to be taken now to prevent today's crisis from becoming tomorrow's catastrophe. The decisions that US leaders make in the next few years will decide whether the future is one of war or peace, poverty or prosperity. If they choose to maintain an inefficient industrial economy, they will be committing the United States to fighting futile wars over diminishing resources; if they instead choose to build an efficient and sustainable economy, they will be breaking free from reliance on foreign imports and laying the foundations for a peaceful and flourishing future.

Inequality is not only increasing between the United States and poor countries, it is also increasing inside the US. The gap between incomes at the top and the bottom has never been wider, and wealth, power and media control have never before been so concentrated in so few hands.[11] Although President Roosevelt's New Deal redistributed wealth and revitalized both economy and democracy in the US, by the 1970s inequality was on the rise and political levers were firmly in the hands of rich individuals and powerful multinational corporations. This has resulted in the loss of both popular and national sovereignty and the creation of a dangerous *democratic deficit.*

Kevin Phillips has extensively studied the relationship between wealth (or the lack of it) and politics. He believes that the history of the rise and decline of empires shows that the imbalance of wealth and democracy in the United States is now unsustainable. "At a certain point in each leading world economic power's history ... some major war proves too burdensome, economic prospects and divisions worsen, and the politics of frustration takes a critical leap forward." The result can be either radical constructive reforms, like the New Deal, or "a new and less democratic regime — plutocracy by some other name."[12]

American political and business leaders have responded to increasing environmental problems and resource shortages in both constructive and destructive ways.[13] The major responses of the federal government under President George W. Bush were destructive: to deny the existence of climate change and other environmental problems and to launch a war to control energy resources in the Middle

East. At the same time, many state and municipal governments led the way in reducing pollution and supporting the development and introduction of renewable technologies.

China provides an example of how leaders in a developing country are likely to respond to growing global crises. While the main concern of US administrations is to maintain the United States' position as an economic and political superpower, the main concern of Chinese governments is to maintain social harmony through providing continuing economic growth. Throughout China's long history there have been many famines and civil wars, and Chinese leaders are very aware of the danger of social disintegration should they fail to meet the basic needs of their large and growing population.

Although the Chinese economy has been expanding rapidly over the last few decades, growth has been unbalanced and uneven.[14] There has been too much dependence on exports, and the gaps are widening between rich and poor, and cities and countryside.[15] As well, standards of living are still relatively low, with high unemployment,[16] poor social services and almost no social safety nets. In addition, China has much less water, land and other resources per person than the United States, and major environmental damage has been caused by the rush to industrialize. As a result China is much more vulnerable to environmental and economic crises.

Top Chinese leaders are quite aware of the risk growing environmental problems pose to the Chinese economy. They frequently stress the need for better environmental management in speeches and are mandating increasing energy efficiency and the introduction of renewable technologies. However, they still have an industrial perspective that values economic expansion over environmental sustainability. Moreover, in China federal environmental initiatives are often blocked or disregarded by lower levels of government. In 2007, President Hu Jintao attempted to include the cost of pollution in the measurement of the Gross Domestic Project. But the early results of *Green GDP* accounting were so grim — with the cost of environmental degradation reducing growth rates in some provinces to zero — that local officials forced the project to be shelved.[17]

China's authoritarian system is opening, but very slowly. Despite censorship and tight controls on the press, information is spreading. Li Datong, a Chinese journalist, believes that these trends are unstoppable. Internet use is growing rapidly, and the press is now publishing opinion columns, some of which are critical of the authorities. Significantly, the leadership recognizes the need for more openness. In a 2007 article in *People's Daily*, China's premier Wen Jiabao said that "democracy, the rule of law, freedom, human rights, equality, and mutual respect are not exclusively capitalist values. They have come about as the result of the gradual advance of history. They are common human values."[18]

As countries industrialize, they require increasingly well-educated and mobile workforces. A better-educated public in turn demands more transparent, accountable and efficient institutions. As a result there is an ongoing tension within industrial societies between the functional need of the system for increasing communication and openness, and the need of ruling elites to limit and distort the flow of information in order to maintain control.

While the United States is a much more open society than China, it is also governed by elites who restrict the flow of information and block changes that could threaten their wealth and power. Orville Schell, a China expert, made the point that "If there is not a censorious state at work where leaders find certain kinds of news and views unwelcome, then there is often a censorious marketplace at work in which advertisers do not welcome the coverage of certain kinds of topics because it disturbs consumers and harms sales."[19] Despite these obstacles, accurate information is spreading and progress is being made in almost every country. However, the pace of constructive change is not keeping up with the rate of environmental destruction. This is due to the resistance of vested interests in combination with the industrial worldview and its consumer values.

The longer environmental issues are avoided, the worse they become. Repairing the damage becomes increasingly difficult and expensive, and the likelihood of environmental and economic collapse increases. For these reasons it is in the interests of business and political elites as much as anyone else to stop pollution, create a sustainable global economy and prevent the catastrophic collapse of the environment and civilization. The danger is that they will not be willing to risk reforms that could cost them their privileged positions. Elizabeth Economy, who studies Chinese environmental issues, said "an authoritarian, decentralized, endemically corrupt political system has little hope of becoming a global environmental leader, much less getting its own environmental house in order. What is missing from China's environmental protection effort to date is ... a system of transparency, accountability and rule of law that rewards officials and businesspeople who do the right thing and punishes those that do not."[20]

China, India, Indonesia, Brazil, Nigeria and every other developing country are facing enormous and growing problems. The costs of food and energy are rising at the same time as the global economy is running out of cheap resources and beginning to stall. Like the United States, they have only two basic choices: to try to maintain and expand unsustainable industrial economies in the face of rising resource costs and degrading environments — or to create sustainable economies. Their challenge will be to maintain social stability (through providing the many poor people in their societies with affordable food and energy), while simultaneously making the transition to a sustainable economy. This will be very difficult for countries that already lack resources, infrastructure and expertise. These problems are compounded by the fact that many developing countries also have governments that are more corrupt and less accountable to the public than those in the developed world. The more governments are controlled by authoritarian elites, the more they are likely to respond to major crises with methods that block change and suppress dissent, and the less they are likely to be interested in solutions that require a redistribution of wealth and power.

As resource shortages grow and the global economy deteriorates, the pressures on developing countries will be extreme. On one hand, this will encourage them to become less dependent on the global economy and to build on their strengths, such as strong community support networks and the retention of (some) traditional ecological knowledge and pre-industrial technologies. Many rural communities that are already living largely outside of the market economy may find ways to survive global economic shocks and create sustainable local economies based on organic farming, renewable energies and a low-consumption lifestyle. On the other hand, as prices rise governments will have immense problems meeting the needs of people living in the megacities of the developing world. Most government resources are likely to be spent on short-term emergency measures to prevent social unrest, such as food and fuel subsidies.[21]

The governments of developing countries are not likely to lead the way in finding solutions to global environmental problems. At this time they are focused on maximizing economic growth in order to raise standards of living; in the near future they will be increasingly focused on the minimal survival needs of their citizens. Moreover, they believe that because people in developed nations pollute more than people in developing nations (for example, people in the US produce more than four times more emissions per capita than the Chinese), rich countries should first reduce their per capita emissions to the levels of developing countries before they ask poorer countries to make cuts.[22] Rajendra Pachauri, head of the Intergovernmental Panel on Climate Change, commented, "Looking at the politics of the situation, I doubt whether any of the developing countries will make any commitments before they have seen the developed countries take a specific stand."[23]

The question of economic justice is central to resolving major global issues and creating a sustainable global system. The poorer four-fifths of the world's population will never agree to any arrangement that leaves them with less than their fair share of the Earth's resources. And yet no meaningful change is possible without their agreement — although the Chinese produce less pollution per person than Americans, their huge population means that China now emits more greenhouse gases each year than the United States.[24]

This problem can be solved using a principle called *contraction and convergence* (C&C). C&C means that developed nations should reduce their emissions while allowing developing nations to increase their emissions until all nations are emitting the same per capita levels of greenhouse gases. Sir Nicholas Stern proposes that average global emissions need to be reduced from their present levels of about 7.7 tons (seven tonnes) per person per year to around 2.2 tons (two tonnes) per person. This would mean that Australia and the US, which produce about 22 tons (20 tonnes) of pollution per head each year, will need to make reductions of 90%; while a developing nation like India, which now produces almost 2.2 tons (two tonnes) of greenhouse gases per person, will have to prevent its per capita emissions from rising.[25]

While C&C is an important principle and one that developing nations are likely to agree to, it is only part of the solution. The challenge is not just to negotiate a fairer way for continuing to pollute and increase global warming but to negotiate agreements that stop further pollution and begin to reduce global warming. Erwin Jackson, policy director of the Australian Climate Institute, said that Sir Nicholas' targets were dangerously conservative. "What he is saying is that it is OK to get on a plane if there is a 50% chance it will crash. The kind of stabilization targets he ... talk[s] about would only give us a 50-50 chance of avoiding dangerous climate change."[26]

The only solution that is both ethical and sustainable is for every person on the planet to be assigned the right to their fair ecological footprint (their *Earthshare*). Any person who consumed and polluted more than their fair share of resources would have to purchase extra resources from other individuals who were consuming less than their share. There are multiple benefits to this approach: it is ethical, logical and practical; it creates a market mechanism for creating a sustainable global economy; it rewards conservation rather than consumption and it provides a simple system for redistributing resources fairly and eliminating the worst poverty on the planet.

Of course, no nation is likely to agree to reduce their consumption of resources and production of waste if this will necessitate a reduction in living standards. It will only be possible to secure international agreements when people can see that their economic security will be improved by the creation of a fair and sustainable global economy. A paradigm shift will need to occur for the nations of the world to change their destructive environmental habits: most people will have to recognize that further economic growth is only possible if it takes place within environmentally sustainable limits; and most people will have to realize that the fastest growing sectors of the economy are those that are reducing their use of increasingly scarce and expensive resources.

Constructive and destructive interventions

While governments are likely to respond to growing crises in many different ways, not every response will be constructive. Constructive interventions reduce imbalances in the system and increase the system's openness and resilience, while destructive responses increase imbalances and make the system more closed, less adaptable and more vulnerable to major shocks.

Constructive interventions, such as restoring damaged ecosystems or developing renewable technologies, address the sources of problems and develop sustainable solutions. Destructive responses, such as censorship or military interventions to secure access to scarce resources, avoid making structural changes. Repressive actions can temporarily prevent social unrest and/or stabilize the existing system. But because they delay needed changes, reduce informed discussion and increase conflict, over the long term they only deepen problems and accelerate the process of collapse.

In every country some leaders will promote structural change once they realize the gravity of the situation; some will reluctantly support change once they realize that the only alternative to creating a sustainable system is environmental, economic and social collapse; and some will oppose any kind of change that threatens their narrow interests. In order to encourage rapid change, leaders and the public need to be educated on the stark choice facing humanity — either transformation or collapse — and a powerful international movement needs to be mobilized in support of constructive action.

At present global leaders are responding far too slowly to growing environmental problems. Only a few currently realize the need for immediate action to avert disaster. However, as scientific evidence mounts, as prices of food and energy rise, as extreme weather events become more frequent and as public awareness and pressure grow, more leaders will begin to speak out and propose constructive policy changes. In the next few years major countries may agree to collaborate in a massive international effort to develop environmental solutions.

> If governments can recognise a cyclical financial emergency and in an instant move heaven and earth (and billions of dollars, pounds sterling and euros) to contain it, why can't they do the same in response to a permanent environmental emergency?
>
> — Andrew Dobson, Professor of Politics, Keele University[27]

One hopes this collaboration will occur, and the rate of global warming and ecological destruction will be slowed. However, while global leaders may agree to efforts to avoid and mitigate the worst consequences of climate change and resource scarcities, they are not likely to agree to the complete economic, political and cultural transformation of their countries and the world system. The objective of international agreements will be to maintain economic growth through reducing environmental damage and increasing the efficient use of resources, not to transform consumer societies into conserver societies. Because the basic worldview, values and structures of the global system will not change, the industrial system will continue to give economic expansion priority over environmental sustainability and will continue to degrade the environment.[28] This process will also continue to increase global inequality and will inevitably provoke conflicts over increasingly scarce resources.

As a result, while efforts to resolve environmental problems within the current paradigm will delay the onset of global crises, they will not be able to prevent them. Nevertheless, we need to encourage politicians and business leaders to reduce pollution, reverse global warming, conserve resources, prevent the loss of biodiversity and support the development and introduction of sustainable technologies. These interventions will not by themselves transform our unsustainable global economy into a sustainable system. But by slowing the pace of environmental degradation and reducing resource consumption, they can lessen the severity of the crises and delay the onset of catastrophic collapse. In addition, constructive responses support the process of societal evolution through accelerating the development and spread of sustainable ideas, values and technologies.

At present global events are dominated by developments leading towards collapse. All countries are racing to accelerate economic growth with little regard for the environmental consequences. Even the most environmentally aware countries are still increasing their annual consumption of resources, and no country has yet mandated that all economic growth must take place within sustainable environmental limits. As a result, many of the efforts to reduce greenhouse gases make more commercial and political sense than scientific sense. For example, growing crops for biofuels on land converted from rainforests or peat bogs produces up to 420 times more carbon dioxide than it saves.[29] Moreover, the cultures and social institutions of all the major countries in the world support consumerism and the exploitation of humans and nature. These views and structures will not change overnight, since they are deeply ingrained not only in laws but also in people's beliefs and behaviors.

The consequences and costs of system failure are all too apparent. States like Somalia, Afghanistan and Iraq have already collapsed, leaving power in the hands of criminals and warring tribes. The question is how well other poor and developing countries will be able to manage severe and growing environmental and economic crises. Food riots have already broken out in many countries, including Egypt, the Philippines and Peru. Will these countries descend into chaos when most people can no longer afford to buy essential supplies of food and energy?[30] When there is not enough to eat, will civil wars break out in deeply divided societies like India, Indonesia or Nigeria?

If global warming is not quickly reversed, and if soils and water supplies continue to degrade, these terrible scenarios are inevitable. We only have to look at examples like the Soviet Union, Yugoslavia, and Iraq to see how quickly even relatively developed societies can break down when faced with severe economic and political crises. The main difference is that in the future these crises will not be localized to one country at a time, and they will not end after a country's economy has been brutally restructured. The crises will be worldwide and they will worsen year by year as both the global environment and the global economy continue to deteriorate.

The challenge for leaders and all citizens in every country is to prevent environmental and economic crises from getting out of control and causing collapse. Proactive measures need to be taken to prevent major crises from developing, then leaping from country to country, interacting and building into perfect storms that destroy economies and governments. If poverty and hunger are allowed to increase in the poorest states, more countries will collapse and become failed states; their failure will increase global instability and accelerate the collapse of the global system.

Political and business leaders in developing countries need to realize that the industrial model of development is no longer viable. Just as there is no longer a need to cover a country with landlines in order to provide people with telephones, there is no longer a need to build a traditional, centralized industrial economy in order to provide economic growth. Leapfrogging the industrial stage of development is now not only possible, but necessary. If they are to avoid economic collapse, developing countries must focus their efforts on creating decentralized, increasingly self-sufficient, sustainable economies.

As a major oil and gas exporter, Russia is unlikely to lead the way towards the creation of a sustainable society.[31] However, Europe, the United States, Japan, India and China are all resource importers, and it is in their interests to become less dependent on increasingly scarce resources. But it will not be easy to make the transition from economic growth based on increasing resource consumption to economic growth based on increasing efficiencies. This massive technological and cultural transformation will require the mobilization of entire nations (similar to efforts made during wartime) and the reorganization of the global economic system. But this type of effort has succeeded before. For example, President Roosevelt mobilized the US public to support rapid economic change with the New Deal and then again in the Second World War. If visionary leaders emerge it is quite possible for rapid transformative change to take place in most of the world's major economies.

Global awareness of the need to create a sustainable system is quickly growing in tandem with the development of practical solutions to environmental, economic and political problems. These synergistic developments are giving birth to powerful progressive movements and informing, encouraging and challenging political and business leaders to make constructive changes. Moreover, the fusion of the new holistic culture with the new technologies is creating the potential for systemic transformation, as it is providing people with a clear vision of a better type of societal system as well as the practical tools for its creation.

Let's examine some of the factors supporting each of the possible future scenarios.

Scenario 1: Business as usual

There is a real danger that the majority of political and business leaders in the world will fail to act quickly enough to preserve major ecosystems and prevent disaster.

Their reluctance to make disruptive changes comes from a combination of ignorance (regarding both the urgency of the problems and the existence of solutions), inertia (the immense difficulty of changing billions of people's beliefs and institutions) and interests (the active obstruction of political and economic elites who fear losing wealth and power).

Most policy makers also believe that the constant expansion of industrial production and consumption is not only beneficial but also necessary for maintaining and improving living standards. The consequence of this worldview is that every country in the world considers economic growth to be more important than ecological conservation. For example, although Canada and Australia have both signed agreements to reduce their production of greenhouse gases, emissions are steadily rising and both are making major investments in expanding fossil fuel exports. Although the Chinese and Indian governments recognize the serious threat of climate change and are concerned about future resource shortages, both are encouraging the rapid growth of their automobile industries. If current growth rates continue, the number of vehicles in China and a number of other Asian countries will increase 15 times within 30 years, dwarfing any fuel efficiency gains.[32]

High commodity prices will make it profitable to exploit previously uneconomic resources such as marginal soils and difficult to access forests and minerals. This may meet growing demand for a number of years and continue the illusion that the global economy does not have to operate within biophysical limits. However, another possibility is that growing food and energy shortages will trigger even sharper price rises, which may in turn provoke a global depression.

The consequences of ignoring the need to change our unsustainable economy will be an accelerating consumption of the planet's remaining natural capital, an accelerating destruction of major ecosystems, and an accelerating rate of global warming. We will not be able to avoid major changes — but if we continue with business as usual we will ensure that when the changes come they will be catastrophic.

Scenario 2: Adjusting the existing system

Most governments in the developed world are becoming increasingly aware of the dangers posed by climate change and environmental degradation.[33] However, it is very difficult for politicians to make significant changes within the present economic and political framework. Their usual response has been to try to gradually reduce the amount of environmental damage done by their economies. Because politicians must try to please voters who are concerned about the environment without alienating environmentally polluting businesses or their employees, policies are often contradictory. For example, the German Chancellor Angela Merkel pushed the European Union to make significant cuts in greenhouse gas emissions while simultaneously opposing efforts to sharply raise fuel efficiency standards for large German luxury vehicles.[34]

The easy way out is to adopt politically expedient solutions. These often have dubious environmental and economic value. Some examples are massive water diversion projects and expensive programs to produce biofuels from food crops, to build new nuclear plants and to capture and store carbon emissions from power plants.[35] The comparative merits of investing the same resources in conservation or the development of renewable and distributed technologies are rarely examined before the projects are approved.[36]

Many analysts believe that it will be very difficult to gather sufficient political support to make the huge investments necessary to transform industrial society, especially in the middle of a worsening global economy. Michael Müller, an official in the German Environment Ministry, commented, "This is a classic situation in which all environmental policy goals take a distant back seat to short-term economic priorities and we drop the ball on important investment decisions."[37]

Attempts to improve the existing system have resulted in some significant achievements, such as the 1989 Montreal Protocol that has been successfully phasing out the production of ozone-destroying chlorofluorocarbons, and the US Clean Air Act of 1990 that curbed acid rain and restored life to many North American lakes and rivers. Many other constructive initiatives are being undertaken now: the European Union is aiming at a 20% reduction in European greenhouse gas emissions by 2020;[38] efforts are being made to develop crops that are more drought and heat resistant;[39] French President Nicholas Sarkozy has announced that France will freeze the building of new motorways and airports and instead invest in rail transport,[40] and Bharrat Jagdeo, President of Guyana, has offered to preserve Guyana's entire rainforest in return for aid in developing a green economy.[41]

These are wonderful initiatives that contribute towards the creation of a sustainable world system. Unfortunately, they will ultimately be futile if the global economy is allowed to continue on its environmentally destructive path. Piecemeal changes and incremental adjustments do not alter the fundamental problem — the industrial system is already consuming and polluting at an unsustainable rate, and almost every government and major corporation is currently planning on expanding their use of natural resources.

Scenario 3: Transformational change

The idea of completely redesigning society in order to preserve the environment strikes many people as both unnecessary and unrealistic. James Rogers, the CEO of Duke Energy, has said that environmentalists' demands that his company stop building coal-fired power plants reflect a "snap-your-fingers, instant transition of the economy" mind-set. "My requirement is to balance reliability, affordability, and clean energy."[42] But what is reliability and affordability? Can we afford to continue using technologies that increase global warming? Can we afford the catastrophic collapse of the environment and civilization?

As long as we have a worldview that believes that human economies exist outside of nature, we won't be able to accurately see either the problems or the solutions. Once we make a paradigm shift and begin to recognize that human economies are completely dependent upon their environments, preserving healthy ecosystems becomes an imperative. The question then is not whether we can afford to create an ecologically sustainable global system — we have no choice if we wish to survive — but what we have to do in order to quickly transform the industrial system and prevent irreversible global warming.

The new worldview makes it possible for the consumer society to be rapidly transformed into a conserver society because it challenges us to evaluate all economic and cultural activities in terms of how they support the transition to a sustainable society. Because the new paradigm is based on systems thinking, it helps people to understand how issues are interconnected, to analyze problems and to develop solutions.

The dynamics of societal change

Figure 2 illustrates the process of societal collapse and transformation (the line represents the global system). Growing environmental problems and resource shortages will provoke cascading crises (this is the point in the diagram at which the solid line representing the global system descends into the red circle). As local crises become generalized (e.g. through a global depression), the global industrial system will become increasingly unstable, chaotic and dysfunctional. While the weakest parts of the system (the poorest countries) are likely to be the first to suffer

Modeling societal change

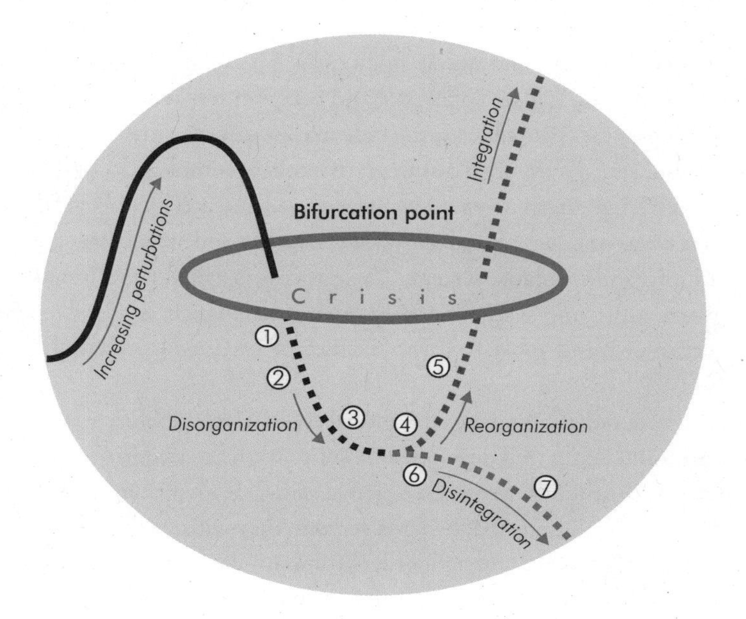

The springboard effect helps systems reorganize.

① People lose faith in the industrial system as crises worsen.

② Human and economic resources are released from the system.

③ Support increases for both inclusive (sustainable) and exclusive (ethnocentric) solutions.

④ If sustainable solutions are supported, constructive reorganization begins.

⑤ The reorganization of the global system accelerates.

or

⑥ If ethnocentric values and structures dominate, conflicts over scarce resources intensify.

⑦ Global civilization disintegrates.

Figure 2: BEST Futures

economic and political collapse, hardship will progressively affect more and more people in developed countries.

The diagram shows the process of increasing crises (the solid line) followed by a period of increasing crisis and system failure (the descending dotted line). As the global system collapses, it loses coherence and becomes less resistant to structural change. At this point it is possible for the industrial system to either reorganize into a sustainable holistic system or to disintegrate into less complex, pre-industrial societies. If constructive alternatives are well developed and are supported by enough people, the process of collapse will allow views, values and social structures to be reorganized into a more viable system. This process is called the springboard effect — since a collapsing system is more open to change, it is possible for it to quickly reorganize into a more functional, efficient and environmentally relevant system.

On the other hand, if unmanageable global crises develop before sustainable alternatives are sufficiently well developed or supported, the majority of people are likely to reorganize around nationalist or tribal values. Rather than trying to create peaceful, sustainable solutions that benefit everyone, they will put the survival of their own group first, resist change and fight other groups for increasingly scarce resources. These conflicts — some of which may be fought with weapons of mass destruction — will exacerbate environmental problems and hasten the catastrophic collapse of major ecosystems. Since international cooperation will have broken down by this stage it will no longer be possible to prevent runaway climate change, which will ruin crops and cause the mass extinction of most species. This disastrous combination is likely to kill off most of the human population and make it almost impossible to ever again establish advanced human economies and civilizations.

Possible time frames for future scenarios

These scenarios are not science fiction, and they are not problems for future generations. The process of global collapse and transformation has already begun. Whether your life is wonderful or terrible is largely a matter of where you live. Some places in the world — like the Lake Chad region in Africa — are already experiencing the devastating consequences of climate change.[43] However, the outlook is far from hopeless. While environmental degradation, resource shortages and conflicts are rapidly developing problems, renewable technologies and whole systems designs are rapidly developing solutions. The question is not whether we can solve the planet's problems — we have the knowledge and the technologies. The question is whether we will solve them in time.

It is impossible to estimate how long it will take global crises to develop or the extent to which national governments and the global economy will be able to manage crises. We can only make very general forecasts because we do not know the precise tipping points at which major biophysical systems will begin to fail or the production of critical resources will begin to decline.[44] Food shortages could develop gradually with rising demand and slowly declining supplies, or develop quickly after another year or two of extreme weather and terrible harvests. Oil and gas shortages could also develop gradually, or suddenly deliver a shock to the global economy if peak production occurs earlier than expected.

It seems likely that major crises will occur sooner rather than later because food and energy prices are rapidly rising, because global temperatures are increasing more quickly than forecast and because thousands of species are on the edge of extinction. While the general consensus is that peak oil and gas production will be reached around 2015, prices are rising because global production is already failing to increase as quickly as demand. Most predictions are that prices will continue to rise over the next few years. This will make energy unaffordable for increasing numbers of people. Global economic crises will occur when major global shortages push oil and gas prices through the roof.

Food prices are also rising because production is not keeping up with demand. If the global production of food actually begins to decline (due to climate change, falling water tables, etc.), prices will skyrocket. If you are one of the billions of poor people in the world, the economic crises have already begun. For those of us who are fortunate to live in the rich world, economic hardship has not yet arrived, but it is closer than we imagine.

The greatest threats are irreversible climate change and species extinction. A study by the United Nations Environment Program shows that glaciers are shrinking at an accelerating rate: the pace doubled between 2006 and 2007.[45] The chairman of the Intergovernmental Panel on Climate Change, Rajendra Pachauri, acknowledged that the situation is much more urgent than previously believed, "If there's no action before 2012, that's too late. What we do in the next two or three years will determine our future. This is the defining moment."[46]

James Hansen and other climate experts have pointed out that there is a high risk of disaster if global temperatures are allowed to rise by even a few degrees. In order to prevent runaway climate change, we will not only have to prevent further greenhouse gas emissions, but also reduce atmospheric carbon dioxide to a level at which the global climate can stabilize. This target will have to be achieved within decades.[47] In the view of Paul Rogers, "It is the period through to around 2015 that is the key. If genuinely major changes are made in that time — perhaps through a unique combination of citizen action and political acumen — then the prospects could be good."[48]

The risks and costs of action versus inaction

Transforming the global economy will be a huge, difficult and enormously expensive task. While environmentalists believe that action is immediately needed to prevent climate change, many other people believe that the threat of global warming is exaggerated. So how do we decide if the potential risks justify the potential costs?

David Spratt and Philip Sutton point out in *Climate Code Red* that the US regulatory standard for nuclear power stations is that there should be not more than a one-in-a-million risk of serious accident. We probably wouldn't fly if the risk of crashing was one-in-a-thousand (in 2004 it was one-in-four-million).[50] But most governments and individuals do not seem to be similarly concerned about the risks of global warming, which scientists warn will likely cause immense economic damage

Figure 3: BEST Futures [49]

The abolition of slavery

A few determined people created a powerful movement that ended thousands of years of legalized slavery.

It is possible to change the world.

Olaudah Equiano
Writer and social activist
(1745-1797)

and kill off a third or more of all the species on Earth by 2050, and possibly cause the deaths of billions of people and the extinction of most species as early as 2100. We now have a double standard — while we are not willing to tolerate much risk in our own lives, we act as if we are willing to accept a very high level of risk for our children and the planet.

The risks and costs of acting or not acting are not comparable. Preventing climate change risks a small, short-term reduction in the rate of economic growth. Not preventing climate change risks the future of humanity.[51]

Constructive change is possible

In the 18th century, slavery — which had existed for thousands of years — was considered normal. But in 1783 a group of British Quakers started a movement to abolish slavery. The movement researched and exposed the atrocities of the slave trade and educated the public with meetings and pamphlets. As the truth about slavery spread, the movement grew in strength and influence. Although many of the richest and most powerful men in Britain were slave owners or slave traders, they were vulnerable to accusations of immoral and unchristian behavior. Slave owners fought hard against the abolitionists, but the public turned against them and in 1807 the slave trade was banned; then in 1834 all slaves in the British Empire were emancipated. The Anti-Slavery Society (which is now called Anti-Slavery International) then worked to outlaw slavery in other countries. Although the struggle to eliminate slavery is not over, legalized slavery no longer exists.

The anti-slavery movement was formed by a few dedicated people at a time when only wealthy white males could vote, most people were illiterate and communications were primitive. Despite this, they succeeded in changing the views, values and laws of Britain in 50 years and creating a movement that spread throughout the world — and won. We don't have much time, but we have the Internet; the process of transformation is well under way. Millions of people are already working for change. If the abolitionists could end slavery two hundred years ago, we can save our planet now.

Never doubt that a small group of thoughtful, committed citizens can change the world. Indeed, it is the only thing that ever has.

— Margaret Mead,
cultural anthropologist,
1901-1978[52]

11 The Design of a Flourishing Earth Community

This new millennium ... should be the millennium of life.

— Evo Morales, the first indigenous President of Bolivia[1]

Health is wholeness

Sustainability is not just a good idea, but a necessity. The global economy will not exist in the future unless it operates within the Earth's carrying capacity. These limits — our planet's annual production of environmental goods and services — define the physical parameters of a sustainable global system. In nature and society, function and form are closely related. As a consequence, the core requirement of a sustainable system is an ecologically relevant worldview that recognizes the interdependence of all life on earth. Besides giving it meaning and direction, a society's view of reality creates a coherent framework for organizing and coordinating social organizations and daily activities. For this reason the development and spread of a holistic paradigm is the key to the constructive transformation of the global economy.

The rationalist worldview of the present Industrial Age is dysfunctional because it primarily supports the exploitation of human and natural environments. It sees reality as being composed of unconnected objects that exist solely for human use. This paradigm facilitates the development of centralized social structures that support political and economic expansion without regard for either human or ecological well-being.[2]

Because the integral, holistic worldview of the emerging Information Age has different goals, it creates different structures. These structures are designed to improve human and natural health and wholeness. Its systems-based perspective sees reality as composed of interdependent processes. This worldview facilitates the self-organization of decentralized social structures that support sustainable development.

Many different elements of a new type of sustainable society are now developing. The integral worldview is also evolving, and with it the systems tools that are needed to successfully integrate emerging values, social structures, economic processes and technologies into a new, functioning societal system.

Conscious self-organization

Although it is impossible to predict the exact design of a future civilization, we can determine the functional requirements of a viable societal system and from these determine its basic structural requirements. A future civilization will only exist if it is sustainable, and it will only be sustainable if it is able to meet essential human and biophysical needs.[4] This means that it must be able to limit the consumption of scarce resources, share these resources more equitably among individuals and regions and ensure that the essential needs of other species are also met. A sustainable system will also need to greatly reduce resource consumption and pollution while simultaneously supporting economic growth.

In order to meet these requirements, new institutions are needed that promote conservation over consumption, cooperation over competition, peace over war. These new social structures are now beginning to develop with the assistance of systems-based views, values and technologies.

> Deep Democracy is our sense that the world is here to help us to become our entire selves, and that we are here to help the world to become whole.
>
> — Arnold Mindell, physicist and psychotherapist[3]

One of the most important new and still emerging properties of the Information Age is system self-awareness — the ability of individuals and organizations to understand how the whole societal system functions. The combination of system theories with system-based technologies allows for a qualitative leap in the ability of individuals and communities to access, create and share knowledge. The emergence of system self-awareness has enormous implications. Millions of people are becoming aware of the need for transformative change. At the same time it is becoming increasingly possible for individuals and communities to autonomously interact with the global network, and to acquire and develop the consciousness and tools they need to organize and govern their own activities.

The advantage of centralized, hierarchical structures is strength and the ability to impose order, but this comes at the cost of flexibility and efficiency. As systems become increasingly centralized and stratified they become less efficient due to the rising costs of distribution, communication, coordination and control. In order for a sustainable global economy to be much more efficient than the industrial economy, it will have to have a relatively egalitarian distribution of power, information and resources. These requirements suggest that it will not be possible to create a sustainable societal system without making a major shift away from centralized, bureaucratic organizations towards more decentralized and self-regulating communities. The shift from a primarily centralized societal system to a primarily decentralized system is the shift from structures that support a deadening process of domination and exploitation to structures that support a flourishing process of environmental and social sustainability. It is the shift from partial democracy to participatory democracy.

However, a decentralized societal network will only function if every part at every level has access to the knowledge and skills needed to appropriately interact with the larger system, to self-regulate and self-organize. The combination of an ecological worldview and systems-based technologies has the potential to empower people with the theoretical and practical tools required to control their own lives, communities and natural environments. An excellent example of this approach is the SEED-SCALE Process for Community Change which teaches people how to ignite community energy, sustain change and expand local successes into positive regional and national projects.[5]

An appropriately decentralized network can improve efficiency by giving all its parts the ability to respond flexibly and autonomously to local conditions. The need for energy and resources can be reduced by having most social and environmental needs met at the local level with local resources. Jeff Vail suggests that it is possible to reduce consumption while improving the quality of life by following three design principles: decentralized production and control; open access to essential knowledge; and environmental and cultural relevance.[6]

The practical structure for a decentralized system is a network of relatively self-sufficient communities that are integrated into wider regional and global networks through the Internet and holarchical social structures.[7] Like some modern European villages, *rurban* communities have the potential to provide the best of rural and urban life. They can be created either through the green redevelopment of existing urban areas, or through creating highly connected and interactive networks of sustainable rural communities.

On an economic level rurban communities have the potential to produce much of their own food and building materials while also making specialized products and services for exchange with the outside economy.[8] On a cultural level they can enjoy the strengths of close family and community ties while still participating in regional and global affairs. The appeal of localized food production, increased self-sufficiency and strong communities is shown by the rapid growth of ideas such as permaculture,[9] vernacular architecture[10] and Slow Food.[11]

Although most needs can be met at a local level, not all functions can or should be devolved: regional issues need to be dealt with at a regional level, and global issues need to be dealt with at a global level. For example, national and international environmental and human rights standards are necessary buffers to guard against any infringement of these rights at the local or regional levels. A decentralized network will require a holarchical structure that supports the appropriate distribution of power and resources and the appropriate self-regulation of each node and level.

While a sustainable global system will need to be primarily decentralized, it will also need to strengthen transnational governance to protect the biosphere, prevent conflicts and ensure the equitable distribution of resources. For this reason we can describe the next stage in societal evolution as the emergence of a polycentric planetary civilization — holarchical networks of human communities organized to live sustainably and preserve the web of life on Earth (the greater Earth community).

Governance and peace

While many positive changes are taking place, the world is still heading towards catastrophic collapse because the dominant values and social structures of the global system promote competition, exploitation and inequality. The challenge is to overcome the resistance of entrenched beliefs and privileged elites and rapidly introduce new laws and social structures that support cooperation, conservation and sharing. A successful example of this is the Sustainable Communities Bill that was introduced in Britain in 2007. It obliges government ministers to work with local authorities and communities on developing local sustainability.[12]

The integral perspective is a powerful tool for changing and aligning global views and values. On one hand, it explains why human societies are dependent on their biophysical environments and why we must create a sustainable global system if we wish to survive. On the other hand, it explains how healthy living systems are organized around the principle of mutual reciprocity: sustainable systems have cooperative designs that ensure that the essential needs of every part are met, thus allowing each part to remain healthy and contribute to the strength and well-being of the whole system.

A holistic perspective is an intrinsically ethical perspective. Once we begin to understand how all physical, biological and human systems are interrelated and mutually dependent, it is then easy to recognize the reciprocal bond between human rights and human responsibility and to accept the right of every species and culture to exist. This worldview produces a value system based on respecting, appreciating and caring for all life — which in turn supports a cooperative and non-violent culture.[14]

Today's competitive world order is unsustainable because the actions of its various parts do not support the well-being of the whole. The fragmented nation-state system was not designed to address global problems; although the world is becoming increasingly integrated, the global economic and political system operates without either systems thinking or adequate global structures.

It will not be possible to create a sustainable global system without more ethical and better global governance.[15] We not only need strong international laws to limit resource consumption and pollution, but we also need new international institutions that have the authority and power to enforce these laws. In order to prevent wars over scarce resources, more failed states and terrorism, countries will also have to agree to share resources more equitably, eliminate weapons of mass destruction and shift funds from war-making to peace-building.

Peace is not the absence of war. Peace is the presence of justice and the absence of fear.

— Ursula Franklin,
physicist and peace activist[13]

While effective planetary governance needs to be established, this does not mean that nation states should be abolished, any more than the existence of national governments eliminates the need for regional or local governments.[16] Moreover, adding a new layer of global government will not be enough to fix the current dysfunctional global system. It will be impossible to end the problems that have plagued humanity for millennia — war, disease, poverty, crime and environmental destruction — without developing more constructive, coordinated, transparent and accountable political institutions.

The keys to better global governance are agreements on goals that benefit all of humanity; the support of all levels of government (from local to global) for these goals;

the ability of all citizens to access information on how various governments are meeting the needs of humans and ecosystems; and the ability of all citizens to be able to directly participate in decisions that affect their own lives. In addition, political participation and power needs to be appropriate: while every level of government should only manage the issues that are relevant to their level and region, at the same time each level needs to have sufficient autonomy and power to effectively manage its own affairs.

The European Union (EU) is an example of how supranational government can evolve. Two major forces have driven the political and economic integration of the continent. First of all, the devastation of two world wars and the dangers of the Cold War persuaded many people that every effort needed to be made to eliminate the causes of conflict in Europe. Secondly, increasing trade and tourism are making European countries progressively more economically interdependent and more appreciative of each other's cultures.

To date national governments in Europe have only given limited authority to the European Union (e.g. to regulate some economic and environmental issues). Nevertheless, the structure of the EU has profound implications. Because the EU effectively reduces political tensions, increases security and raises living standards, it is a very attractive model — in fact, it is the only great power whose power the majority of people in every continent would like to see increase rather than decrease.[17]

The European Union shows that it is possible to create cooperative solutions in which all parties win. The strongest and richest nations in Europe have agreed to share some of their power and wealth with their poorer neighbors in exchange for sustainable security (i.e. reducing the causes of conflict) and expanding markets.[18] At the same time, poorer countries have agreed to engage in major political, economic and environmental reforms in order to secure membership in the European Union. Their elites have either voluntarily supported the reform process or been forced from power by popular movements.

European political and business leaders supported the establishment of the EU in order to accelerate the modernization of Europe. However, industrialization is a two-edged sword. While it has raised educational levels and general living standards, it has also destroyed many traditional industries. Market-driven globalization means that Europe cannot escape the international *race to the bottom* — the flight of industry to low-wage countries and the competition of cheap imported goods. As a result many European countries suffer from major social dislocation and high unemployment.

The major historical forces that are shaping global events are very visible in European politics: the dominant trend towards increasing industrialization and globalization (supported by most center-right and center-left governments), the reaction against industrialization (supported by right-wing nationalist parties on one side and fundamentalist immigrant groups on the other), and the post-industrial trend towards a sustainable global system (supported by green parties).

The problem for European elites is that they can neither separate European developments from global events nor stop the evolution of industrial Europe into a more democratic and egalitarian post-industrial civilization. The European Union needs more power to be able to effectively respond to growing international problems. Individual countries are not well equipped to deal with growing global issues like climate change, energy shortages and economic inequality.[19] Although national elites and national governments do not want to surrender any power, they are under increasing pressure to give more authority to the European parliament. This trend will only be reversed if a catastrophic global environmental and economic collapse creates the conditions for the rise of powerful nationalistic and fascist movements that oppose peaceful and cooperative solutions.

The same factors that created and are changing the European Union — the need for security and globalization — are also creating pressures for stronger international institutions. Most national governments and multinational corporations are likely to continue to oppose stronger international governance. However, it will not be possible for one country or one region to create a sustainable economy in isolation, since all countries will be affected by climate change and other growing global crises. Only two options exist: either nations will place self-interest first and the global environment and economy will collapse, or nations will cooperate and create a sustainable planetary civilization. As people become aware that the survival of humanity is at stake, the pressure on governments to create a sustainable global system will be immense.

Individual and societal health

Although societies are living systems with dynamics and structures that individuals do not have, because they are collections of individuals they exist to serve human needs. When we talk about societal change we are talking about changes in the views, values and behaviors of many individuals.

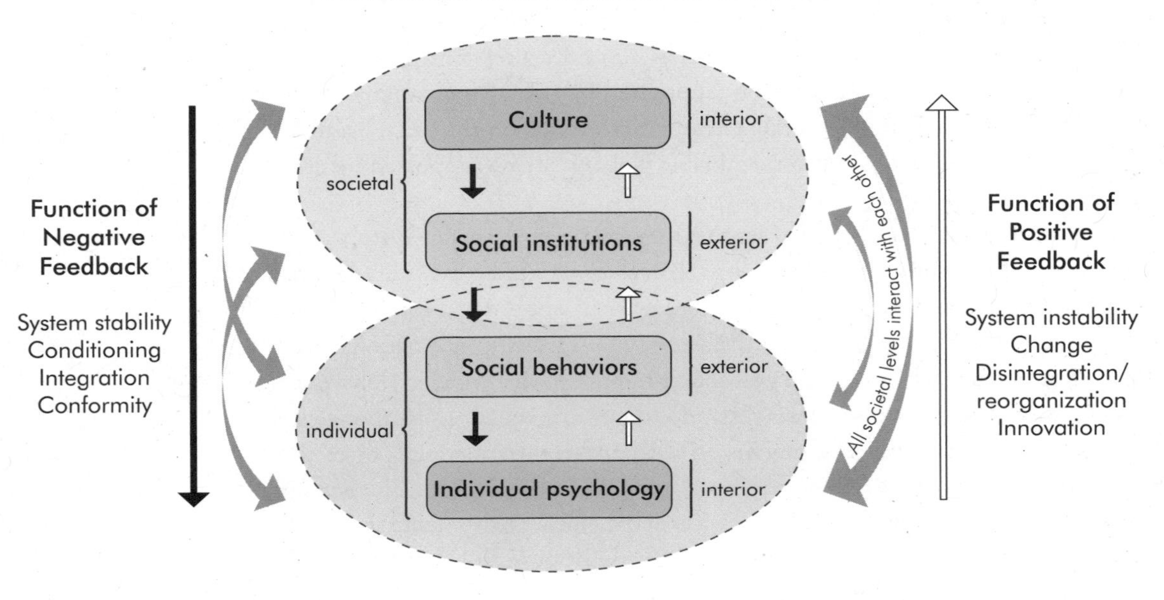

Figure 1: BEST Futures

Figure 1 shows the interactive relationships between individuals and the societies they live in. There are four holarchical (nested) levels to these relationships: the worldview and culture that pattern, direct and coordinate the whole society; the institutions that physically organize it and give it functionality; the social behaviors that regulate and coordinate individual activities and the individual psychological structures that enable people to interpret reality and act in the world.

In general, culture shapes social institutions which in turn shape individual behaviors. Both culture and the behaviors of others help to shape an individual's intellectual and emotional outlook on life. However, societies are interactive systems. When we are children, we have very little influence on society: we learn our views, values and behaviors from our parents, peers and community. But as we grow older and gain knowledge and influence, our words and actions have a stronger affect. They then either reinforce our society's culture and institutions or help to change them.

Because a society's worldview and culture shapes not only its institutions but also the psychological structures of its members, people inevitably share many if not most of the beliefs and behaviors of their societies. We cannot separate our lives from our social and natural environments. It is not possible to have a healthy economy on a dying planet, and it is not possible for people to be completely healthy when they live in a dysfunctional and destructive culture.

The world system is destructive and violent because its structures are designed to maximize the exploitation of both humans and the natural world. John Galtung, a leading peace and conflict theorist, said structural violence occurs in situations where social structures prevent individuals from achieving their full human potentials.[20] Structural violence occurs whenever institutions prevent real needs from being met — for example, through denying access to food or medicine, through physical violence, through emotional neglect or trauma or through social exclusion from education or participation in public affairs.

Industrial civilization values the objective over the subjective (things over human relationships, thoughts over emotions, wealth over happiness, the material over the spiritual); its goal is to maximize the production and accumulation of goods. To do this it must create *free markets* in which workers can be easily moved to wherever they are needed to maximize production, and in which goods, land and other resources can be easily exchanged and converted into financial capital. Since family, community and cultural bonds are a barrier to the movement and commodification of people and natural resources, the process of industrialization always involves the break-up and destruction of traditional family and community ties and traditions.[21] All forms of industrial societies, whether capitalist or socialist, weaken community bonds and relationships between humans and nature.

The psychologist Erik Erikson described mental health as a state of *psychosocial integration* in which people experience themselves as both complete individuals and full members of their communities and culture.[22] Individuals that are unable to develop constructive bonds with their families, peers and communities become psychosocially *dislocated* and dysfunctional. A lack of nurturing and social support (in combination with other factors such as genetics) can result in various forms of mental illness, addictions, anxiety, hyperactivity and emotional trauma. Because these problems are more likely to emerge when people are under stress, they are on the rise in our increasingly fast-paced and competitive world.

Bruce Alexander, an addictions researcher, has pointed out that the consumer society with its individualistic values and demands for mobility constantly undermines people's community bonds and cultural traditions. "Addiction in the modern world can be best understood as a compulsive lifestyle that people adopt as a desperate substitute when they are dislocated from the myriad intimate ties between people and groups — from the family to the spiritual community — that are essential for every person in every type of society."[24]

People often suffer emotional trauma when they are subject to overwhelming stress (either suddenly or over an extended period of time). Trauma can result in crippling, long-term disabilities, such as depressions, anxieties and inappropriate feelings of fear and anger. It is not only individuals that are traumatized by accidents, stress or violence. Many experts argue that families and entire cultures can suffer emotional trauma when exposed to unbearable conditions such as extreme poverty, disease or war.[25] Because these traumas are important parts of a group's experience of reality, they begin to shape its members' views, values and behaviors. This contributes to the development of paranoid and aggressive cultures that pass their dysfunctional behaviors from one generation to the next.

Trauma specialist Peter Levine stated, "Trauma is among the most important root causes for the form modern warfare has taken. The perpetuation, escalation, and violence of war can be attributed in part to post-traumatic stress. Our past encounters with one another have generated a legacy of fear, separation, prejudice and hostility. This legacy is a legacy of trauma fundamentally no different from that experienced by individuals — except in its scale."[26]

It is very difficult for a child growing up in a destructive culture to learn constructive behaviors and to lead an emotionally healthy life. A sustainable society must have a culture and institutions that meet all essential human needs, including our needs for close and constructive relationships with our families, communities and natural environments.

In his 2000 study *Bowling Alone*, Robert Putnam catalogued the increasing loss in the United States of all forms of connections between individuals and their friends, families and communities. This is shown by steep declines in civility, social trust and participation in community organizations and the sharp rise of aggressive interpersonal behaviors. In his view the physical basis for civic engagement is disappearing as commercial malls and spread-out suburbs replace traditional neighborhoods, and televisions and computer monitors replace interpersonal relationships.[27]

Reciprocity and trust — the bonds that create social cohesion — will continue to erode as long as our society values profits over relationships. The American cultural

We must all learn to live together as brothers or we will all perish together as fools. We are tied together in the single garment of destiny, caught in an inescapable network of mutuality. And whatever affects one directly affects all indirectly. For some strange reason I can never be what I ought to be until you are what you ought to be. And you can never be what you ought to be until I am what I ought to be. This is the way God's universe is made; this is the way it is structured.

— Rev. Martin Luther King, Jr., civil rights leader[23]

historian Morris Berman said, "The point is that the rebuilding of social capital cannot occur in a context in which power, money, celebrity and the like have become the key values of the dominent culture."[28] We will only be able to create a sustainable system if we have social as well as environmental health and wholeness, and we will only have social health if we redesign not only our culture but also the physical layout of our cities to support strong families and communities.

The culture pattern of a sustainable civilization

Societal evolution occurs in response to human and societal needs for growth and survival.[29] It involves the emergence of progressively more complex societal systems with new and better environmental control capabilities. New structures will only evolve and be adopted if they are more efficient, adaptable and environmentally relevant, and if they have improved abilities to acquire and utilize energy, resources and information.

We can model the requirements of a sustainable society using both deduction (starting with its biophysical parameters and the essential needs of humans and ecosystems) and induction (studying the factors that are already supporting the transformation to a sustainable system). Since all the major components of a new type of global system will have to support ecologically sustainable outcomes, we are also able to define many of its key environmental, economic, cultural and structural requirements.

Because a new type of societal system must be more functional than its predecessor (i.e. it must be able to meet human and social needs more effectively and efficiently), evolution involves the congruent development of interdependent and mutually reinforcing views, values, social structures, economic processes and technologies. For this reason it is possible to outline the essential culture pattern of a sustainable planetary civilization. Figure 2 summarizes the shifts in the Universal Culture Pattern that will have to occur in order for the current industrial system to be transformed into a holistic system.

Although we can describe the critical requirements of a sustainable planetary civilization, at this point we can only give a very general description of what it will look like and how it will work. This is partly because a future planetary civilization will be composed of economically and culturally diverse regions, nations and communities. While every part of the world will have to develop a sustainable economy in order to avoid collapse, every country, culture and bioregion is likely to develop its own approach.

Also, we can only create a general model because a new type of living social system is not a machine that can be invented and assembled — it has to evolve. And a key element of evolution is the complex process of trial and error, disassembly and reassembly, in which new structures emerge with previously unknown capabilities. Because we can determine the fundamental design requirements of a sustainable system, we can create conditions which will help it to develop — but we can not foresee structures and functions that do not yet exist. The next stage of evolution will always be a surprise.

Figure 2:

BEST Futures

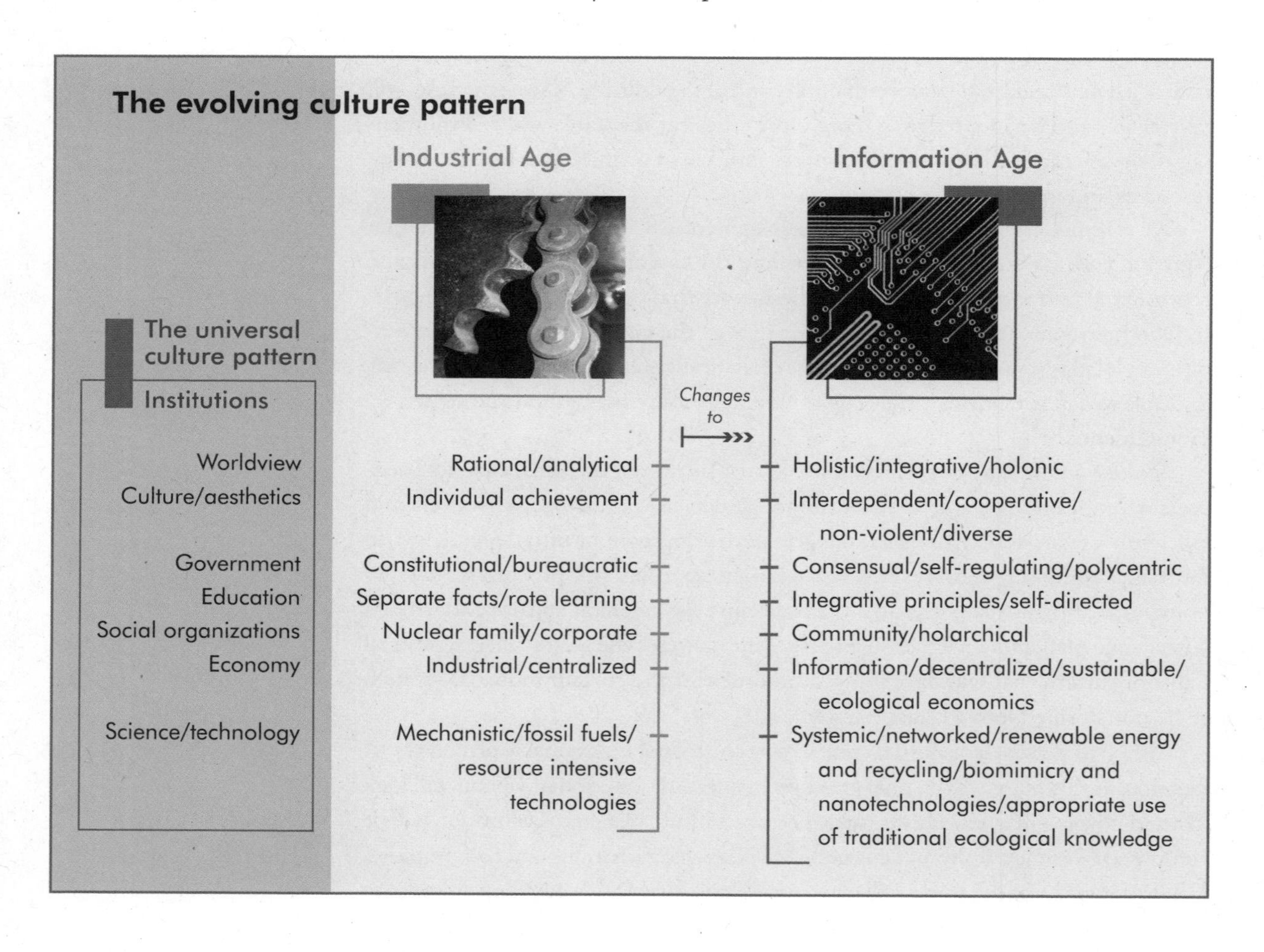

12 Tools for Transformation

> Love and compassion are necessities, not luxuries. Without them humanity cannot survive.
>
> — The Dalai Lama, head of Tibetan Buddhism[1]

The power of love

I REMEMBER A SUMMER AFTERNOON when I was driving an ambulance to a hospital and feeling depressed about all the misery in the world. My partner was in the back bandaging up the victim of a senseless assault, and I was thinking about how stupid and cruel people can be.

Just then I turned a bend in the road and passed a corner store. A group of teenagers were standing outside laughing, and down the street the sun was shining on a row of neat houses, each with flowers lining the paths to their front doors. I saw how much love each family put into their homes, and realized that the world is made up of hundreds of millions of caring families. Our planet has serious problems, but these problems can be solved because most people are intelligent, hard-working — and caring.

The coming decades will not only be a time of great crises, but also a time of great opportunities. For the first time in history tens of millions of people are working for constructive change. The strengths of this movement are that it is enormous and diverse, organic and self-organizing. It is composed of many of the brightest, most creative and most courageous people on the planet. It brings together modern science, ancient wisdom, love and faith. It is driven by both the need for humanity to survive and the desire for a better future.

But the global movement has serious weaknesses. It is largely uncoordinated and still lacks the political and economic power to prevent the destruction of nature and civilization. Our task is to unite and give it the tools it needs to successfully transform the world.

The moral imperative

As environmental, economic and social crises multiply, they will threaten our standards of living first, and then our very survival. Sooner or later everyone — families, communities, businesses and governments — will be forced to act. The question then is not whether we should act, but when we will act. Will we act while constructive change is still possible? Or will we deny the reality of the dangers and avoid acting until the problems become unmanageable and disaster becomes inevitable? The survival of life on Earth is not a problem for someone else in some other place at some other time — it is a problem for each of us right now. It is not only a global issue, but also a personal and local issue.

> Before it's too late, we need to make courageous choices that will recreate a strong alliance between man and Earth. We need a decisive "yes" to care for creation and a strong commitment to reverse those trends that risk making the situation of decay irreversible.
>
> — Pope Benedict XVI, head of the Roman Catholic Church[2]

Preserving life on Earth is not only the greatest practical challenge that humanity has ever faced, it is also the greatest moral challenge. Pope Benedict has said that polluting is a sin and declared it the moral obligation of all Roman Catholics to protect the environment.[3] Anglicans, US Southern Baptists[4] and other Christian churches, as well as many other religious leaders such as the Dalai Lama, the head of Tibetan Buddhism, and Mata Amritanandamayi, a Hindu saint, also emphasize that it is our duty to preserve the environment and all the species on our planet.

Our moral responsibilities and our personal interests are interconnected. Some people have already begun to move back to the land in order to survive the failure of the global economy. Although producing our own energy and growing our own food are good ideas, by themselves they will not keep us or our children safe from the catastrophic collapse of nature and society. There is no safe place to hide on a dying planet. Our only option is to work together to restore the biophysical systems that maintain life on Earth.

Our species has reached the edge of destruction because individual interest has been made more important than the collective good, ownership more important

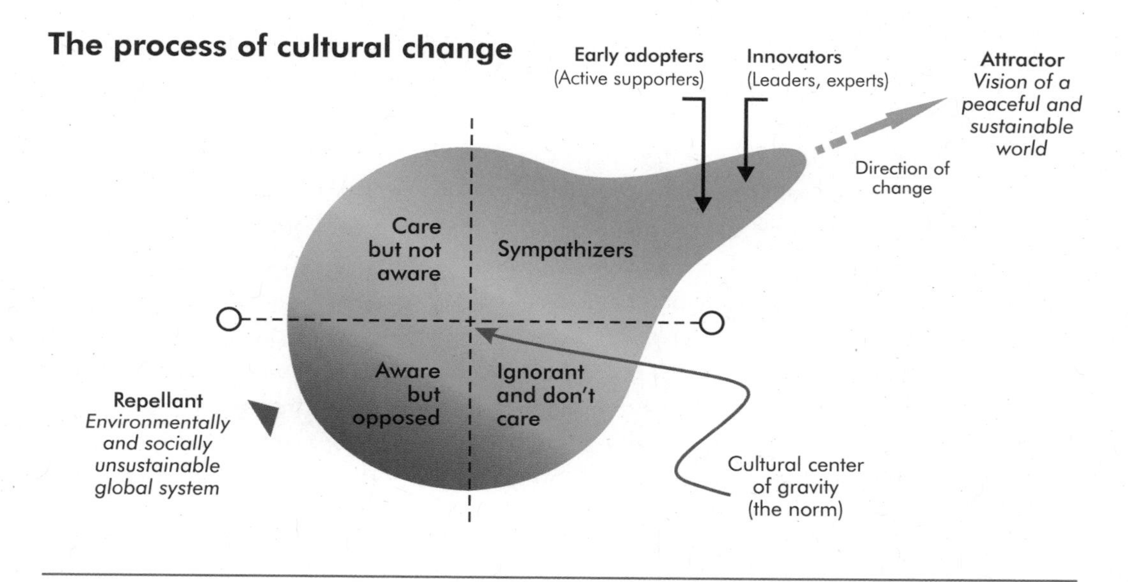

Figure 1: N. Davidson and G. Taylor [5]

than relationship and money more important than morality. In order to avoid personal and collective disaster we must once again put our common interests at the center of our lives. This is a time for commitment, courage and sacrifice. Now, more than ever, humanity needs heroes.

Helping people become engaged

Figure 1 illustrates the process of cultural and technological change. Social change involves a wide spectrum of people, from experts and activists who introduce new ideas and lead organizations, to individuals who have recently become aware of the issues and just begun to change their own views and actions. Innovators initiate and lead change, followed by early adopters and then by people who are attracted to the new ideas. At the opposite end of the process are people who actively oppose the change, either because they disagree with the new ideas and values or because the change is not in their interests.[6]

The goal is to shift the cultural center of gravity of the global system from its current unsustainable position to a sustainable position — a position now held by early adopters of sustainable values and technologies. This massive change in global values and behaviors will require the active participation of not only the majority of people who are already aware of the need for change, but also the involvement of most of the people who currently don't know or don't care.

We will need different types of approaches and organizations to engage people who have different levels of awareness, who come from different cultures and who are involved in different types of activity. For this reason social activists will need to assist the formation of multiple levels of organizations from local single-issue groups to global networks of experts.[7]

While the key elements of a sustainable system have begun to emerge, they are still very fragmented. We can help develop and integrate the different elements through presenting a clear and unifying vision of a sustainable alternative. The only way for hundreds of thousands of autonomous organizations to coordinate their activities with each other is to be aligned around a common vision, to have common values and to interact in ways that promote trust, participation and democratic decision making.

As Donella Meadows has explained, the quickest way to transform a system is to change the dominant paradigm.[8] Rapid change is possible once people understand that the existing world system is failing and can see that a sustainable alternative is possible. Our priority is to clarify this alternative and unite people around it.

The Earth Charter

The ethical requirements of a sustainable global system have already been developed and agreed to by most of the leading environmental organizations in the world. This vision was articulated in 2000 in The Earth Charter,[9] a document developed under the auspices of the United Nations to provide an ethical foundation for the emerging world community.[10] The Earth Charter is the result of a broad international consultation process in the 1990s that involved experts in law, scientists, theologians, indigenous peoples and various civil groups from diverse cultures.

The Earth Charter describes the social and environmental principles that are essential for the design of a sustainable future. Its fundamental concepts can be summarized as:

> The interdependence of all beings must be recognized.
>
> Humanity has a moral responsibility to care for all life on Earth.
>
> Life cannot be sustained without preserving and restoring ecological health and wholeness.

Essential human needs cannot be met without social and economic justice.

Environmental and social sustainability cannot exist in the absence of peace and democratic governance.

The Charter identifies sixteen principles that are necessary components of a sustainable global community. Because economic sustainability is impossible without environmental sustainability, and environmental sustainability is impossible without peace, social and economic justice and democratic government, all of these requirements must be met in order for a successful societal transformation to take place. The Earth Charter forms the cornerstone of a common vision of a sustainable world. For this reason all individuals, organizations and governments should be encouraged to read the Charter, adopt its principles and put them into practice.

Nevertheless, it is not necessary for every individual, business or government to agree to all sixteen principles before they can start to take constructive action. People can support the need for a sustainable solution to a particular local or global issue without either understanding or agreeing to the need for a sustainable global system.

Moreover, while many environmental organizations can easily agree to all the principles of the Earth Charter, at this time it is not realistic to expect everyone on the planet to have a similar awareness of environmental and social issues. It will not be possible for most people and most governments to reach agreement on a detailed plan of action because every community and nation will have to develop their own culturally and environmentally appropriate solutions. If we wish to involve as many people as possible in the task of creating a sustainable world, we must find ways to help people take constructive action on issues that concern them.

Building global agreement

At this time there are three major obstacles standing in the way of global agreement on the need to create a sustainable world system: only a minority of the world's people and governments understand the urgent need to stop environmental destruction; many powerful interests actively oppose constructive change; and the hundreds of thousands of organizations working to build a better world lack a common vision and strategy.

The practical challenge is to find ways to rapidly raise global awareness of the critical issues, and then to have agreements that support sustainable outcomes signed into local, national and international laws. Unless clear agreements are made and written into enforceable law we can be sure that many businesses and governments will continue to claim that they are promoting sustainability when in fact they are continuing to degrade the environment.

> As things stand, the movement remains disparate, atomised and marginalised.
>
> This frees politicians to expand airports and increase road capacity. Parliaments fiddle while the planet burns, and individuals are pressured to take responsibility for global climate change by "switching off at the wall". And so, inevitably, the Titanic's deck chairs are rearranged — and energy use goes up, rather than down, on Energy Saving Day.
>
> — Ann Pettifor, executive director of Advocacy International[11]

It may prove easiest to build local and global awareness and agreements in a step-by-step manner. The starting point is to first develop the broadest possible agreement around the most fundamental concept — the need to avoid catastrophic climate change — and then to use this as a basis for establishing further agreements. The assumption behind this approach is that will be easier for the nations of the world to reach agreement on the need to prevent the common threat of runaway global warming than it will be to reach agreement on the need to redistribute resources. After the need for urgent transformative change has been accepted, then people and governments may be more willing to tackle difficult issues such as reducing consumption, sharing resources and abolishing weapons of mass destruction.

A step-by-step approach could involve getting progressive organizations to agree to launch a series of world-wide campaigns designed to mobilize global opinion. Each campaign would build support for developing international agreement on one of the essential needs of a sustainable system. A logical order for the campaigns would be:

1. Humanity will only be able to prevent the catastrophic collapse of major ecosystems and human societies if we immediately act to stop runaway climate change and the further loss of biodiversity.
2. The global economy is unsustainable because it is constantly degrading the environment. Our collective survival depends on human economies operating within the carrying capacity of the Earth's ecosystems.
3. Human economies will only be sustainable when they meet essential human and biophysical needs for health and wholeness. Cultural and genetic diversity is necessary for ecological and social health, wholeness and resilience. It is not only a moral responsibility of humanity to preserve all life on Earth, but also a practical necessity.
4. Power and resources will have to be redistributed in order to meet essential human and biophysical needs.
5. A sustainable global system is not possible without peace and accountable governance.

Focusing attention on global warming and limits to growth will clarify the importance of systemic change and the urgency for local and global action on all types of issues. This growing awareness will then support major shifts in views and values, which will in turn support policy shifts on the part of governments and businesses. This will create a *virtuous cycle* of accelerating change in which structural change is possible.

Building global agreement one step at a time does not mean that everyone should only work at first on raising awareness of the dangers of environmental collapse and leave social justice, peace and other issues for later. The point is rather to put all issues within a clear ecological context, so that everything we do raises awareness of the need to prevent catastrophic collapse of nature and civilization. If the world community first reaches agreement on the need to create an environmentally sustainable economy, this will not end the process of social change. Instead it will focus global attention on the fact that it will be impossible to achieve ecological and economic sustainability without cultural and social transformation.

Figure 2: BEST Futures

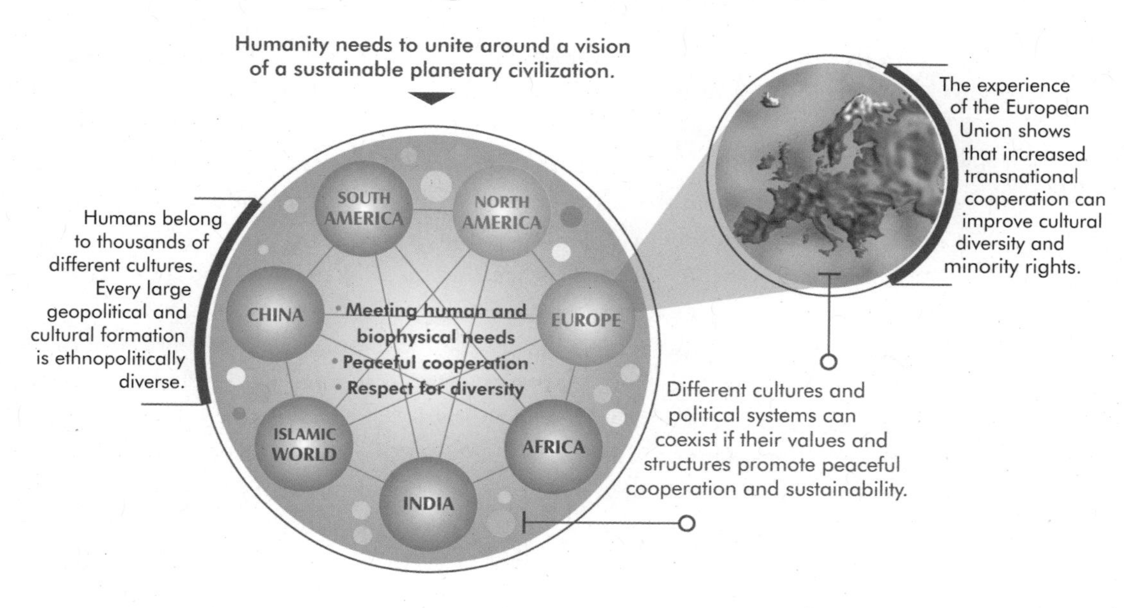

Just as we should not ignore other issues in order to raise awareness around the need to prevent runaway global warming, we cannot sacrifice one principle in order to reach agreement on another.[12] It would be meaningless to try to reach local, national and international agreements on the need to preserve life on Earth through trade-offs that sacrifice social health and wholeness. It is of no long term benefit to workers to save jobs at the expense of the environment, and it is of no long term benefit to the environment to try to reach agreements that are economically and socially unsustainable. We cannot solve global issues by win-lose approaches that benefit one group at the expense of another, or that solve one problem by creating another. Sustainable solutions work over the long-term precisely because they create win-win situations that meet the real interests of all parties.

Figure 2 illustrates how a sustainable planetary civilization can be established on the basis of a few fundamental agreements. The European Union serves as a model for this process: there has been no need to abolish national governments in order to increase international cooperation and security.

Supporting constructive change

Rapid positive transformation will only take place if powerful civil movements encourage and support the process of change. For popular movements to flourish people must know that solutions are possible and practical, or they will feel intimidated by the size and complexity of the problems they are facing. We can empower people by providing them with useful theoretical and practical tools for educating and organizing their communities. These tools include information on how local and global issues affect them; skills and resources to help their communities develop sustainable solutions that are relevant to their own cultures and environments and examples of constructive leadership and successful changes.

Since people learn the most when they are involved in solving problems, activists need to encourage the formation of many different types of local, regional and global coalitions. Figure 3 illustrates how to build coalitions around shared values. The principle is to create organizations that enable people with diverse views to work together in areas where there is agreement on the need for sustainable solutions. This does not mean that other issues and disagreement should be ignored, but that unconnected issues should be dealt with in separate settings. For instance, people with opposing beliefs on whether abortion should be legal can work together in a coalition formed to preserve the diversity of life on Earth.

Because the global fault line is the issue of ecological sustainability, the criteria for determining whether or not to support a particular policy is whether or not it contributes to the creation of a sustainable global system. Some examples: in order for economic production to be constructive it must take place within ecologically sustainable limits; in order for political policies to be constructive they must support ecological and social health and wholeness.

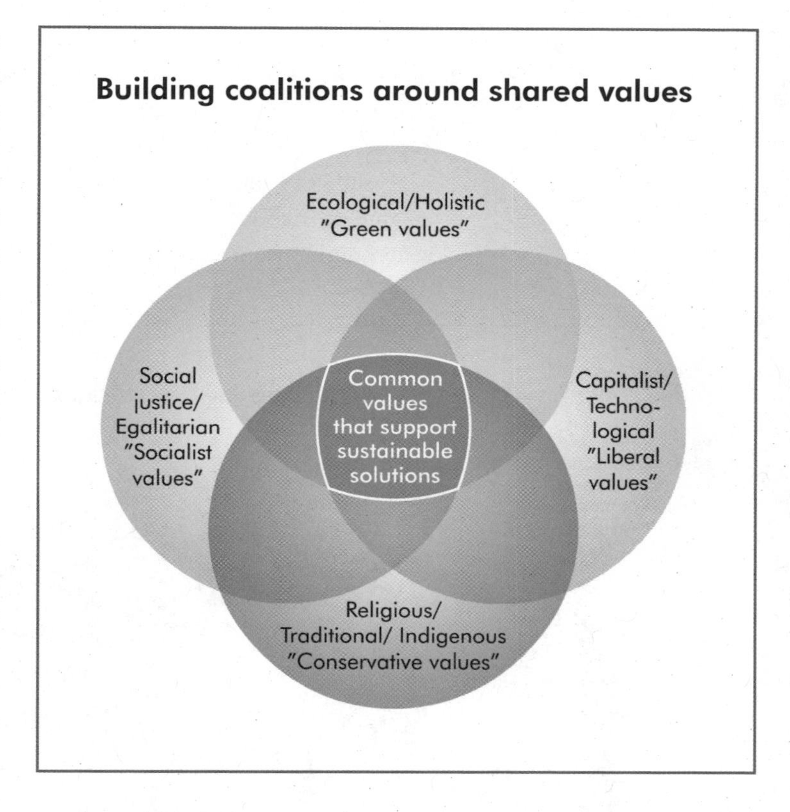

Figure 3: BEST Futures

Include and transcend

While shared values and common dangers will begin the process of dialogue, they will not be enough to overcome the deep differences in values and interests that are now blocking peace and cooperation. Many of these conflicts cannot be resolved by the current world system because its views, values and structures promote competition and win-lose outcomes. Structural change is necessary to resolve so-called intractable interpersonal and international conflicts.

Biological and social evolution builds new types of systems out of existing structures through combining them in more functional ways. In this process outdated structures and functions are eliminated or put to better uses (as in the case of our tailbones), while new structures and abilities are added. This process can be termed *include and transcend*, as existing components become the building blocks for new and more complex evolutionary levels. We can use this method to integrate current views into a systems perspective. All perspectives are valuable to the extent that they contain part of the truth. When the relevant parts of different perspectives are combined, they can provide a new and more complete view of reality.

However, while different organizations and governments can be brought together around shared views and values and while a systems approach can be used to reframe issues and develop workable solutions, major differences will still exist. Since conflicting views and interests will be the main obstacles to constructive action, the process of building local and global agreements is likely to be a complex and extended process of conflict resolution.

Conflict resolution is not about negotiating compromises that do not satisfy anyone. It is about changing destructive relationships (in which one group prevents another from meeting its needs) into constructive relationships (in which the

Figure 4: BEST Futures

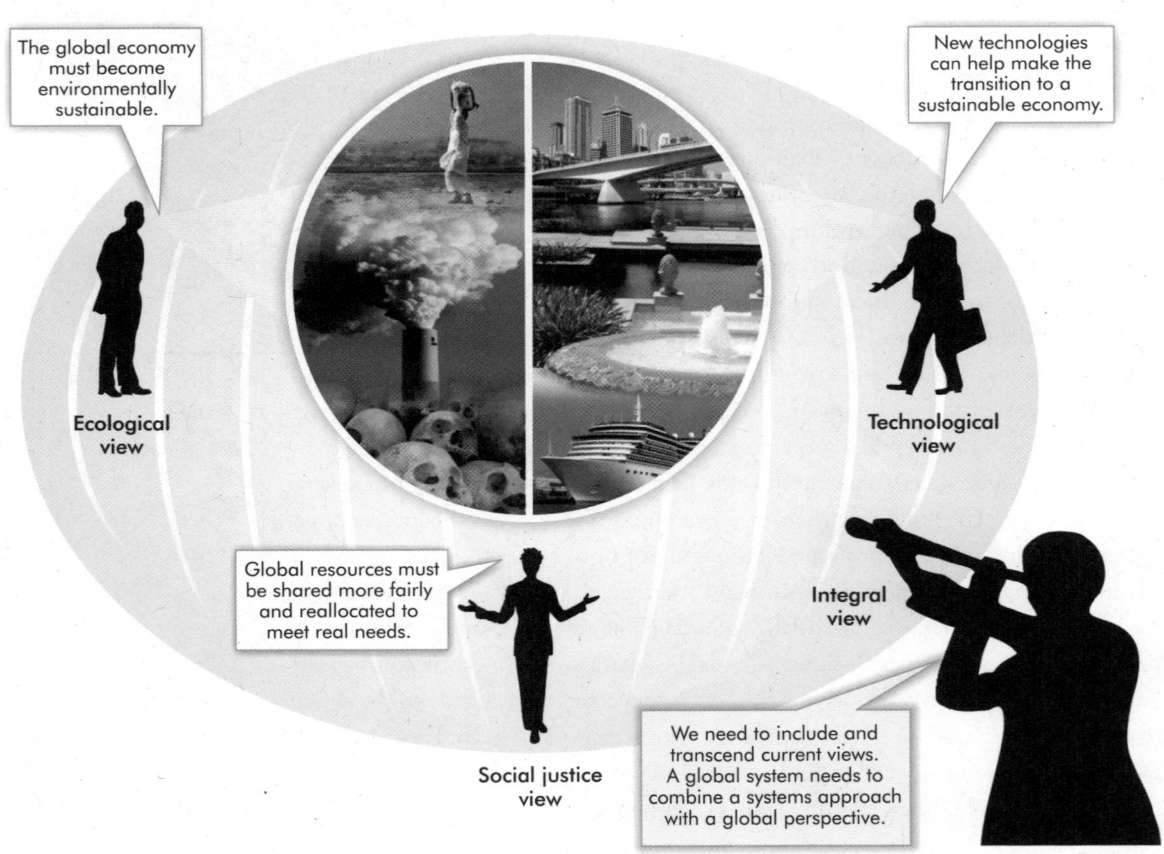

different parties work together to achieve mutually beneficial objectives). This process involves discovering common interests that transcend the dispute. Once real interests are separated from negotiating positions, and essential needs from wants, it is possible to make the shift from competition to cooperation.[13]

Because it is in everyone's interests to survive, the survival of humanity is the one common goal that has the potential to unite all the people and nations of the world. But more than negative reasons will be needed to develop long-term trust and cooperation. It will not be possible to end wasteful and destructive conflicts without eliminating their causes, which are usually rooted in fears about physical safety, material security and cultural identity. Sustainable security can only be achieved when people live in healthy environments and are able to meet their real needs. As a result, the focus of conflict resolution must be to create comprehensive agreements that meet the essential needs of all parties, including their needs for peace, material security and respect. This involves replacing structures that encourage violence, exploitation and inequality with structures that support peace, cooperation and sharing.

Most efforts to resolve international conflicts fail because of a lack of vision and integrity. Negotiators are frequently told not to sign agreements unless they secure strategic advantages that strengthen their side and weaken their opponents. Governments need to understand that sustainable security cannot be achieved through the use of oppressive force or be built on inequality; it will only exist when people have no reasons to mistrust, fear or fight others. Genuine conflict resolution starts with the desire for peace — it is a transformational process that builds trust and overcomes hatred through creating solutions that benefit everyone.[14]

> The choice between love and fear is made every moment in our hearts and minds. That is where the peace process begins. Without peace within, peace in the world is an empty wish. Like love, peace is extended. It cannot be brought from the world to the heart. It must be brought from each heart to another, and thus to all mankind.
>
> — Paul Ferrini, minister and writer[15]

Truth, trust and transformation

Because the industrial system supports the exploitation of nature and society, it destroys and wastes not only natural resources but also human lives, health and energy. The global system is environmentally and socially unsustainable because it is extremely inefficient and dysfunctional. Tremendous resources are devoted to managing conflicts in all their external and internal forms — from warfare, to legal disputes, to family breakdowns, to mental illnesses.

The industrial system creates an impoverished world where people in developing countries suffer from material scarcity while people in developed countries suffer from emotional scarcity. Feelings of neediness, anxiety and fear are at the roots of the conflicts and destructive behaviors that permeate our societies.

It will be impossible to create a sustainable system without developing more functional and constructive views, values and institutions. Transformational change is about the shift from violent to peaceful, from fearful to trusting, from defensive to open, from competitive to cooperative, from scarcity to abundance, from illness to wellness. It is about healing our planet, healing our civilization and healing ourselves. The trauma expert Peter Levine said: "Transformation is the process of changing something in relation to its polar opposite. In the transformation between a traumatic state and a peaceful state, there are fundamental changes in our nervous systems, feelings and perceptions .... Through transformation, the nervous system regains its capacity for self-regulation. Our emotions begin to lift us up rather than bring us down .... We begin to face life with a developing sense of courage and trust. The world becomes a place where bad things may happen but they can be overcome. Trust, rather than anxiety, forms the field in which all experience occurs."[16]

The transformation of society starts from the bottom up with individuals realizing the necessity of creating more authentic and healthy relationships. Our views and values gradually change as we begin to understand the difference between constructive and destructive values and behaviors. The next step is to bring our actions in line with our new understandings — to interact with other people, institutions and the environment in ways that support healthy relationships and sustainable outcomes. In our interactions we not only begin to heal the world around us, but we also heal ourselves.

Marshall Rosenberg, the founder of the Center for Nonviolent Communication, believes that we will only be able to overcome the alienation and violence in the world if we develop more compassionate ways of talking to each other.[17] Our society does not teach us to take responsibility for our feelings and to express our emotional needs in constructive ways. As a result, people frequently communicate their needs through criticisms, demands and moralistic judgements. These provoke defensive and aggressive responses that hinder healthy relationships and block conflict resolution. Changing old emotional habits takes a lot of study and practice. But we need to learn better ways of relating to one another because we can't force people to be loving and caring. We will only be able to transform our violent, destructive global system into a peaceful and constructive civilization through nonviolent means. As Mahatma Gandhi said, "We must be the change we want to see in the world."[18]

However, nonviolence is not the same as passivity or pacifism.[19] Nonviolence is an attitude of nonaggression — the desire not to do harm to others and to find peaceful solutions to problems. A nonviolent approach enables us to replace hatred with compassion, and to view people who are destructive as ignorant or mentally ill rather than as evildoers who deserve punishment. But this does not mean that we do not have the right to self-defense, or that destructive people should not be stopped from doing violence, held accountable for their actions and if necessary locked up in prisons or mental hospitals where they can do no harm.

If aggressors cannot be persuaded to desist from destructive acts, then other people have no choice but to use (minimal) force to stop the abuse and establish peace and security. The primary difference between nonviolence and violence is not the use of force but its objectives. While a murderer and a surgeon may both use knives, one harms and the other heals. Making and keeping peace is different from making war because it has as its goal protection rather than punishment and liberation rather than oppression. At this stage in history it is not possible to do without soldiers and police, but it is possible to define the behaviors that distinguish cops from criminals.

Courage and commitment

Constructive change will only occur if we are courageous enough to act on our convictions. We not only have to change our own behaviors, but also challenge others to change their destructive values and behaviors. We do this through engaging in what Susan Scott calls *fierce conversations*. She pointed out that, "While many are afraid of 'real', it is the unreal conversation that should scare us to death."[20]

Change will occur to the extent that support for constructive solutions increases and resistance to change decreases. Figure 5 illustrates where we are likely to find supporters and opponents. The first group that we should reach out to are ethical people who are already aware of the issues (blue), as they will probably be the most willing to commit themselves to constructive action. The next (and much larger) group that we should educate and work with are the many honest people who care about the planet and their children's futures but who have only a partial awareness of the issues (purple). Most people in the world fit into these ethical categories, including most political, business, community and religious leaders.

Potential opponents are unethical people who are willing to harm others to advance their personal interests. We can put these into two categories: the large group that don't know the issues and don't care because they are narrowly focused

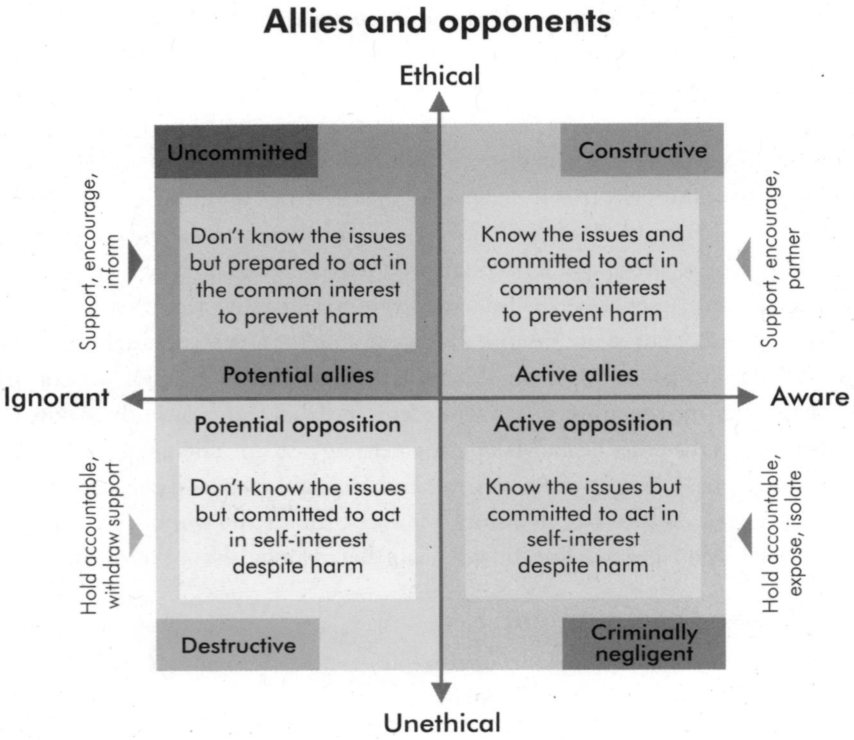

Allies and opponents

Ethical

Uncommitted

Constructive

Support, encourage, inform

Don't know the issues but prepared to act in the common interest to prevent harm

Know the issues and committed to act in common interest to prevent harm

Support, encourage, partner

Potential allies

Active allies

Ignorant

Aware

Potential opposition

Active opposition

Hold accountable, withdraw support

Don't know the issues but committed to act in self-interest despite harm

Know the issues but committed to act in self-interest despite harm

Hold accountable, expose, isolate

Destructive

Criminally negligent

Unethical

Figure 5: N. Davidson and G. Taylor[21]

on their immediate lives (yellow), and the smaller, but more dangerous group that know about the issues and are actively working to prevent change (red). Their destructive actions need to be exposed and neutralized.

Although the number of people who understand the need for urgent action is growing, the majority of the world's population (including most political and business leaders) are still largely unaware of the growing dangers. As a result, people who call out warnings are still frequently ignored and sometimes silenced. All this is about to change.

Supporting evolution: the magical formula

At this time most people are too busy with their lives to pay much attention to national or global issues. However, we have already begun to enter the phase of growing global crises. As rising energy and food prices send shocks through the global economy, more and more people will begin to ask questions. As the crises worsen people will increasingly look for alternative political and economic solutions. At this point, if positive solutions and good leaders exist, rapid constructive change is possible.

The situation is similar to that of Europe in the late 1930s. Before the Second World War, people were busy working, fixing up their houses, planning vacations and saving for their children's education and their retirement. There were many warnings that war was coming, but they were dismissed by mainstream politicians. As a result, when war broke out in 1939, it came as a shock to most people.

The mobilization for war that took place in the United States after it was attacked by Japan is an example of how rapid change is possible when people unite behind a common cause. Priorities changed overnight. Defeating fascism was suddenly more important than paying off the mortgage or buying new furniture. The entire economy was changed over from consumer to military production within a few months. Men went to war, women went into the factories and the country went from high unemployment to a labor shortage. Although the United States was only at war for three and a half years, during this time enormous quantities of military goods were produced and scientific breakthroughs were made in many areas including radar, jet aircraft and nuclear energy.

The example of the Second World War shows that once the majority of people and governments in the world reach agreement on the urgent need to create a sustainable global economy, economic structures and technologies can be rapidly transformed. As the old saying goes, where there is a will, there is a way.

Skeptics argue that attempts to create a sustainable economy will result in economic ruin. In contrast, studies by the Organization for Economic Co-operation and Development (OECD) suggest that greenhouse gas emissions could be cut by a third by 2030 at a cost of just 0.03% of the annual growth of the world GDP.[22] The complete transformation of the global economy will be much more expensive. However, it is not likely to cost more than the Second World War, when American military expenditures rose from 1% of the national income in 1939 to 42% in 1944 — an enormous effort that did not destroy the American economy.[23]

It is estimated that the Iraq war will cost the United States between $1 trillion and $4 trillion.[24] If this money had instead been invested in developing renewable energies, by now the US would be well on the way to creating a more efficient,

Rosie the Riveter (1942)

Uniting for constructive change

The Allies united in WWII to defeat fascism. Now humanity must unite to create a peaceful and sustainable world.

When the will exists, rapid change is possible.

Millions of women workers contributed to the war effort.

Figure 6: BEST Futures [25]

prosperous and sustainable economy and freeing itself from dependence on expensive energy imports.

People need to become aware that the transformation of the global system is a requirement, not a choice. Living in a continually degrading world where the climate becomes hotter and more erratic each year, and where global crises grow progressively worse is not an option. This future would be a nightmare from which we and our children could not awake.

The problem is that most people and most governments will not become convinced of the need for change until the global economy has begun to collapse. This process will cause immense misery and damage. The danger is that the downward spiral of poverty, hunger, wars and environmental destruction will become unmanageable and humanity will cross one ecological and social tipping point after another.

The hope is that as the global environment and economy begin to collapse, the majority of people will wake up to the urgent need for constructive change. But for this to happen a clear vision of a sustainable alternative must already exist and be supported by a critical mass of well-organized people. Progressives will also need to give more and more leadership as the crises mount. We will need to intervene to stop the descent towards disaster, chart a constructive course of action and guide the process of global reorganization.

Millions of people and thousands of organizations are already working towards a better future. The momentum for change is further supported by the emergence of systems-based societal and material technologies. But our trump card is the transformative power of societal evolution. Not only is knowledge spreading around the world, but a global consciousness is developing — an awareness that humanity is interconnected and that the fate of our species is inextricably bound up with the fate of all life on Earth. With this awareness come new views, new values and new relationships. A sustainable global society has already begun to emerge, and unless there is a catastrophic collapse, its evolution cannot be stopped.

Our role is to accelerate the process through helping the emerging elements of the new sustainable system come together. We can do this through:

- Developing a common vision of a sustainable future.
- Networking knowledge of local and global problems and solutions.
- Helping to develop coalitions in support of constructive initiatives.
- Helping communities develop sustainable solutions through empowering them with systems tools and other resources.

Societal evolution is magical because it is a self-organizing process involving billions of people. It is too vast and complex a process to be either directed or controlled, but it can be supported. And at this critical juncture, the evolution of our species needs all the protection and support that we can give.

Love and faith: the gift of our ancestors

In 1950 few people could imagine how electronics would change the world. In 1985 few people could imagine the rapid and relatively non-violent collapse of the Soviet Union. Few people can now imagine that industrial civilization will either collapse or be transformed within their lifetimes. But it will. We are living in a time of great changes, great fears and great opportunities.

It is our destiny to save life on Earth. Although we are the best educated generations in history and although we have tools that our forebears could never have dreamed of possessing, still the challenges are daunting. Will we have the wisdom, the courage and the commitment to avoid disaster and create a sustainable future?

If we look to the past, we can have hope for the future. Everyone that is alive today is here because of the love and commitment of thousands of previous generations.

You wouldn't exist if any one of your ancestors had not been protected and cared for as a child. No generation in the past ever neglected or failed in their duty to future generations, so why should we? We are the products of a long evolutionary journey that has been sustained by constant love and faith. The same love will give us the courage and strength to create a better world for the countless generations that are still to come.

Endnotes

Introduction

1. This saying is attributed to Albert Einstein at The Quotation Page. [cited May 5, 2008]. quotationspage.com/quotes/Albert Einstein/31.
2. Although this present period is often referred to as the Information Age, the world system is still dominated by the worldview, values, social institutions and economic structures of the Industrial Age. However, the Industrial Age is drawing to a close. Systems-based electronic technologies such as the Internet are part of a larger paradigm shift to a new societal system.
3. For more information see *The BEST Model* Graphic Presentation at http://www.bestfutures.org/content/view/17/35/.
4. Vast quantities of resources remain to be discovered, but they are in polar regions, deep underground, under the ocean or in space. Most of the major sources of easily accessed and utilized fresh water, topsoil, forests, fossil fuels and minerals have already been discovered. See Chapters 1 and 2.
5. The eco-philosopher Joanna Macy (joannamacy.net) believes that we are living in a time of irrevocable choice: we must support the transition from a destructive way of life to a life-sustaining society or face environmental collapse and the dieback of human populations. She calls this shift *The Great Turning*. David Korten's book *The Great Turning: From Empire to Earth Community* (Berrett-Koehler, 2006) outlines a strategy for developing an *Earth Community* in support of spiritual and social change.
6. These scenarios are discussed in detail in Chapter 10. Critics will of course argue that another option is the continuing successful expansion of the current global system. We have yet to hear any arguments in favor of this option that address and solve the problem of biophysical limits to growth. If you know any, please send them to us!

7. Attributed to Buckminster Fuller at the CreatingMinds.org website: creatingminds.org/quotes/change.htm.
8. Evolutionary systems theory has been developed from the application of general systems theory to the study of living biological and social systems. Its particular strength is its ability to explain how living systems self-organize, adapt to change and evolve. See BEST Futures at bestfutures.org.
9. The shift to a sustainable system is the shift from a fragmented, mechanistic view of reality (a world of disconnected parts) to an integrated, holistic view (a world of interconnected and interacting systems). This new paradigm began to emerge with field theories in the 1800s (the work of Faraday, Maxwell and others) which laid the basis for the unified universe of modern physics. Integral systems perspectives are increasingly influencing worldviews and values and supporting the emergence of a sustainable planetary civilization. This is explained further in Chapters 7 and 8.
10. Alexis Madrigal. "Scientists Build First Man-Made Genome; Synthetic Life Comes Next." *Wired*, January 24, 2008. [cited May 11, 2008]. wired.com/science/discoveries/news/2008/01/synthetic_genome
11. The original version of the diagram appeared in Alastair M. Taylor, "Integrative Principles in Human Societies", in Henry Margenau, ed., *Integrative Principles of Modern Thought*, Gordon and Breach, 1972, p. 238. Revised version: [cited May 19, 2008]. bestfutures.org/content/view/16/34.
12. See Thomas S. Kuhn. *The Structure of Scientific Revolutions*, 3rd ed. University of Chicago Press, 1996. Paradigm changing innovation can occur anywhere in the societal system — with new technologies (e.g. the Internet), with new ideas and values (e.g. feminism) and with new forms of social organization (e.g. international non-governmental organizations).
13. The Kennedy administration was worried that the Soviets had a lead in the space race that would give them the military advantage in information gathering, communications and the command and control of weapons systems. The moon mission was launched as a way of mobilizing public support behind the US space program.

Part 1: Collapse — the dominant trend

1. The Earth Charter Initiative website: earthcharter.org.

Chapter 1

1. Quoted on the Climate Protection Campaign website: climateprotectioncampaign.org/quotes.php.
2. Paul Erhlich is well known for his 1968 book, *The Population Bomb*, which stated that the continuous growth of the world's population was unsustainable because the planet has finite resources. He predicted that food supplies would be unable to keep up with demand, causing global famines by the end of the century. Julian Simon took the opposite position, arguing in his 1981 book, *The Ultimate Resource*, that population growth is good for the economy and sustainable because new resources and resource substitutes can always be found.

3. Limiting growth does not mean forbidding the production of more goods and services, but ensuring that all resource consumption is sustainable (i.e. that it takes place within the Earth's biophysical limits). Since global consumption already exceeds sustainable biophysical limits, this means that all further growth will have to be achieved through increasing the efficiency of resource use, increasing the production of renewable resources and energy and decreasing waste and pollution. See Chapter 7.
4. Quoted by Lindsey Grant in "Too Many Old People or Too Many Americans? Thoughts About the Pension Panic." NPG Forum, July 1988. [cited March 18, 2008]. npg.org/forum_series/pension_panic.htm.
5. Donella H. Meadows et al. *The Limits to Growth.* Universe, 1972.
6. I describe the debate as taking place between two sides because there are only two possible positions: either continuous physical growth is sustainable because there are no biophysical limits to growth, or continuous growth is unsustainable because the planet's resources are limited. While there may be a range of positions (e.g. most growth is possible but with inevitable environmental costs) they ultimately fall within the limitless-growth-is-sustainable or limitless-growth-is-unsustainable camps.
   I call the two sides orthodox economists and holistic ecologists, not because everyone on one side is an economist and everyone on the other an ecologist, but because pro-growth positions support mainstream economic theories, while pro-limits positions support holistic ecological theories.
7. Ancient and modern world populations are estimated. Numbers are based on data from US Census Bureau. *Historical Estimates of World Populations.* 2006. [cited March 18, 2008]. census.gov/ipc/www/worldhis.html. Data for the period 1950 to 2050 is from the United Nations *World Population Prospects. The 2006 Revision Population Database.* [cited March 18, 2008]. esa.un.org/unpp.
8. Science Blog. "United Nations Issues Study on Women's Education and Fertility." Press Release DEV/2092 POP/605, February 7, 1996. [cited March 18, 2008]. scienceblog.com/community/older/archives/L/1996/A/un960224.html.
9. See note 7.
10. International Institute for Applied Systems Analysis. Probabilistic Population "Projections for the Millennium Ecosystem Assessment." Updated January 30, 2007. [cited March 18, 2008]. iiasa.ac.at/Research/POP/millennium/index.html#fig1.
11. United Nations Population Fund. *Population, Environment and Poverty Linkages, Operational Challenges.* Population and Development Strategies Series #1, 2001, p. xi. [cited March 18, 2008]. unfpa.org/upload/lib_pub_file/81_filename_pop_env_pov.pdf.
12. Norbert Pintsch. "Rural-urban migration." May 6, 2007. thattakedona.blogspot.com/2006/06/rural-urban-migration.html.
13. World consumption of private and public goods and services is estimated based on UNDP Human Development Report 1998. [cited May 15, 2008]. hdr.undp.org/reports/global/1998/en/) and Sandra Poncet. *The Long Term Growth Prospects of the World Economy: Horizon 2050.* CEPII Working Paper #2006- 16, October 2006. [cited

March 18, 2008]. cepii.fr/anglaisgraph/workpap/summaries/ 2006/wp06-16.htm. The limits of environmental sustainability are also an estimate based on *WWF Living Planet Report 2002*, p. 4. [cited May 15, 2008]. panda.org/news_facts/publications/living_planet_report/lpr02/index.cfm.

14. Poncet, *The Long Term Growth Prospects of the World Economy*. Projections are for an increase in the world Gross Domestic Product from $32 trillion in 2000 to $118 trillion at constant US$2000 exchange rates. If expected exchange rate movements are taken into account, the world GDP in 2050 will be $142 trillion (in 2000 C.E. US$). The other economic projections in this section are also from this study and also adjusted to reflect projected changes in the purchasing power of different currencies. The projections do not take into account recent increases in global growth rates, which averaged 3.4% per year between 2002 and 2007.
15. Alice Calaprice. The New Quotable Einstein. Princeton, 2005, pp. 294-295.
16. Global growth was 3.9% in 2006 and 3.9% in 2007. See World Bank. *Global Economic Prospects 2008: Technology Diffusion in the Developing World*. [cited March 18, 2008]. econ.worldbank.org. Since global consumption involves both goods and services, there is not a direct relationship between economic growth and the increased consumption of renewable and non-renewable resources. As modern economies develop they normally expand services and use resources more efficiently. Nevertheless, increasing production and consumption in the current industrial system usually involves increasing both resource inputs and waste outputs. Chapter 7 describes ways in which economic growth can be decoupled from increased resource consumption.
17. World Wildlife Fund. *WWF Living Planet Report 2002*, p. 4.
18. Diagram adapted from WWF. World Wildlife Fund. "The Living Planet Through Time, 1961-2003." *Living Planet Report 2006*, p. 21. [cited April 7, 2008]. panda.org/news_facts/publications/living_planet_report/footprint/index.cfm. All statistics in Chapter 1 on biologically productive areas and ecological footprints are based on Table 3 of this report, p. 38. Data on growth rates: World Bank."Global Economic Prospects 2008: Global Growth." [cited May 22, 2008]. web.worldbank.org/WBSITE/EXTERNAL/EXTDEC/EXTDECPROSPECTS/EXTGBLPROSPECTS/0,,contentMDK:21539434~menuPK:612510~pagePK:2904583~piPK:2904598~theSitePK:612501,00.html.
19. Some plans are being made to mine Helium 3 on the Moon. Because it is an extremely rare source of energy, it is worth several billion dollars a ton, which may mean that it is profitable to strip mine it from the Moon's surface and transport it back to Earth: Nick Davidson. "Making a mint out of the Moon." *BBC News*, April 11, 2007. [cited March 18, 2008]. news.bbc.co.uk/2/hi/science/nature/6533169.stm.
20. UN Food and Agricultural Organization. "Making every drop count." News release, February 14, 2007. [cited March 18, 2008]. fao.org/newsroom/en/news/2007/1000494/index.html.
21. World Wildlife Fund, "Table 3: The Living Planet Through Time, 1961-2003." Living Planet Report 2006.

22. Living Planet Report 2006 (note 21).
23. See note 18. Definitions of natural capital and income: William E. Rees."Revisiting Carrying Capacity: Area-Based Indicators of Sustainability." 1995. [cited May 22, 2008]. dieoff.org/page110.htm.
24. See note 18.
25. See note 18.
26. "Considering the almost unavoidable growth of the world population, the desired growth of welfare per capita in North and South (RIO 21) and the desired (or "necessary") reduction of environmental pressure from local up to global scales, sustainable development to fulfil peoples needs requires radical improvement of the eco-efficiency (depending on assumptions and on specific need ranging from a factor five up to fifty).": Leo Jansen. Technology in Sustainable Development: The Challenge of evolutionary change! Version 18-03-2008, p. 2. [cited March 18, 2008]. indecol.ntnu.no/indecolwebnew/events/conferences/061000/061000_jansen.doc.
27. See note 18.
28. David R. Klein. "The Introduction, Increase, and Crash of Reindeer on St. Matthew Island." Journal of Wildlife Management, Vol. 32#2 (1968), pp. 350-367. [cited March 18, 2008]. greatchange.org/footnotes-overshoot-st_matthew_island.html.
29. Donella H. Meadows et al. *Limits to Growth: The 30-Year Update.* Chelsea Green, 2004, p. 1.

Chapter 2

1. I suspect that the person who called for help told the dispatchers that a woman was in labor in order to have an ambulance sent without attracting the notice of the police.
2. An energy slave is the amount of energy required to replace the work of one person in the non-human infrastructure (e.g. machines, roads, draft animals). Calculations by Walter L. Youngquist. *GeoDestinies: The Inevitable Control of Earth Resources over Nations and Individuals.* National Book Company, 1997, p. 32.
3. Electric Power Research Institute. *Electricity, the economy and environmental sustainability: Environmental Issues.* [cited March 19, 2008]. epriweb.com/public/000000000001013114.pdf.
4. International Energy Agency. *World energy outlook 2007.* [cited March 19, 2008]. worldenergyoutlook.org/.
5. Adapted from a chart by Donella Meadows et al. *Limits to Growth: the 30-Year Update.* Chelsea Green, 2004, p. 95. Reproduced by courtesy of the publisher.
6. World oil demand is steadily rising. The average American now uses approximately 25 barrels of oil a year and will use more with rising consumption. In China, the average is 1.3 barrels per year per person, while in India the average is less than one. But for China and India to reach even one-quarter of the per capita level of US oil consumption, world output would have to rise by 44%: David R. Francis. "China's Risky Scramble for Oil." Christian Science Monitor, January 20, 2005. [cited March 19, 2008]. csmonitor.com/2005/0120/p16s01-cogn.html.

7. Adapted from a chart by Kjell Aleklett. "Oil: A Bumpy Road Ahead." World Watch Magazine, Vol.19#1 (January/February 2006). [cited May 15, 2008]. peakoil.net/uhdsg/WorldWatch_2006_Jan.doc. Note: the data does not take into account recent deep-water discoveries made in the Gulf of Mexico and off the coast of Brazil. While these are very large oil fields, some experts question whether technical difficulties will permit high levels of production.
8. Economics dictates that the highest grades of easily available non-renewable resources (oil, gas, copper, silver, coal) are extracted first, with the result that the quality of available resources tends to decline over time. This means that in the long term, the tendency is for both costs and waste output to rise, as companies have to find and utilize less accessible supplies as well as extract and process lower grades of resources.
9. Even the oil industry optimists warn that "[M]any complex challenges could keep the world's diverse energy resources from becoming the sufficient, reliable, and economic energy supplies upon which people depend. These challenges are compounded by emerging uncertainties: geopolitical influences on energy development, trade, and security; and increasing constraints on carbon dioxide emissions that could impose changes in future energy use. While risks have always typified the energy business, they are now accumulating and converging in new ways." "Hard Truths" about Global Energy Detailed in New Study by the National Petroleum Council. Press Release, July 18, 2007. [cited March 19, 2008]. npc.org/7-18_Press_rls-post.pdf.
10. ExxonMobil. *A Report on Energy Trends, Greenhouse Gas Emissions and Alternative Energy.* February, 2004. [cited March 19, 2008]. esd.lbl.gov/SECUREarth/presentations/Energy_Brochure.pdf.
11. Kjell Aleklett. "Oil: A Bumpy Road Ahead." See also Fredrik Robelius. Giant Oil Fields — Highway to Oil. 2007. [cited March 19, 2008]. publications.uu.se/abstract.xsql?dbid=7625. His diagrams can be seen at peakoil.net/GiantOilFields.html.
12. For example, it is unlikely that production from the Canadian tar sands will exceed 3 million barrels per day by 2025. This will not begin to compensate for falling reserves of conventional oil: Byron W. King. "2006 Boston ASPO: The Canadian Tar Sands." Whiskey and Gunpowder, November 13, 2006. [cited March 19, 2008]. whiskeyandgunpowder.com/Archives/2006/20061113.html.
13. Colin J. Campbell and Jean H. Laherrère. "The End Of Cheap Oil." *Scientific American,* March 1998. [cited March 19, 2008]. dieoff.org/page140.htm.
14. Ian Traynor and David Gow. "Europe sets benchmark for tackling climate change." *The Guardian,* March 10, 2007. guardian.co.uk/world/2007/mar/10/eu.greenpolitics.
15. The Swedish government is planning major initiatives to reduce oil consumption. However, these plans have not been made within the context of developing a sustainable economy. Commission on Oil Independence. *Making Sweden an Oil-Free Society.* June 21, 2006. sweden.gov.se/content/1/c6/06/70/96/7f04f437.pdf.
16. For example, see Startech Environmental Corporation. Startech Environmental successfully completes first phase of "hydrogen from coal program" for Department of

Energy. Press Release, May 1, 2006. [cited March 19, 2008]. prnewswire.com/gh/cnoc/comp/113537.html. Both nuclear fusion and gas hydrates have the potential to be major future sources of energy, but they are not yet proven either economically viable or safe. For example, see Richard Embleton. "Methane Hydrates: the next great energy source?" December 19, 2006. [cited March 19, 2008]. oilbeseeingyou.blogspot.com/2006/12/methane-hydrates-next-great-energy.html.

17. International Energy Agency. *World energy outlook 2006: Summary and Conclusions*, p. 4. [cited April 14, 2008]. iea.org/Textbase/npsum/WEO2006SUM.pdf.
18. Richard Wachman. "IEA probes fears that oil will run out." *The Guardian*, May 25, 2008. [cited May 27, 2008]. guardian.co.uk/business/2008/may/25/oil.
19. David Strahan. "End of oil heralds climate pain." *BBC News*, March 29, 2007. [cited March 19, 2008]. news.bbc.co.uk/2/hi/science/nature/6505127.stm.
20. Soot from burning coal, diesel, wood and dung is another major factor contributing to global warming: James Randerson. "Scientists warn of soot effect on climate." *The Guardian*, March 24, 2008. [cited May 15, 2008]. guardian.co.uk/environment/2008/mar/24/climatechange.fossilfuels.
21. Andrew C. Revkin. "Skeptics on Human Climate Impact Seize on Cold Spell." *New York Times*, March 2, 2008. [cited May 15, 2008]. nytimes.com/2008/03/02/science/02cold.html?_r=1&scp=1&sq=Skeptics+on+human+climate+impact+seize+on+cold+spell&st=nyt&oref=slogin. A longer weather oscillation may help to cool the planet over the next decade. However, temperatures will again be rapidly rising by about 2020: Richard Black, "Next decade 'may see no warming'." *BBC News*, May 1, 2008. [cited May 15, 2008]. ews.bbc.co.uk/2/hi/science/nature/7376301.stm.
22. Intergovernmental Panel on Climate Change. "Summary for Policymakers" in *Climate Change 2007: The Physical Science Basis. Contribution of Working Group I to the Fourth Assessment Report of the Intergovernmental Panel on Climate Change*. Cambridge, 2007. [cited March 19, 2008]. ipcc.ch/pdf/assessment-report/ar4/wg1/ar4-wg1-spm.pdf.
23. Photos courtesy of NASA. For 2013 prediction, see Richard Black, "Arctic sea ice melt 'even faster'". *BBC News*, June 18, 2008. [cited July 17, 2008]. bbc.co.uk/2/hi/science/nature/7461707.stm.
24. Data: HM Treasury. *Stern Review on the economics of climate change*. 2006. [cited April 14, 2008]. hm-treasury.gov.uk/independent_reviews/stern_review_economics_climate_change/stern_review_report.cfm; Intergovernmental Panel on Climate Change. "Summary for Policymakers" in *Climate Change 2007: Impacts, Adaptation and Vulnerability. Contribution of Working Group II to the Fourth Assessment Report of the Intergovernmental Panel on Climate Change*. Cambridge, 2007. [cited March 19, 2008]. ipcc.ch/pdf/assessment-report/ar4/wg2/ar4-wg2-spm.pdf.; David Spratt and Phillip Sutton. *Climate code red: the case for sustainability emergency*. [cited March 19, 2008]. climatecodered.net.
25. An example of major climate change was the 2005 drought in the tropical rainforests of the Amazon. Scientists warn that a combination of forest fires, droughts, and

deforestation will combine to create a "cycle of destruction" in the Amazon. See James Painter. "Amazon 'faces more deadly droughts'." *BBC News*, March 23, 2007. [cited March 19, 2008]. news.bbc.co.uk/2/hi/americas/6484073.stm. The World Bank estimates that the number of natural disasters has quadrupled from 100 per year in 1975 to 400 per year in 2005. Desmond Tutu. "This fatal complacency." *The Guardian*, May 5, 2007. [cited March 19, 2008]. guardian.co.uk/commentisfree/story/0,,2073021,00.html.

26. *IPCC Fourth Assessment Report 2007.*
27. Mark Lynas. *Six Degrees: Our Future on a Hotter Planet.* National Geographic, 2008.
28. Alison Benjamin. "Stern: Climate change a 'market failure'." *The Guardian*, November 29, 2007. [cited May 15, 2008]. guardian.co.uk/environment/2007/nov/29/climatechange.carbonemissions.
29. HM Treasury. *Stern Review on the economics of climate change.* 2006. For example, the United Nations Environmental Program estimates that it is realistic to expect that losses from droughts, storm surges, hurricanes and floods alone will double every 12 years, with a cost to the global economy of $1 trillion per year by 2040: Daniel Wallis. "Extreme weather could cost $US1 trillion." The Courier Mail, November 15, 2006. [cited March 19, 2008]. news.com.au/couriermail/story/0,,20761001-462,00.html?from=rss.
30. Sam Carana, "Cut carbon to 350 ppm, says James Hansen." Gather, April 8, 2008. [cited April 14, 2008]. gather.com/viewArticlePF.jsp?articleId=281474977304582.
31. David Spratt and Phillip Sutton. *Climate code red: the case for sustainability emergency.*
32. "Earth's carbon dioxide, methane soars." *The Australian*, April 24, 2008. [cited May 15, 2008]. theaustralian.news.com.au/story/0,25197,23591089-11949,00.html.
33. Jonathan Amos. "Methane ice poses climate puzzle." *BBC News*, December 13, 2006. news.bbc.co.uk/2/hi/science/nature/6166011.stm. Also see University of Wyoming. "Ocean Floor Reveals Clues To Global Warming." Science Daily, January 13, 2004. [cited March 19, 2008]. sciencedaily.com/releases/2004/01/040113080810.htm.
34. Ian Sample. "Warming hits 'tipping point'." *The Guardian*, August 11, 2005. [cited March 19, 2008]. guardian.co.uk/environment/2005/aug/11/science.climatechange1.
35. Edward Hill. "Calm before the storm." Inroads Earth Day Newsletter, 2007. [cited March 19, 2008]. royalroads.ca/about-rru/the-university/news-events/inroads-newsletter/2007-EarthDay/climate_research.htm. For further information see Audrey Dallimore. Paleo-environmental records of climate change. Natural Resources Canada. [cited March 19, 2008]. ess.nrcan.gc.ca/2002_2006/rcvcc/j27/2_2_e.php.
36. Sepp Hasslberger. Global Warming: Methane Could Be Far Worse Than Carbon Dioxide. February 1, 2005. [cited March 19, 2008]. newmediaexplorer.org/sepp/2005/02/01/global_warming_methane_could_be_far_worse_than_carbon_dioxide.htm.
37. "Australia 'must commit to 25 pc emissions cut'." ABC News, March 13, 2008. [cited April 14, 2008]. abc.net.au/news/stories/2008/03/13/2188315.htm.
38. International Energy Agency. *World energy outlook 2006: Summary and Conclusions*, p. 10. [cited April 14, 2008]. iea.org/Textbase/npsum/WEO2006SUM.pdf.

39. UNDP *Human Development Report 2007/2008*. Fighting climate change: human solidarity in a divided world. [cited March 19, 2008]. hdr.undp.org/en/reports/global/hdr2007-2008/.
40. For example see Geoffrey Lean. "Global warming 'Is Three Times Faster Than Worst Predictions.'" *The Independent*, June 3, 2007. [cited March 19, 2008]. countercurrents.org/lean030607.htm; Deborah Zabarenko. "Arctic ice cap melting 30 years ahead of forecast." Yahoo! News, May 1, 2007. [online – link expired]. The Southern Ocean is also absorbing carbon dioxide more slowly than predicted: Michael McCarthy. "Earth's natural defences against climate change 'beginning to fail'." *The Independent*, May 18, 2007. [cited March 19, 2008]. news.independent.co.uk/environment/climate_change/article2556466.ece.
41. Steve Connor. "The Earth today stands in imminent peril." *The Independent*, June 19, 2007. [cited March 19, 2008]. environment.independent.co.uk/climate_change/article2675747.ece.
42. The most vociferous opponents of limits to growth are corporations who make money from polluting. George Monbiot points out that much of the financing for discrediting climate change science came from ExxonMobil and Phillip Morris (the tobacco giant that spent millions arguing that there is no link between smoking and cancer): "The denial industry." *The Guardian*, September 19, 2006. [cited March 20, 2008]. environment.guardian.co.uk/climatechange/story/0,,1875762,00.html. Also see Sharon Begley, "The truth about denial." *Newsweek*, August 13, 2007. [cited March 20, 2008]. newsweek.com/id/32482.
43. For example, the US government spent 20 times as much in 2006 on military research as on energy research: Andrew C. Revkin. "Budgets Falling in Race to Fight Global Warming." *New York Times*, October 30, 2006. [cited March 20, 1988]. nytimes.com/2006/10/30/business/worldbusiness/30energy.html?ex=1177128000&en=bdff07eb7b69be91&ei=5070.
44. Dr. Caroline Lucas, a Member of the European Parliament, pointed out that the European Union saw no contradiction between signing an agreement to reduce CO_2 emissions in April, 2007, and a few weeks later signing another agreement with the US which will increase the number of trans-Atlantic flights by 25 million per year and increase CO_2 emissions by 3.8 million tons. In her view "the EU's policy priorities remain focused on the objectives of greater free trade and international competitiveness rather than becoming a leader in the global fight against climate change ..." Caroline Lucas. "EU 'must walk its climate talk'." *BBC News*, June 1, 2007. [cited March 20, 2008]. news.bbc.co.uk/2/hi/science/nature/6711245.stm.
45. "Canada facing Kyoto probe." Courier Mail, May 8, 2008. [cited May 15, 2008]. news.com.au/couriermail/story/0,23739,23663811-5012763,00.html.
46. Daniel Howden. "No Blair! America's parting gift to Britain's PM." *The Independent*, May 19, 2007. [cited March 20, 2008]. news.independent.co.uk/world/americas/article2559994.ece.

47. Peter Walker. "China speeds towards 'biggest greenhouse gas producer' title." *The Guardian*, April 24, 2007. [cited March 20, 2008]. environment.guardian.co.uk/climatechange/story/0,,2064485,00.html. However, while not agreeing to mandatory limits on emissions, the Chinese government is planning on making "significant achievements in controlling greenhouse gas emissions": Audra Ang. "China Won't Adopt Caps on Gas Emissions." *Washington Post*, June 4, 2007. [cited March 20, 2008]. washingtonpost.com/wp-dyn/content/article/2007/06/04/AR2007060400167.html.
48. The 2008 meeting of the G-8 industrial powers — which account for almost 60% of the world economy — also failed to set binding targets, the leaders instead agreeing to "seriously consider" emission cuts of 50% by 2050. See "G-8 urged to do more for climate". *BBC News*, July 8, 2008. [cited July 10, 2008]. news.bbc.co.uk/2/hi/asia-pacific/7495641.stm.
49. "UN warning over global ice loss." *BBC News*, June 4, 2007. [cited March 20, 2008]. news.bbc.co.uk/2/hi/science/nature/6713139.stm.
50. Alex Kirby. "Water scarcity: A looming crisis?" *BBC News*, October 19, 2004. [cited March 20, 2008]. news.bbc.co.uk/2/hi/science/nature/3747724.stm.
51. Adapted from the diagram "Worldwide Supply" in Alex Kirby's article, note 50.
52. See note 50.
53. Millennium Ecosystem Assessment. *Ecosystems and Human Well-being: Synthesis.* Island Press, 2005, p. 31.
54. UN-Water. *Coping with Water Scarcity.* August 2006, p. 2. [cited March 20, 20008]. unwater.org/downloads/waterscarcity.pdf.
55. Sylvia Westall. "Melting mountains a 'time bomb'." *Courier Mail*, April 15, 2008. [cited May 15, 2008]. news.com.au/couriermail/story/0,23739,23545407-5012763,00.html.
56. Intergovernmental Panel on Climate Change. "Summary for Policymakers" in *Climate Change 2007.*
57. Jon Gertner. "The Future Is Drying Up." *New York Times*, October 21, 2007. [cited March 20, 2008]. nytimes.com/2007/10/21/magazine/21water-t.html?scp=2&sq=the+future+is+drying+up&st=nyt.
58. New Agriculturalist. "Thirst threatens with impending water crisis". 2004. [cited March 21, 2008]. new-agri.co.uk/00-4/develop/dev06.html.
59. For example, water tables are falling by 7-10 feet (2-3 meters) per year under the North China Plain, by up to 26 feet (8 meters) per year under the Chenaran Plain in northeastern Iran, and by 3-10 feet (1-3 meters) per year in many key agricultural areas in India (including the Punjab, India's breadbasket): Lester Brown. "Water deficits growing in many countries." Great Lakes, September 8, 2002. [cited March 21, 2008]. greatlakesdirectory.org/zarticles/080902_water_shortages.htm.
60. Unfortunately, "Although overdrafting and contamination of groundwater aquifers are known to be widespread and growing problems (UNEP 1996), comprehensive data on groundwater resources and pollution trends are not available at the global level.": Carmen Revenga. "Will there be enough water?" EarthTrends, October, 2000. [cited

March 21, 2008]. earthtrends.wri.org/features/view_feature.php?theme=1&fid=17. For example, around 140 million people are being poisoned by arsenic (a metal naturally found in soils) in their drinking water: Richard Black. "World facing 'arsenic timebomb'." *BBC News*, August 30, 2007. [cited March 21, 2008]. news.bbc.co.uk/2/hi/science/nature/6968574.stm. Information on the state of the Yangtze: "Pollution risks Yangtze's 'death'." *BBC News*, May 30, 2006. [cited March 21, 2008]. news.bbc.co.uk/2/hi/asia-pacific/5029136.stm.

61. A geography professor at the University of Manitoba has calculated that the world is rapidly approaching the maximum level of sustainable water withdrawals: Vaclav Smil. *Feeding the World: A Challenge for the Twenty-First Century*. MIT, 2001, pp. 43-44.
62. UNESCO. Water: a crisis of governance says second UN *World Water Development Report*. March 9, 2006. [cited March 21, 2008]. portal.unesco.org/en/ev.php-URL_ID=32057&URL_DO=DO_TOPIC&URL_SECTION=201.html.
63. World Hunger Information Service. *World Hunger Facts 2008*. [cited March 21, 2008]. worldhunger.org/articles/Learn/world%20hunger%20facts%202002.htm.
64. Hannah Belcher. "Asian rice crisis starts to bite." *Al Jazeera*, March 28, 2008. [cited April 15, 2008]. english.aljazeera.net/NR/exeres/CB6E8E48-C288-4066-90B8-8F23DFDCDFEE.htm?FRAMELESS=true&NRNODEGUID=%7bCB6E8E48-C288-4066-90B8-8F23DFDCDFEE%7d.
65. UNFAO. "Global cereal supply and demand brief." *Crop Prospects and Food Situation*. (No. 2, April 2008). [cited May 15, 2008]. fao.org/docrep/010/ai465e/ai465e04.htm.
66. Rooster photo courtesy of Nora Arias Loftis.
67. Melissa Schober. "US Responds to Worsening Global Food Crisis." Green Daily, May 1, 2008. [cited May 14, 2008]. greendaily.com/2008/05/01/us-responds-to-worsening-global-food-crisis.
68. Lester R. Brown, *Plan B: Rescuing a Planet under Stress and a Civilization in Trouble*. Norton, 2003, p.138.
69. Another problem with meat-based diets is that the world's 1.5 billion cattle already produce 18% of global carbon dioxide emissions — more than cars, planes and all other forms of transport put together: Geoffrey Lean. "Cow 'emissions' more damaging to planet than CO_2 from cars." *The Independent*, December 10, 2006. [cited March 21, 2008]. news.independent.co.uk/environment/article2062484.ece. Factory farming is also a major source of both pollution and disease: Danielle Nierenberg. "Rethinking the Global Meat Industry" in *State of the World 2006*. Norton, 2006, p. 24.
70. George Monbiot. "Credit crunch? The real crisis is global hunger. And if you care, eat less meat." *The Guardian*, April 15, 2008. [cited May 27, 2008]. guardian.co.uk/commentisfree/2008/apr/15/food.biofuels.
71. James Randerson. "Food crisis will take hold before climate change, warns chief scientist." *The Guardian*, March 7, 2008. [cited March 21, 2008]. guardian.co.uk/science/2008/mar/07/scienceofclimatechange.food.
72. Lester R. Brown et al. "Feeding Nine Billion" in *State of the World 1999*. Norton, 1999, p. 118.

73. Geoffrey Lean. "Exposed: the great GM crops myth." *The Independent*, April 20, 2008. [cited May 15, 2008]. independent.co.uk/environment/green-living/exposed-the-great-gm-crops-myth-812179.html.
74. Lester Brown. *Outgrowing the Earth: The Food Security Challenge in an Age of Falling Water Tables and Rising Temperatures*. Norton, 2005. [cited March 21, 2008]. earth-policy.org/Books/Out/Ote4_4.htm. He quoted USDA plant scientist Thomas R. Sinclair, who stated that "except for a few options which allow small increases in the yield ceiling, the physiological limit to crop yields may well have been reached under experimental conditions." The Indian biologist Vandana Shiva also doubts that a second green revolution is possible. In her view the Green Revolution in India's Punjab State has been a failure. "It has led to reduced genetic diversity, increased vulnerability to pests, soil erosion, water shortages, reduced soil fertility, micronutrient deficiencies, soil contamination, reduced availability of nutritious food crops for the local population, the displacement of vast numbers of small farmers from their land, rural impoverishment and increased tensions and conflicts.": Research on the Scientific Basis for Sustainability. 2-4-9 How Can We Enable Increased Food Production? [cited March 21, 2008]. sos2006.jp/english/rsbs_summary_e/2-4-food-and-water.html.
75. Dipanjan Ghosh. "Search for Future Viands: Algae and Fungi as Food." *Resonance*, May 2004. [cited March 21, 2008]. ias.ac.in/resonance/May2004/pdf/May2004p33-40.pdf.
76. At the time of writing, the last data was available for the harvest of 2006/07. US Foreign Agricultural Service. *Grain: World Markets and Trade.* [cited March 21, 2008]. Circular FG 05-06, May 2006. fas.usda.gov/grain/circular/2006/05-06/graintoc.htm.
77. Raja M. "India grows a grain crisis." Asia Times Online, July 21, 2006. [cited March 21, 2008]. atimes.com/atimes/South_Asia/HG21Df01.html.
78. "Rice prices 'to keep on rising'." *BBC News*, April 11, 2008. [cited April 15, 2008]. news.bbc.co.uk/2/hi/business/7341978.stm.
79. Lester Brown. "Exploding U.S. Grain Demand for Automotive Fuel Threatens World Food Security and Political Stability". Earth Policy Institute, November 3, 2006. [cited March 21, 2008]. earth-policy.org/Updates/2006/Update60.htm. Brown also pointed out that, "From an agricultural vantage point, the automotive demand for fuel is insatiable. The grain it takes to fill a 25-gallon tank with ethanol just once will feed one person for a whole year. Converting the entire U.S. grain harvest to ethanol would satisfy only 16 percent of U.S. auto fuel needs .... There are alternatives to creating a crop-based automotive fuel economy. The equivalent of the 2 percent of U.S. automotive fuel supplies now coming from ethanol could be achieved several times over, and at a fraction of the cost, by raising auto fuel efficiency standards by 20 percent.": Lester Brown. "Distillery demand for grain to fuel cars vastly underestimated." Earth Policy Institute, January 4, 2007. [cited March 21, 2008]. earth-policy.org/Updates/2007/ Update63.htm. See also "EU biofuels policy is a 'mistake'." *BBC News*, August 17, 2007. [cited March 21, 2008]. news.bbc.co.uk/2/hi/science/nature/6949861.stm.

80. "UN warns on soaring food prices." *BBC News*, December 17, 2007. [cited March 21, 2008]. news.bbc.co.uk/2/hi/in_depth/7148880.stm.
81. Emily Buchanan. "Assessing the global food crisis." *BBC News*, April 22, 2008. [cited May 27, 2008]. news.bbc.co.uk/2/hi/7361945.stm.
82. Ten to twenty percent of the world's 3.7 billion acres (1.5 billion hectares) of cropland is now degraded to some extent: Brian Halweil. "Farmland Quality Deteriorating" in *Vital Signs 2002*, Norton, 2002, p. 102.
83. Lester R. Brown. *Plan B: Rescuing a Planet under Stress and a Civilization in Trouble.* Norton, 2003, p. 47.
84. Chris Hawley. "World's Land Turning to Desert at Alarming Speed." Associated Press, June 15, 2004. [cited March 21, 2008]. commondreams.org/headlines04/0615-10.htm.
85. The Chinese government reported that pollution destroyed 760,000 acres (307,000 hectares) of arable land in the first 10 months of 2006 alone. The Chinese government is worried about the continuing loss of farm land. Land and Resources Minister Sun Wensheng said that the amount of agricultural land in China must not be allowed to fall below 297million acres (120 million hectares): "Pollution 'hits China's farmland'." *BBC News*, April 23, 2007. [cited March 21, 2008]. news.bbc.co.uk/2/hi/asia-pacific/6582571.stm.
86. Intergovernmental Panel on Climate Change. "Summary for Policymakers" in Climate Change 2007: Impacts, Adaptation and Vulnerability. *Contribution of Working Group II to the Fourth Assessment Report of the Intergovernmental Panel on Climate Change.* Cambridge, 2007, p. 11. [cited March 19, 2008]. ipcc.ch/pdf/assessment-report/ar4/wg2/ar4-wg2-spm.pdf. Also see Fred Pearce. "Climate change warning over food production." New Scientist 18:17 (April 26, 2005). [cited March 21, 2008]. newscientist.com/article.ns?id=dn7310; Lester Brown. World grain stocks fall to 57 days of consumption: grain prices starting to rise. Earth Policy Institute, June 15, 2006. [cited March 21, 2008]. earth-policy.org/Indicators/Grain/2006.htm.
87. Adapted from a *BBC News* chart: Richard Black "'Only 50 years left' for sea fish." *BBC News*, November 2, 2006. [cited March 21, 2008]. news.bbc.co.uk/2/hi/science/nature/6108414.stm. Data from FAO. See Boris Worm et al. "Impacts of Biodiversity Loss on Ocean Ecosystem Services." Science 314 (November 3, 2006), p. 787.
88. "China sees climate impacts ahead." *BBC News*, April 23, 2007. [cited March 21, 2008]. news.bbc.co.uk/2/hi/science/nature/6585775.stm. In response to these grim predictions the Chinese government has requested international help in adapting to global warming: "Help Needed to Prepare for Climate Change." *China Daily*, April 12, 2007. [cited March 21, 2008]. china.org.cn/english/environment/206980.htm.
89. Gwynne Dyer. "Don't mention the warming." June 4, 2007. [cited March 21, 2008]. gwynnedyer.com/articles/Gwynne%20Dyer%20article_%20%20G8%20and%20Climate%20Change.txt.
90. Black, "Only 50 years left for sea fish."
91. Brown. *Plan B*, pp. 137-141.

92. Smil. *Feeding the World,* p. 315.
93. United Nations University. Looming desertification could spawn millions of environmental refugees. *Mongabay.com,* December 14, 2006. [cited March 24, 2008]. news.mongabay.com/2006/1214-unu.html.
94. Zoe Cormier. "Climate refugees" in The New Climate Almanac. GlobeandMail.com, February 17, 2007. [cited March 24, 2008]. theglobeandmail.com/servlet/story/RTGAM. 20070216.wclimatealmanac/BNStory/ClimateChange/home/?pageRequested=3.
95. Attributed to Chief Seattle's "Letter to President Franklin Pierce," 1854. *The Columbia World of Quotations,* 1996. [cited May 14, 2008]. bartleby.com/66/10/48910.html.
96. Julia Whitty. "Animal extinction — the greatest threat to mankind." *The Independent,* April 30, 2007. [cited March 24, 2008]. news.independent.co.uk/environment/article2494659.ece.
97. Sigmar Gabriel. "Biodiversity 'fundamental' to economics." *BBC News,* March 9, 2007. [cited March 24, 2008]. news.bbc.co.uk/2/hi/science/nature/6432217.stm.
98. The mean estimate is 10 million: Edward O. Wilson. *The Future of Life.* Vintage, 2002, p. 14. The precise number of species disappearing each day is unknown because scientists don't know exactly how many species exist and estimates of the rate of extinction vary from one thousand to ten thousand per million per year (pp. 99-101).
99. The applied scientific knowledge of the natural world far exceeds the scientific knowledge of humanity. We still do not understand and cannot replicate countless natural processes.
100. From 1970 to 2003 humans wiped out an average of 1% of the populations of all other species each year. World Wildlife Fund. "Living Planet Index." Living Planet Report 2006. [cited May 27, 2008]. panda.org/news_facts/publications/living_planet_report/index.cfm.
101. Ian Sample. "World's wildlife and environment already hit by climate change, major study shows." *The Guardian,* May 15, 2008. [cited May 27, 2008]. guardian.co.uk/environment/2008/may/15/climatechange.scienceofclimatechange.
102. John Roach. "By 2050 Warming to Doom Million Species, Study Says." *National Geographic News,* July 12, 2004. [cited March 24, 2008]. news.nationalgeographic.com/news/2004/01/0107_040107_extinction.html.
103. United Nations Secretariat of the Convention on Biological Diversity. "Elevating climate change and biodiversity to the top of the political agenda." Press release, April 13, 2007. [cited March 24, 2008]. biodiv.org/doc/press/2007/pr-2007-04-13-potsdam-en.pdf.
104. "Without protection, Amazon rainforest faces huge loss: scientists." *CBC News,* March 22, 2006. [cited March 24, 2008]. cbc.ca/health/story/2006/03/22/amazon-rainforest060322.html. Each year forests covering an area the size of Great Britain are being destroyed, primarily in tropical areas. In temperate zones reforestation is expanding forested areas, but tree plantations lack the biodiversity of old growth forests. Because trees in temperate areas have darker foliage than tropical trees, they tend to absorb rather than reflect heat, thus contributing to global warming.

105. Daniel Howden. "Deforestation: the hidden cause of global warming". *The Independent,* May 14, 2007. [cited April 15, 2008]. http://news.independent.co.uk/environment/climate_change/article2539349.ece.
106. Steve Connor. "Loss of coral may cause food supply crisis." *The Independent,* April 4, 2007. [cited March 24, 2008]. independent.co.uk/environment/nature/loss-of-coral-may-cause-food-supply-crisis-443253.html.
107. James Randerson. "Sealife at risk from rapid acidification." *The Guardian,* May 23, 2008. [cited May 28, 2008]. guardian.co.uk/environment/2008/may/23/climatechange.water.
108. Andy Coghlan. "Pollution triggers bizarre behaviour in animals." NewScientist.com, September 3, 2004. [cited March 24, 2008]. http://www.newscientist.com/article.ns?id=dn6343.
109. World Wildlife Fund. "Toxic pollution in the Arctic." July 13, 2006. [cited March 24, 2008]. panda.org/about_wwf/where_we_work/europe/what_we_do/arctic/polar_bear/threats/toxic_pollution/index.cfm.
110. See note 96.
111. Rhett A. Buttler. "Global warming may cause biodiversity extinction." *Mongabay.com,* March 21, 2007. [cited March 24, 2008]. news.mongabay.com/2007/0322-extinction.html.
112. Data: Worldwatch Institute. *State of the World 2000.* Norton, 2000; John Carey."China's Air Pollution Gorilla". NowPublic.com, July 6, 2007. [cited March 24, 2008]. nowpublic.com/china_s_air_pollution_gorilla; Robert F. Kennedy Jr."Bush's Crimes Against Nature." [cited March 24, 2008]. AlterNet, October 7, 2004. alternet.org/election04/20124.
113. James Lovelock. *Gaia and the Theory of the Living Planet,* 2nd rev. ed. Gaia Books, 2005.
114. Ecosystems provide us with"provisioning services such as food, water, timber, and fiber; regulating services that affect climate, floods, disease, wastes, and water quality; cultural services that provide recreational, aesthetic, and spiritual benefits; and supporting services such as soil formation, photosynthesis, and nutrient cycling. The human species, while buffered against environmental changes by culture and technology, is fundamentally dependent on the flow of ecosystem services." Millennium Ecosystem Assessment. *Ecosystems and Human Well-being: Synthesis.* Island Press, 2005, p. 15.
115. See note 97.
116. Millenium Ecosystem Assessment. *Ecosystems and Human Well-Being: Synthesis.* p. 1. [cited May 13, 2008]. millenniumassessment.org/documents/document.356.aspx.pdf.
117. Janet Larsen. "Wild fish catch hits limits: oceanic decline offset by increased fish farming." Earth Policy Institute, 2005. [cited March 24, 2008]. earthpolicy.org/Indicators/Fish/2005.htm.
118. "Orangutans 'face greater threat'." *BBC News,* February 6, 2007. [cited March 24, 2008]. news.bbc.co.uk/2/hi/science/nature/6337107.stm.
119. Richard Black. "Cash row at wildlife trade forum." *BBC News,* June 16, 2007. [cited March 24, 2008]. news.bbc.co.uk/2/hi/science/nature/6758215.stm.

120. Africa Rice Center. "Nerica — a technology from Africa for Africa." [cited March 24, 2008]. warda.org/NERICA%20flyer/technology.htm.
121. Tony Paterson. "Norway's 'Doomsday Vault' holds seeds of survival." *The Independent*, February 25, 2008. [cited March 24, 2008]. independent.co.uk/news/europe/norways-doomsday-vault-holds-seeds-of-survival-786773.html.
122. A July, 2007 report from the International Energy Agency warns that consumers will see the beginnings of a serious oil and gas shortage within two years. "Not only does oil look extremely tight in five years time, but this coincides with the prospect of even tighter natural gas markets at the turn of the decade.": "Oil and gas shortages likely within 5 years: report." *CBC News*, July 9, 2007. [cited March 24, 2008]. cbc.ca/consumer/story/2007/07/09/oilshortage070709.html.
123. Courtesy of Millenium Ecosystem Assessment. The original may be seen at maweb.org/documents/document.356.aspx.pdf, p. 26. For tonne equivalents, multiply vertical axis figures by 0.9.
124. John W. Miller. "Global fishing trade depletes African waters." *Wall Street Journal*, July 23, 2007. [cited May 27, 2008]. icsf.net/icsf2006/ControllerServlet?handler=EXTERNALNEWS&code=getDetails&id=34873&userType=&fromPage=.
125. See note 89.
126. Ibid.
127. Jonathan Fildes. "Call to halt deep sea 'plunder'." *BBC News*, February 19, 2007. [cited March 24, 2008]. news.bbc.co.uk/2/hi/science/nature/6374971.stm.
128. Data on child mortality: United Nations Environment Programme. *Vital Water Graphics Executive Summary*, 2002. [cited March 24, 2008]. unep.org/dewa/assessments/ecosystems/water/vitalwater/summary.htm.
129. *United Nations. Human Development Report 2005*, p. 52. [cited March 24, 2008]. hdr.undp.org/en/media/hdr05_complete.pdf.
130. Lester Brown. *Plan B 3.0: Mobilizing to Save Civilization*. Norton, 2008, p. 282.
131. Data: Lester Brown. *Plan B 3.0: Mobilizing to Save Civilization*; Tim Jackson. "The Challenge of Sustainable Lifestyles" in Worldwatch Institute. *State of the World 2008*. Norton, 2008, p. 59; Alex Kirby. "Fossil fuel subsidies 'must end'." *BBC News*, June 21, 2004. [cited March 24, 2008]. news.bbc.co.uk/2/hi/science/nature/3818995.stm.
132. In 2006 global GDP was over $US 48 Trillion: World Bank. [cited March 28, 2008]. siteresources.worldbank.org/DATASTATISTICS/Resources/GDP.pdf.
133. Catastrophic climate change is likely even if 50% reductions are made in greenhouse gas emissions by 2050. A chart showing temperature probabilities and their predicted consequences at each level of atmospheric carbon dioxide equivalent is available in Philip Sutton, "Zero-minus-fast: the best target for a safe planet?" Friends of the Earth Australia, March 2007. [cited July 16, 2008]. foe.org.au/resources/chain-reaction/chain-reaction-editions/chain-reaction-99-march-2007/zero-minus-fast-the-best-target-for-a-safe-planet.

Chapter 3

1. This quote has been attributed to various sources. See The Quotations Page at quotationspage.com/forum/viewtopic.php?t=35.
2. For a list of positive developments, see "Somewhere Over the Rainbow." *The Economist*, January 24, 2008. [cited May 16, 2008]. economist.com/world/international/displaystory.cfm?story_id=10564141.
3. UNDP. *Human Development Report 2003*. [cited April 24, 2008]. hdr.undp.org/en/reports/global/hdr2003/.
4. Mohandas K. Gandhi (1869-1948) said, "I suggest that we are thieves in a way. If I take anything that I do not need for my own immediate use and keep it I thieve it from somebody else. I venture to suggest that it is the fundamental law of Nature, without exception, that Nature produces enough for our wants from day to day, and if only everybody took enough for himself and nothing more, there would be no pauperism in this world, there would be no more dying of starvation in this world. But so long as we have got this inequality, so long we are thieving." M. K. Gandhi. Trusteeship. Navjeevan Trust, 1960, p. 3. [cited March 25, 2008]. gandhi-manibhavan.org/main/q4.htm.
5. Franklin D. Roosevelt. Second Inaugural Address. January 20, 1937, para. 29. [cited March 25, 2008]. bartleby.com/124/pres50.html.
6. 87% of the games were played for wagers that averaged $589. At the same time, 99% of those questioned said that they are honest in business. "Survey: CEO golfers cheat." CNNMoney.com, June 26, 2006. [cited March 25, 2008]. money.cnn.com/2002/06/26/news/ceos/fore/index.htm. Dishonesty isn't confined to the USA. In a British survey 61% of people interviewed admitted committing petty crimes such as avoiding taxes, stealing from work, submitting false insurance claims and keeping extra money when given too much change. One of the lead researchers, Dr. Stephen Farrell, blamed the crime on selfish attitudes, "The values and the behaviour of those at the centre of society are indicative of the moral state of our society, perhaps more so than violent and other street crimes.": Nigel Morris. "Criminal nation: Two-thirds break law regularly." *The Independent*, June 25, 2007. [cited March 25, 2008]. news.independent.co.uk/uk/crime/article2705315.ece.
7. "Wealth of UK richest 'rises 20%'." *BBC News*, April 28, 2007. [cited March 25, 2008]. news.bbc.co.uk/2/hi/business/6603893.stm.
8. Real human needs and their costs are discussed in Chapter 6.
9. Information on British workers: "Bus drivers' 'early death risk'." *BBC News*, November 29, 2007. [cited March 25, 2008]. news.bbc.co.uk/2/hi/health/7116720.stm. Data on US mortality rates: UNDP. Human Development Report 2005, p. 59. [cited March 25, 2008]. hdr.undp.org/en/media/hdr05_complete.pdf.
10. Kjelle Arne Brekke and Desmond McNeill. "Identity Signaling in Consumption: A Case for Provision of More Public Goods." Background paper for World Development Report 2003: Dynamic Development in a Sustainable World. 2003.
11. Courtesy of David G. Myers. *From Psychology 8th edition*. Worth Publishers, 2007.

12. Yuri Dikhanov. "Trends in Global Income Distribution, 1970-2000, and Scenarios for 2015." UNDP *Human Development Report 2005* Occasional Paper 2005/8, p. 12. [cited March 25, 2008]. hdr.undp.org/en/media/globalincometrends.pdf.
13. David G. Myers. "The secret of happiness." *Yes!* Magazine, Summer 2004. [cited March 25, 2008]. yesmagazine.org/article.asp?id=866; Mike Rudin." The science of happiness." *BBC News*, April 30, 2006. [cited March 25, 2008]. news.bbc.co.uk/2/hi/programmes/happiness_formula/4783836.stm.
14. Steve Schifferes. "Richer, healthier but not happier." *BBC News*, April 8, 2008. [cited April 24, 2008]. news.bbc.co.uk/2/hi/business/7336336.stm.
15. Tom Kelly and Tristan Swanwick. "Friends worth weight in cash." *The Courier-Mail* (Brisbane, Australia), May 2, 2007, p. 11.
16. In 1949 the median size of a new American house was 1,100 square feet; by 2005 it was 2,227 square feet: Alan Durning. "Redesigning the forest economy." in Lester Brown et al., *State of the World 1994*. Norton, 1994; Sam Roberts. "Who Americans Are and What They Do, in Census Data." *New York Times*, December 15, 2006. [cited March 25, 2008]. nytimes.com/2006/12/15/us/15census.html?ex=1323838800&en=0854df3ef0203103&ei=5090&partner=rssuserland&emc=rss.
17. See Roberts, note 16.
18. John Schumaker. "An economy built on overconsumption cannot be sustained." *CCPA Monitor*, June 2002, pp. 34-35.
19. David R. Butcher. "Sleep Deprivation and Workplace Riskzzzzz." ThomasNet, August 1, 2006. cited March 25, 2008]. news.thomasnet.com/IMT/archives/2006/08/sleep_deprivation_workplace_injuries.html?t=archive.
20. Rates of mental illness in the United States were higher than in any other country surveyed. Nigeria and China had the lowest rates. The WHO World Mental Health Survey Consortium. "Prevalence, Severity, and Unmet Need for Treatment of Mental Disorders in the World Health Organization World Mental Health Surveys." *Journal of the American Medical Association*, Vol. 291 #21 (June 2, 2004). [cited March 25, 2008]. jama.ama-assn.org/cgi/reprint/291/21/2581.
21. N.R. Kleinfield. "Diabetes and Its Awful Toll Quietly Emerge as a Crisis." *New York Times*, January 9, 2006. [cited March 25, 2008]. query.nytimes.com/gst/fullpage.html?sec=health&res=9907E2DA1F30F93AA35752C0A9609C8B63.
22. Jeremy Laurance. "Children at risk of early death as obesity rises." *The Independent*, February 28, 2006. [cited March 25, 2008]. news.independent.co.uk/uk/health_medical/article348191.ece.
23. Roma Luciw. "One in three Canadians is a workaholic." *Globe and Mail*, May 15, 2007. [cited March 25, 2008]. theglobeandmail.com/servlet/story/RTGAM.20070515.wworkaholic0515/BNStory/Front.
24. "Advertising & youth/body image." YouthXchange, July 22, 2005. [cited March 25, 2008]. youthxchange.net/main/b262_advertising_youth-d.asp.
25. "Sexualisation 'harms' young girls." *BBC News*, February 20, 2007. [cited March 25, 2008]. news.bbc.co.uk/2/hi/health/6376421.stm.

26. "STDs rife among US teenage girls." *BBC News*, March 12, 2008. [cited March 25, 2008]. news.bbc.co.uk/2/hi/americas/7290088.stm.
27. Data: Worldwatch Institute. *State of the World 2004*. Norton, 2004; Lester R. Brown. Plan B Budget for Saving Civilization. Earth Policy Institute, 2007. [cited March 25, 2008]. earth-policy.org/Books/Seg/PB2ch13_ss4.htm.
28. Data: UNDP. *Human Development Report 2005*, p. 37. [cited March 25, 2008]. hdr.undp.org/en/media/hdr05; Millennium Ecosystem Assessment Reports. [cited March 25, 2008]. maweb.org/en/index.aspx.
29. Attributed to Mahatma Gandhi in BrainyQuote. [cited May 14, 2008]. brainyquote.com/quotes/authors/m/mohandas_gandhi.html.
30. Howard French. "Lives of Poverty, Untouched by China's Boom." *New York Times*, January 13, 2008. [cited April 24, 2008]. nytimes.com/2008/01/13/world/asia/13china.html?scp=1&sq=lives+of+grinding+poverty%2C+untouched+by+china%27s+boom&st=nyt.
31. Branko Milanovic. *Worlds Apart: Measuring International and Global Inequality*. Princeton University Press, 2005, pp.140-144.
32. Per capita incomes are measured in terms of purchasing power parity (i.e. with incomes adjusted among countries to reflect their real purchasing ability). Milanovic. *Worlds Apart*, pp. 129-130.
33. See note 32.
34. If anyone has access to more recent data on global incomes, poverty and inequality, I would appreciate receiving it. Please contact: graeme@bestfutures.org.
35. For example, see the website of Anti-Slavery International: antislavery.org.
36. Wealth by itself is not a problem, other than that it promotes unsustainable patterns of consumption. The main problem is that the political and economic power of wealthy people distorts the democratic process and resource allocation.
37. Data: World Institute for Development Economics Research of the United Nations University. "Pioneering Study Shows Richest Two Percent Own Half World Wealth." Press Release, December 5, 2006. [cited March 25, 2008]. wider.unu.edu/events/past-events/2006-events/en_GB/05-12-2006/.
38. Central Intelligence Agency. "Mozambique" and "United Kingdom" in *The 2008 World Factbook*. [cited April 19, 2008]. cia.gov/library/publications/the-world-factbook/geos/mz.html and cia.gov/library/publications/the-world-factbook/geos/uk.html.
39. UNDP. *Human Development Report 2005*, p. 24. [cited March 25, 2008]. hdr.undp.org/en/media/hdr05. Although many children are still dying from preventable causes, immunization campaigns have succeeded in reducing mortality in children under five to record lows.
40. These figures are approximate. The World Bank revised its estimates in late 2007 to take into account purchasing power. The new figures show that many more people in the world live in extreme poverty than had been previously estimated: Branko Milanovic. "Developing countries worse off than once thought." Yale Global, February

11, 2008. [cited March 25, 2008]. carnegieendowment.org/publications/index.cfm?fa=view&id=19907&prog=zch,zgp&proj=zsa,zted.

41. "Bangladesh children toil to survive." *Al Jazeera*, April 14, 2008. (online). [cited April, 24, 2008]. english.aljazeera.net/NR/exeres/F1AE7554-93D8-4786-BA8D-33E60D2F8965.htm?FRAMELESS=true&NRNODEGUID=%7bF1AE7554-93D8-4786-BA8D-33E60D2F8965%7d.
42. "Two of every 3 children physically abused: study." *The Hindu*, April 10, 2007. [cited March 25, 2008]. hindu.com/2007/04/10/stories/2007041003601100.htm.
43. "Around the world, at least one in every three women has been beaten, coerced into sex or otherwise abused during her lifetime. Nearly one-third of American women (31 percent) report being physically or sexually abused by a husband or boyfriend at some point in their lives.": Family Violence Prevention Fund. "Domestic Violence is a Serious, Widespread Social Problem in America: The Facts". [cited March 25, 2008]. endabuse.org/resources/facts.
44. Psychologists interviewed female and male prostitutes in five countries. 90% of the prostitutes "wanted out" of their way of life: Abigail Zuger. "Prostitution more stressful than combat, study finds." *Globe and Mail*, August 19, 1998, p. A11.
45. For example, the European Union's Common Agricultural Policy levies tariffs of up to 324% on farm imports from poor countries: Maxine Frith. "EU subsidies deny Africa's farmers of their livelihood." *The Independent*, May 16, 2006. [cited May 16, 2008]. independent.co.uk/news/world/politics/eu-subsidies-deny-africas-farmers-of-their-livelihood-478419.html.
46. In 2007, four countries had gender equality at the cabinet level: Chile, Spain, Sweden and Finland. However, at current annual rates of growth in the proportion of women members of national parliaments, global gender parity in national legislatures will not be reached until 2068: UNICEF. "Inequality in politics." in State of the World's Children 2007. [cited March 25, 2008]. unicef.org/sowc07/profiles/inequality_politics.php.
47. Maxine Frith. "UN: Women denied representation, making war on poverty hard to win." *The Independent*, March 8, 2006. [cited March 25, 2008]. news.independent.co.uk/world/politics/article349915.ece.
48. UNDP. *Human Development Report 2005*, p. 61. [cited May 13, 2008]. http://hdr.undp.org/en/reports/global/hdr1995/.
49. See note 47.
50. UNDP. Human Development Report 2005, p. 61. [cited March 25, 2008]. hdr.undp.org/en/media/hdr05.
51. Jerome Singer quoted by Newton Minow in "The Stranger in the Living Room." *Kidspeak*, Vol. 3#1 (Fall 2006), p.3. [cited May 14, 2008]. aap.org/sections/media/KidspeakFall06.pdf.
52. American Psychological Association. "Childhood Exposure to Media Violence Predicts Young Adult Aggressive Behavior, According to a New 15-Year Study". Press Release March 9, 2003. [cited March 31, 2008]. apa.org/releases/media_violence.html.

53. Derrick Jensen. *The Culture of Make Believe.* Context Books, 2002, p. 205.
54. Data from James Gasana. "Remember Rwanda?" *World Watch* Magazine, Vol. 15#5 (September/October 2002). [cited March 31, 2008]. worldwatch.org/node/524.
55. Anup Shah. "World military spending." Global Issues webpage. [cited April 25, 2008]. globalissues.org/Geopolitics/ArmsTrade/Spending.asp.
56. "Statement by Ambassador Mark. D. Wallace on the 2008/2009 U.N. budget." USUN press release, December 11, 2007. [cited April 25, 2008]. usunnewyork.usmission.gov/press_releases/20071211_367.html.
57. Dan Plesch. "Disarmament: the forgotten issue." *openDemocracy*, December 12, 2007. [cited May 16, 2008]. opendemocracy.net/article/globalisation/disarmament_the_forgotten_issue.
58. Benny Peiser. "Existential risk and democratic peace." *BBC News*, November 15, 2007. [cited April 25, 2008]. news.bbc.co.uk/2/hi/science/nature/7081804.stm.
59. For example: Bob Rigg. "The evisceration of a disarmament body." *openDemocracy*, April 27, 2007. [cited March 31, 2008]. opendemocracy.net/globalization-wmd/multilateral_disarmament_4567.jsp.
60. The space-based laser (ABL) systems will be able to strike anywhere on the Earth's surface. Another new weapons systems being developed by the United States is called prompt global strike. It will enable ballistic missiles armed with conventional warheads to hit any place on earth within minutes or hours of a decision to attack. The US control paradigm and the alternative — sustainable security — are discussed in: Paul Rogers. "Global security: a vision for change." *openDemocracy*, April 12, 2007. [cited March 31, 2008]. opendemocracy.net/conflict/security_vision_4520.jsp.
61. Luke Harding. "Russian missile test adds to arms race fears." *The Guardian*, May 30, 2007. [cited March 31, 2008]. guardian.co.uk/frontpage/story/0,,2091131,00.html.
62. Richard Fisher, Jr. "Developing US-Chinese Nuclear Naval Competition In Asia". International Assessment and Strategy Center, January 16, 2005. [cited March 31, 2008]. strategycenter.net/research/pubID.60/pub_detail.asp.
63. Subir Bhaumik. "India launches ballistic missile." *BBC News*, May 7, 2008. [cited May 16, 2008]. news.bbc.co.uk/2/hi/south_asia/7387082.stm.
64. Paul Rogers. "Britain's 21st-century defence." *openDemocracy*, February 15, 2007. [cited March 31, 2008]. opendemocracy.net/conflict/britain_defence_4352.jsp.
65. Dan Plesch. "Disarmament: the forgotten issue".
66. The International Monetary Fund estimates that around $1.8 trillion (US) is laundered annually, most of it derived from corrupt practices. Most of this money is transferred from developing countries to banks in developed countries. Without the cooperation of the world's largest financial centers, drug lords, warlords and corrupt officials could not find safe havens for their plunder: Akere Muna. "The man with the monogrammed clothes." *Globe and Mail*, June 7, 2007. [cited March 31, 2008]. theglobeandmail.com/servlet/story/RTGAM.20070607.wcomment0607/BNStory/International.
67. Small Arms Working Group. "Small Arms Facts Sheets 2006." [cited March 31, 2008]. fas.org/asmp/campaigns/smallarms/sawg.htm#Facts. The international arms

trade fuels wars around the world. For example civil war in the Congo has killed 5.4 million people since 1998: Joe Bavier. "Congo war-driven crisis kills 45,000 a month-study." Reuters AlterNet, January 22, 2008. [cited April 25, 2008]. alertnet.org/thenews/newsdesk/L22802012.htm.

68. "Iraq 'is university of terrorism'." *Al Jazeera* English, April 17, 2007. [cited March 31, 2008]. english.aljazeera.net/NR/exeres/B3470AA6-FC96-487C-8D8E-E23BB70E604F.htm.
69. Most Muslims have a very different view of the world than most Americans. They believe that the main reason the United States invaded Iraq was to steal the oil. In their view American policy is hypocritical: if the United States was really concerned about democracy and the welfare of the people it wouldn't be supporting Arab dictators and the Israeli occupation of Palestine.
70. Anatol Lieven. "At the Red Mosque in Islamabad." *openDemocracy*, June 4, 2007. [cited March 31, 2008]. opendemocracy.net/at_the_red_mosque_in_islamabad.jsp.
71. The strength of al Qaeda lies in its ability to simultaneously draw on the two causes of conflict — identity conflicts (over religion, values, and ethnic identities) and resource conflicts (over oil, the unequal distribution of resources). Surveys indicate that the majority of Muslims resent the West for both abusing their culture and exploiting their resources. However, support for al Qaeda's is falling because the majority of Muslims disagree with their fundamentalist values and brutal tactics.
72. Afghanistan now produces 93% of the world's opiates. The war-torn province of Helmand is now the single biggest drug-producing area in the world, surpassing whole countries like Colombia: Alastair Leithead. "Afghanistan opium at record high." *BBC News*, August 27, 2007. [cited March 31, 2008]. news.bbc.co.uk/2/hi/south_asia/6965115.stm.
73. At the end of 2000, 791,600 black American men were behind bars and 603,032 were enrolled in colleges or universities: StoptheDrugWar.org. "More Black Men in Prison Than College, Study Finds". August 30, 2002. [cited March 31, 2008]. stopthedrugwar.org/chronicle-old/252/jpistudy.shtml.
74. UNDP. *Human Development Report 2005*, pp. 151-152. [cited March 31, 2008]. hdr.undp.org/en/media/hdr05_complete.pdf.
75. Benjamin Barber. *Jihad vs. McWorld: How Globalism and Tribalism Are Reshaping the World.* Ballantine Books, 1996.
76. Wade Davis. *Light at the Edge of the World: A Journey Through the Realm of Vanishing Cultures.* Douglas & McIntyre, 2001, p. 13.
77. Drawing of the Mayan nobles from the Codex Nutall. Photo of shaman courtesy Wade Davis from *Light at the Edge of the World.* Douglas & McIntyre, 2001.
78. Quoted by Donnamarie in "Elegent Simplicity." A Garden of Roses and Lilies, July 13, 2006. [cited May 16, 2008]. donnamarie.wordpress.com/2006/07/13/elegant-simplicity/.
79. "The average American's day includes six minutes playing sports, five minutes reading books, one minute making music, thirty seconds attending a play or concert,

twenty-five seconds making or viewing art, and four hours watching television.": George Scialabba. "How Bad Is It?" Review of *Dark Ages America* by Morris Berman in the New Haven Review of Books, June 2007. [cited March 31, 2008]. morrisberman.blogspot.com/search?q=scialabba.

80. Bill Gates is optimistic that "more smart people" on the planet and new innovations will be able to improve global living standards and narrow the gap between rich and poor nations: Heather Scoffield. "To hell in a handbasket? No way, Bill Gates says." *Globe and Mail*, February 21, 2007. [cited May19, 2008]. theglobeandmail.com/servlet/story/RTGAM.20070221.wrgates21/BNStory/Business/. While we agree that human ingenuity can solve global problems, we don't agree that the problems can be solved within the existing world system which is based on constant material growth and the exploitation of both nature and humans for power and profit. Influential people like Bill Gates and Jimmy Carter are doing wonderful work for the benefit of humanity now that they have retired. When they headed Microsoft and the US government, institutional goals and structures restricted their ability to promote global values. The primary goal of Microsoft is to make profits; the primary goal of the US government is to promote American interests.
81. Comments made during a lecture hosted by the Ethos Foundation in Brisbane, Australia on May 5, 2007.

Chapter 4

1. The size of the global footprints from 1650 to 1950 are estimates based on data from Angus Maddison. *The World Economy*, Vols. 1 and 2. OECD, 2006.
2. In fascism, all individual and social interests are subordinate to the needs of the nation. In socialism, the economy is collectively controlled either by the state or the local community and operated for the general benefit. Production and prices are planned. In capitalism, most of the economy is privately owned and operated for profit. Production and prices are determined through a free market. Although there are still fascist and socialist states in the world (e.g. Burma and Cuba), at present the dominant global economic system is capitalism. China may call itself socialist, but its economic system is predominantly capitalist. Social democratic countries also have predominantly capitalist economies, with the government owning and controlling only a few sectors such as public communications and transport. In reality no economies are completely capitalist or socialist: in the United States various levels of government own roads, schools, fire departments, military installations and regulate large sections of the economy.
3. The global footprint in 2000 is estimated based on data from World Wildlife Fund. *Living Planet Report 2002*. [cited April 1, 2008]. panda.org/news_facts/publications/living_planet_report/lpr02/index.cfm.
4. Jeffrey Sachs estimates that total economic activity in the world (gross world product or GWP) rose 49 times between 1820 and 2000: *The End of Poverty: Economic Possibilities for Our Time*. Penguin, 2005, p. 28. Since all data regarding historical global

consumption are estimates, different sources give different figures. Because the consumption of goods and services (economic growth) expands faster than the consumption of renewable resources, the global consumption of renewable resources may have increased six times from 1900 to 2000.

5. Until 2005, most of the world's population lived in rural areas: David Whitehouse. "Half of humanity set to go urban." *BBC News*, May 19, 2005. news.bbc.co.uk/2/hi/science/nature/4561183.stm.
6. The global footprint for 2050 has been projected by the World Wildlife Fund. *Living Planet Report 2006*. [cited April 1, 2008]. panda.org/news_facts/publications/living_planet_report/lp_2006/index.cfm.
7. World Wildlife Fund. "Human footprint too big for nature." October 24, 2006. [cited April 1, 2008]. panda.org/index.cfm?uNewsID=83520. There is not a direct relationship between economic growth and the consumption of renewable resources: efficiencies tend to increase over time and economic growth involves many factors that do not require large renewable inputs (e.g. the growth of services and the increased consumption of non-renewable resources and recycled materials).
8. Societal dynamics and evolution is explained in more detail in BEST Futures. Introduction to *Time-Space-Technics*, Part 2. [cited May 8, 2008]. bestfutures.org/media/documents/TST_Presentation_14-36.pdf.
9. For an overview of the historical evolution of worldviews and societal systems, see Alastair M. Taylor. *Time-Space-Technics: An Evolutionary Model of Societies and Worldviews*. 1999. [cited April 1, 2008]. bestfutures.org/content/view/16/34/.
10. Fritjof Capra provides an excellent description of the development of the industrial worldview ("the Newtonian world-machine"): *The Turning Point: Science, Society, and the Rising Culture*. Simon and Schuster, 1982. He also describes the development of the systems-based worldview of the emerging Information Age.
11. H. H. Gerth and C. Wright Mills, eds. *From Max Weber: Essays in Sociology*. Oxford University Press, 1946, p. 155.
12. "Consumer culture no accident". David Suzuki Foundation, March 7, 2003. [cited April 25, 2008]. davidsuzuki.org/about_us/Dr_David_Suzuki/Article_Archives/weekly03070301.asp.
13. Marx stated that exchange in traditional societies took the form of a producer taking a commodity (C) to the market and selling in for money (M) in order to be able to purchase another commodity. This exchange can be written as C-M-C. In capitalist societies, the capitalist takes money to the market and purchases a commodity in order to be able to sell it at a profit for more money. The exchange is now M-C-M. While in traditional societies the only purpose of the money was to be a medium of exchange, in capitalist societies the only purpose of commodities is to make money. Because the capitalist is not interested in the social usefulness of the goods being traded but rather in their potential to make profits, the entire process is profoundly alienating. The alienation does not just involve the capitalist, who may be buying shares in unseen products or loaning money to unseen investors, but also to employees,

who frequently only work because of the paycheck and never meet the people that are consuming the products they make. See Dino Felluga."Modules on Marx: On Capital." *Introductory Guide to Critical Theory.* Purdue University, November 28, 2003. [cited April 1, 2008]. cla.purdue.edu/english/theory/marxism/modules/marxcapitalismmainframe.html.

14. Richard Sanders. "Sustainability — implications for growth, employment and consumption." *International Journal of Environment, Workplace and Employment,* Vol. 2 #4 (2006), p. 389.
15. Sir Peter Ustinov. UNICEF Goodwill Ambassador Archive, section 6. [cited April 1, 2008]. unicef.org/videoaudio/PDFs/gwa_sirpeterustinov_brollscript.pdf.
16. Joel Bakan. *The Corporation: The Pathological Pursuit of Profit and Power.* Viking Canada, 2004, pp. 75-76.
17. For example: Pew Global Attitudes Project. "Global Unease With Major World Powers". June 27, 2007 [cited April 1, 2008]. pewglobal.org/reports/display.php?ReportID=256.
18. Jeff Vail. *A Theory of Power.* iUniverse, 2004, p. 34. [cited April 1, 2008]. jeffvail.net.

Chapter 5

1. The Khmer (Cambodian) Empire flourished between the 9th and 14th centuries C.E. At its height Angkor had a population of 500,000 people, and with its supporting farms and irrigation canals covered an area of 1,158 square miles (3,000 square kilometers). The collapse was probably caused by deforestation, over population, topsoil erosion and the silting up of irrigation canals: "Map reveals ancient urban sprawl." *BBC News,* August 14, 2007. [cited April 2, 2008]. news.bbc.co.uk/2/hi/science/nature/6945574.stm.
2. Jared Diamond. *Collapse: How Societies Choose to Fail or Succeed.* Viking, 2005, pp. 13-15.
3. Julian Borger. "Darfur conflict heralds era of wars triggered by climate change, UN report warns." *The Guardian,* June 23, 2007. [cited April 2, 2008]. environment.guardian.co.uk/climatechange/story/0,,2109490,00.html.
4. Joseph Tainter. *The Collapse of Complex Societies.* Cambridge University Press, 1988.
5. Suk H. Kim and Mahfuzal Haque. "The Asian financial crisis of 1997: Causes and policy responses." *Multinational Business Review,* Spring 2002. [cited April 2, 2008]. findarticles.com/p/articles/mi_qa3674/is_200204/ai_n9026596.
6. Jack Goldstone. *Revolution and Rebellion in the Early Modern World.* University of California Press, 1991, p. 469 cited in Thomas Homer-Dixon. *The Upside of Down: Catastrophe, Creativity, and the Renewal of Civilization.* Island Press, 2006.
7. The danger of pandemics (world-wide diseases) is growing. Factory farms add antibiotics to feed to prevent outbreaks of diseases among overcrowded animals. These atrocious conditions support the development of diseases that are resistant to drugs. Drug resistant forms of diseases like tuberculosis are also breeding among poorly fed and housed prison populations in developing countries and among homeless people in the developed world. Threats like the spread of nuclear weapons are also growing.

Because the technology for making atomic bombs is more than 60 years old, every developing country sooner or later acquires the capability to build nuclear weapons as well as chemical and biological weapons.

8. For example, the OECD-FAO assumed that "average weather conditions" (p. 3) would continue between 2007-2016 with world oil prices declining to US$55 by 2012 and then rising to US$60 by 2016 (p.15): OECD/FAO. *2007 OECD-FAO Agricultural Outlook 2007-2016.* [cited April 25, 2008]. oecd.org/dataoecd/6/10/38893266.pdf.
9. Probable future scenarios are discussed in Chapter 10.
10. See the BEST Futures. Introduction to *Time-Space-Technics*, Part 2. [cited May 8, 2008]. bestfutures.org/media/documents/TST_Presentation_14-36.pdf for a more detailed explanation of societal dynamics.
11. Adapted from Richard Heinberg. *The Party's Over: Oil, War and the Fate of Industrial Societies.* New Society, 2003, p. 30. Updated with data from BP *Statistical Review of World Energy 2008.* [cited July 16, 2008]. bp.com/productlanding.do?categoryId=6929&contentId=7044622.
12. For example see Heinberg. *The Party's Over* and Paul Roberts. *The End of Oil: On the Edge of a Perilous New World.* Mariner, 2005.
13. One reason why conflicts and drugs do not destabilize the global system is because war and addictions are an integral part of it — for example drug money is frequently reinvested in legitimate businesses and used to finance the campaigns of politicians in many countries.
14. "EU must halt climate's doomsday clock, Prince Charles tells heads of Europe." *Daily Mail*, February 14, 2008. [cited March 19, 2008]. dailymail.co.uk/pages/live/articles/news/news.html?in_article_id=514345&in_page_id=1770.
15. "Doomsday Clock' Moves Two Minutes Closer To Midnight." Bulletin of the Atomic Scientists, January 17, 2007. [cited May 19, 2008]. thebulletin.org/media-center/announcements/20070117.html. "The Bulletin of the Atomic Scientists was founded in 1945 by University of Chicago scientists who had developed the atomic bomb and were deeply concerned about the use of nuclear weapons and the danger of nuclear war. In 1947 the Bulletin introduced its clock to convey the perils posed by nuclear weapons. The Doomsday Clock evokes both the imagery of apocalypse (midnight) and the contemporary idiom of nuclear explosion (countdown to zero)." Movement of the minute hand signals the organization's assessment of world events.
16. However, awareness is rapidly growing around the world. For example, polls show that 72% of the people living in British Columbia, Canada, fear that the world will end in two to three generations unless concerted action is taken on global warming: Geoff Olsen. "The future isn't what it used to be." *Common Ground*, July 2007. [cited April 2, 2008]. commonground.ca/iss/0707192/cg192_future.shtml.
17. Thomas Homer-Dixon. *The Ingenuity Gap: How Can We Solve the Problems of the Future?* Vintage Canada, 2001.
18. Albert Bartlett. "Arithmetic, Population and Energy." Global PublicMedia, 2004. [cited April 2, 2008]. globalpublicmedia.com/transcripts/645.

Part 2: Transformation — the emerging trend

1. Paul Hawken. *Blessed Unrest: How the Largest Movement in History is Restoring Grace, Justice and Beauty to the World.* Viking, 2007. [cited May 20, 2008]. gristmill.grist.org/story/2007/5/15/16245/1159.

Chapter 6

1. *Eco, ecological* and *green* have also become buzzwords used to improve corporate images. *Environmentally friendly* often means less destructive (at the best) rather than non-destructive and non-polluting. Such labels are only genuine when a company's productive processes and products have been inspected and certified by environmental organizations.
2. Of course sustainable is a relative term, since nothing lasts forever.
3. Adapted from Ervin Laszlo. *Evolution: The Grand Synthesis.* Shambhala, 1987, p. 55.
4. UN General Assembly. *Report of the World Commission on Environment and Development.* Resolution A/RES/42/187, December 11, 1987. [cited April 2, 2008]. un.org/documents/ga/res/42/ares42-187.htm.
5. ICUN/UNEP/WWF. *Caring for the Earth: A Strategy for Sustainable Living.* 1991. [cited April 2, 2008]. coombs.anu.edu.au/~vern/caring/caring.html.
6. Health and wholeness can also be described as functional integrity: David Pimentel et al. *Ecological Integrity: Integrating Environment, Conservation, and Health.* Island Press, 2000.
7. Indigenous communities were sustainable for thousands of years because they had developed economies and cultures that were appropriate to their environments. In parts of the world where these environments have not been destroyed (e.g. in the Amazon rainforest), indigenous societies that still retain their traditional ecological knowledge and traditional economies remain sustainable.
8. The triune brain is described in Thomas Lewis et al. *A General Theory of Love.* Vintage, 2000. The human needs hierarchy is adapted from W. Huitt. "Maslow's Hierarchy of Needs." *Educational Psychology Interactive.* Valdosta State University, 2004. [cited April 2, 2008]. chiron.valdosta.edu/whuitt/col/regsys/maslow.html.
9. See Figure 1 in the Introduction.
10. *Spiritual needs* refers to our need to understand reality and lead fully integrated lives. People satisfy these needs in many different ways: through devotional practices, meditation, art, community service, scientific studies and so on.
11. See Huitt, note 8.
12. People are happy when they feel able to satisfy their needs and reach their goals — i.e. when they are in control of their environments: F. Heylighen. "Happiness." *Principia Cybernetica Web,* 1999. [cited April 2, 2008]. pespmc1.vub.ac.be/HAPPINES.html.
13. Steve Crabtree. "The economics of happiness." *Gallup,* January 10, 2008. [cited April 2, 2008]. gmj.gallup.com/content/103549/Economics-Happiness.aspx.
14. Fred R. Shapiro, ed. *The Yale Book of Quotations.* Yale, 2006, p. 819.

15. Of course the military-industrial complexes of the world will never admit this, but it is easy to see that lobbyists for the arms industry have much more influence over foreign policy in industrialized countries than peace organizations. The Iraq war is a prime example: US military contractors were some of the largest contributors to the campaigns of the politicians that launched the war and have been the major beneficiaries of the conflict. The 2008 US government decision to sell tens of billions of dollars of high-tech weapons to Saudi Arabia, Egypt and Israel (two dictatorships and a country engaged in the illegal occupation of another nation) will promote conflicts and wars in the region for years to come.
16. Steve Bass believes that the design of the global economy prevents constructive change: "Economic growth is considered an inviolable principle, rather than people's rights and welfare, or environmental processes and thresholds; Environmental benefits and costs are externalised; Poor people are marginalised, and inequities entrenched; Governance regimes are not designed to internalise environmental factors, to iron out social inequities, or to develop better economic models; Therefore unsustainable behaviour has not been substantially challenged." Steve Bass. *A new era in sustainable development.* International Institute for Sustainable Development, 2007, p. 2. [cited April 2, 2008]. iied.org/mediaroom/docs/new_era.pdf.
17. Company executives must produce quarterly or annual profits if they want to keep their jobs, and politicians must show that they are producing results as they must run for re-election every few years. These institutional pressures make it difficult for decision-makers to engage in long-term strategic planning or to make major reforms that have high initial costs that will not pay off for many years — especially in the absence of a public debate on the need for fundamental changes. Because fundamental reforms are rarely made, businesses and governments spend much of their time to crisis managing crises.
18. For examples of how the market system distorts the economy: Jack Manno."Commodity Potential: An Approach to Understanding the Ecological Consequences of Markets" in Pimentel et al., *Ecological Integrity,* note 6.
19. National and global accounting usually measures income and productivity with Gross National Product (GNP), Gross Domestic Product (GDP), Gross National Income (GNI), Net National Product (NNP) and Net National Income (NNI). These are all variations of the same system, which tracks financial transactions such as wages, rents, profits and interest.
20. Mark Anielski. *The Economics of Happiness.* New Society Publishers, 2007. Various measures for calculating environmental and social welfare are being developed by NGOs, governments, the World Bank and others. These include Ecological Footprints, the Human Development Index (HDI), the Index of Sustainable Economic Welfare (ISEW), the Genuine Progress Indicator (GPI), the Environmental Performance Indicator (EPI), Green Net Product (GNP), Genuine Savings (or Genuine Investment) indicators, Environment and Sustainable Development Indicators (ESDI), the Sustainable Net Benefit Index (SNBI) and

Sustainable National Income (SNI). For example, the relationship between environment, economics and subjective well-being are examined in *The Happy Planet Index: An index of human well-being and environmental impact.* New Economics Foundation, 2006. [cited April 2, 2008]. neweconomics.org/gen/z_sys_publicationdetail.aspx?pid=225.

21. Robert F. Kennedy. *Remarks of Robert F. Kennedy at the University of Kansas, March 18, 1968.* [cited April 2, 2008]. jfklibrary.org/Historical+Resources/Archives/Reference+Desk/Speeches/RFK/RFKSpeech68Mar18UKansas.htm.
22. For example, Adam Cresswell. "The preventable things making us sick." *The Australian,* May 25, 2007, p. 7.
23. Although most social scientists agree that all societies require similar institutions to function, they define the Universal Culture Pattern in different ways. In Figure 6 we use Alastair Taylor's model which adds science and technology to the usual six institutions of the UCP. This is done to emphasize every society's need for environmental relevance.
24. The linkages between ecosystem services and human well-being are illustrated in Millennium Ecosystem Assessment. *Ecosystems and Human Well-being: Synthesis.* Island Press, 2005, p. 50. [cited April 2, 2008]. maweb.org/documents/document.356.aspx.pdf.

Chapter 7

1. The evolution of societal systems and worldviews is explained in more detail in the Time-Space-Technics area of our website: bestfutures.org/content/view/16/34/.
2. The process of paradigm change in the sciences is described in Thomas Kuhn. *The Structure of Scientific Revolutions,* 3rd ed. University of Chicago Press, 1996.
3. Fritjof Capra. *The Turning Point: Science, Society, and the Rising Culture.* Simon and Schuster, 1986, p. 72.
4. Courtesy of The Natural Edge Project: naturaledgeproject.net/About.aspx.
5. Quoted in Peter Maass. "The Breaking Point." *New York Times,* August 21, 2005. [cited April 3, 2008]. nytimes.com/2005/08/21/magazine/21OIL.html?pagewanted=6.
6. Natural Edge Project. Chapter 1 (Part 4) – "A critical mass of enabling technologies" in *The Natural Advantage of Nations* (Vol. I): *Business Opportunities, Innovation and Governance in the 21st Century.* 2007. [cited April 3, 2008]. naturaledgeproject.net/NAONChapter1.4.aspx.
7. Leo Jansen. *Technology in Sustainable Development.* Version 3-4-2008. [cited April 3, 2008]. indecol.ntnu.no/indecolwebnew/events/conferences/061000/061000_jansen.doc.
8. Matt Richtel. "Start-Up Fervor Shifts to Energy in Silicon Valley." *New York Times,* March 14, 2007. [cited April 3, 2008]. nytimes.com/2007/03/14/technology/14valley.html?ex=1186027200&en=b1d13f892772934d&ei=5070.
9. John Vidal. "Hologram prince hails new money for alternative energy." *The Guardian,* January 21, 2008. [cited April 3, 2008]. guardian.co.uk/environment/2008/jan/21/energy.renewableenergy.

10. Donella Meadows et al. *Limits to Growth: The 30-Year Update.* Chelsea Green, 2004, p. 97.
11. European Renewable Energy Council and Greenpeace. *energy [r]evolution: A Sustainable World Energy Outlook.* January 2007, p. 60. [cited April 3, 2008]. energyblueprint.info/home.0.html. For example, all US electrical energy needs could be met by covering 100 miles x 100 miles (160 kilometers x 160 kilometers) of desert with solar panels (based on 1kw/m² x 5 hours per day).
12. Janet L. Sawin. *Are renewables approaching a tipping point? Highlights from the REN21 renewables 2007 Global Status Report.* Worldwatch, March 3, 2008. [cited April 3, 2008]. worldwatch.org/node/5629.
13. For example, solar power plants can heat saline liquids which can then run turbine generators 24 hours a day: Peter Popham, "Sicily to build world's first solar power plant." *The Independent,* March 28, 2007. [cited April 3, 2008]. news.independent.co.uk/europe/article2398895.ece.
14. David Sassoon. *Fabled Oilman Sinking $10 Billion into World's Biggest Wind Farm.* Solve Climate website, February 23, 2008. [cited April 3, 2008]. solveclimate.com/blog/20080223/fabled-oilman-sinking-10-billion-worlds-biggest-wind-farm.
15. "Google's cheaper-than-coal target." *BBC News,* November 27, 2007. [cited April 3, 2008]. news.bbc.co.uk/2/hi/science/nature/7115786.stm.
16. European Renewable Energy Council and Greenpeace. *energy [r]evolution,* p. 83.
17. "The List: The World's Largest Solar Energy Projects." *Foreign Policy,* March, 2008. http://www.foreignpolicy.com/story/cms.php?story_id=4239.
18. "A solar powered world?" *Green Energy News,* Vol. 12#48 (February 22, 2008). [cited April 3, 2008]. green-energy-news.com/arch/nrgs2008/20080016.html.
19. Paul Rodgers. "Wind-fuelled 'supergrid' offers clean power to Europe." *The Independent,* November 25, 2007. [cited April 3, 2008]. environment.independent.co.uk/climate_change/article3194088.ece.
20. Ed Pilkington. "Big oil to big wind: Texas veteran sets up $10bn clean energy project." *The Guardian* April 14, 2008. [cited April 29, 2008]. guardian.co.uk/environment/2008/apr/14/windpower.energy.
21. Associated Press. "Algae Emerges as a Potential Fuel Source." *New York Times,* December 2, 2007. [cited April 3, 2008]. nytimes.com/2007/12/02/us/02algae.html?st=cse&sq=algae+emerges+as+a+potential+fuel+source&scp=1.
22. Anne Trafton. "'Major discovery' from MIT primed to unleash solar revolution." *MIT News,* July 31, 2008. (cited August 4, 2008] web.mit.edu/newsoffice/2008/0xygen-0731.html.
23. Jeremy Rifkin. *The Hydrogen Economy: The Creation of the World-Wide Energy Web and the Redistribution of Power on Earth.* Tarcher/Putnam, 2002, p. 184.
24. In order to calculate the environmental impact of a product or service it must be tracked throughout its entire life-cycle (through extraction, manufacture, use, waste and recycling). The Material Input per Service Unit (MIPS) concept has been designed to assist with these measurements.

25. Christopher Flavin. "Building a Low-Carbon Economy" in World Watch Institute. *State of the World 2008*. Norton, 2008, p. 88.
26. "The cost of lighting the world." *BBC News*, October 23, 2006. [cited April 3, 2008. news.bbc.co.uk/2/hi/technology/6067900.stm; "New efficient bulb sees the light." *BBC News*, December 28, 2007. [cited April 3, 2008]. news.bbc.co.uk/2/hi/uk_news/scotland/glasgow_and_west/7162606.stm.
27. Doug Struck. "A Big Drop in Emissions Is Possible With Today's Technology." *Washington Post*, January 21, 2008. [cited April 29, 2008]. washingtonpost.com/wp-dyn/content/article/2008/01/20/AR2008012001171_pf.html.
28. Sherrill Nixon. "Green offices that slash absenteeism." *Sydney Morning Herald*, January 16, 2008. [cited April 3, 2008]. smh.com.au/articles/2008/01/15/1200159449417.html.
29. Daniel Esty and Andrew Winston. *Green to Gold: How Smart Companies Use Environmental Strategy to Innovate, Create Value, and Build Competitive Advantage*. Yale University Press, 2006.
30. Rajendra Singh has brought water back to more than 1,000 villages in Rajastan, India by restoring and building traditional earth dams: "50 people who could save the planet." *The Guardian*, January 5, 2008. [cited May 29, 2008]. guardian.co.uk/environment/2008/jan/05/activists.ethicalliving.
31. Envirofit International. "Clean Cookstoves." 2008. [cited May 20, 2008]. envirofit.org/clean_cookstoves.html.
32. *What is Biomimicry?* Biomimicry Institute website. [cited April 29, 2008]. biomimicryinstitute.org/about-us/what-is-biomimicry.html.
33. Stephen Gillett. *Nanotechnology: clean energy and resources for the future*. Foresight Institute, October 2002, p. 72. [cited April 3, 2008]. foresight.org/impact/whitepaper_illos_rev3.pdf.
34. Janine Benyus. *Biomimicry: Innovation Inspired by Nature*. Harper, 2002.
35. Data: E. Cook. "The flow of energy in an industrial society." *Scientific American*, Vol. 224 (September 1971).
36. Jim Goldman. "Big opportunity in 'Nano'solar." *CNBC*, October 27, 2007. [cited April 4, 2008]. nanosolar.com/cache/cnbc.htm.
37. Rebecca Morelle. "Creating life in the laboratory." *BBC News*, October 19, 2007. [cited April 4, 2008]. news.bbc.co.uk/2/hi/science/nature/7041353.stm.
38. Stephen Gillett points out that nanotechnology will increasingly remove our dependence on metals: "As nanoscale fabrication makes accessible the ultimate materials strengths set by covalent chemical bonds, the structural metals that dominate present technology will become obsolete. If carbon becomes the 'ultimate material,' the carbonate rock that forms the bulk of the crustal carbon reservoir becomes an important backstop resource. Indeed, the very silicates that make up most of a rocky planet become a valuable feedstock for a mature nanotechnology.": Gillet, p. 7.
39. Diagram by Charles Blake and BEST Futures 2008. Further information: charlesblake@qldnet.com.au. In July 2008, the US National Science Foundation

brought together leading chemists to examine how carbon dioxide could be used to produce fuels, pharmaceuticals, plastics and other carbon-based products. "Experts to Discuss Recycling Carbon Dioxide." National Science Foundation press release. [cited July 17, 2008]. nsf.gov/news_summ.jsp?cntn_id=111815.

40. For example, Coskata, a startup biofuel company, believes it can make ethanol for under $1 a gallon. Any form of carbon-rich sources (e.g. old tires) can be baked to break their molecular bonds and release carbon monoxide and hydrogen gases. The gases are then sent to a bioreactor where anaerobic organisms digest them and secrete ethanol. [cited July 17, 2008]. coskata.com.
41. Elisabeth Rosenthal. "Environmental Cost of Shipping Groceries Around the World." *New York Times,* April 26, 2008. [cited May 20, 2008]. nytimes.com/2008/04/26/business/worldbusiness/26food.html?_r=1&scp=1&sq=Environmental%20cost%20of%20shipping%20groceries%20around%20the%20world&st=cse&oref=slogin.
42. Peter Bijur. "Global Energy Address to the 17th Congress of the World Energy Council." September 14, 1998 cited in *Limits to Growth: The 30-Year Update,* p. 97.
43. John Perry Barlow. "The Economy of Ideas." *Wired,* 2003. [cited May 21, 2008]. wired.com/wired/archive/2.03/economy.ideas_pr.html.
44. "Seven Questions: The Silent Tsunami." *Foreign Policy,* April, 2008. [cited May 21, 2008]. foreignpolicy.com/story/cms.php?story_id=4296.
45. "Global food system 'must change'." *BBC News,* April 15, 2008. [cited May 21, 2008]. news.bbc.co.uk/2/hi/science/nature/7347239.stm.
46. Consumers would be better able to compare the real price of making major purchases like appliances, cars and houses, if they were told the estimated cost of owning the product over its entire life cycle. Such costs include original purchase price plus interest charges, power and fuel costs, maintenance and insurance. Taxes should also be added to the purchase price of every product to cover the costs of cleaning up the pollution caused in its manufacture, operation and disposal.
47. Sir John Sorrell. "Time to leave the comfort zone." *BBC News,* May 20, 2008. [cited May 29, 2008]. news.bbc.co.uk/2/hi/science/nature/7410305.stm.
48. Ashley Seager. "Task force gives housing the green light." *The Guardian,* May 12, 2008. [cited May 21, 2008]. guardian.co.uk/environment/2008/may/12/greenbuilding.carbonemissions.
49. Collaborative Innovation Systems, Brisbane, Australia: neil.davidson6n@optusnet.com.au.
50. "Prescription for a healthy lifestyle, Australia." *Medical News Today,* October 3, 2005. [cited April 4, 2008]. medicalnewstoday.com/articles/31349.php; Jill Stark. "Children's obesity rates keep rising." *The Age,* July 20, 2007. [cited April 4, 2008]. theage.com.au/news/national/childrens-obesity-rates-keep-rising/2007/07/19/1184559956554.html.
51. Weight gains are not the result of more food being available — the majority of people in developed countries have been able to eat as much as they wanted for 50 years. They are mostly the result of lifestyle changes such as driving more and walking less;

eating more fast food and more junk food; watching more television; playing and exercising less and eating as a way of managing stress. The increasing pressure and pace of the consumer culture encourages people to eat and drink too much, work too hard and live too fast. The design of our cities, which are built for driving, not walking, adds to our stressful, sedentary lifestyle. As a result many people feel that they are too busy, too tired and too out of shape to get regular exercise.

52. Prevention is not given priority over treatment for many interconnected reasons: health care systems are designed to treat illness, not maintain health; health problems are often ignored until they become full-blown illnesses; there is no proper accounting system tracking personal and public health costs and benefits; maximum resources are often devoted for minimum objectives (such as prolonging the length of a person's life at all costs regardless of the quality of that life); there is not a whole-systems approach to the maintenance of health and treatment of illness; individuals, businesses and governments are not held responsible for making choices, producing products or designing systems that damage health. For all these reasons health care systems are so busy responding to catastrophic medical crises that they lack the resources to engage in prevention.
53. "Call for rethink in obesity fight." *BBC News*, December 14, 2007. [cited April 4, 2008]. news.bbc.co.uk/2/hi/health/7142176.stm.
54. Fritjof Capra. *The Hidden Connections: a Science for Sustainable Living*. Anchor, 2004, pp. 233-234.
55. William McDonough and Michael Braungart. *Cradle to Cradle: Remaking the Way We Make Things*. North Point, 2002.
56. See Zero Emissions Research and Initiatives (ZERI) at zeri.org.
57. Paul Hawken, Amory Lovins and L. Hunter Lovins. *Natural Capitalism: Creating the Next Industrial Revolution*. Back Bay, 2000.
58. Michael Grumwald. "The Clean Energy Scam." *Time*, March 27, 2008. [cited May 21, 2008]. time.com/time/magazine/article/0,9171,1725975,00.html.
59. Susie Mesure. "The £20bn food mountain: Britons throw away half of the food produced each year." *The Independent*, March 2, 2008. [cited April 8, 2008]. independent.co.uk/life-style/food-and-drink/news/the-16320bn-food-mountain-britons-throw-away-half-of-the-food-produced-each-year-790318.html.
60. Lev Grossman. "Bill Gates goes back to school." *Time*, June 7, 2007. [cited April 8, 2008]. time.com/time/health/article/0,8599,1630188,00.html.
61. Jorn Madslien. "Carmakers face obesity challenge." *BBC News*, July 6, 2007. [cited April 8, 2008]. news.bbc.co.uk/2/hi/business/6277636.stm.
62. It is also estimated that 1.4% of all the power produced in Britain in 2020 will be used by televisions on standby — that is, by televisions that no one is even watching: "Gadgets 'threaten energy savings'." *BBC News*, July 4, 2007. [cited April 8, 2008]. news.bbc.co.uk/2/hi/science/nature/6266082.stm.
63. Richard Sanders. "Sustainability — implications for growth, employment and consumption." *International Journal of Environment, Workplace and Employment*, Vol. 2 #4 (2006), p. 396.

64. Herman Daly. *Beyond Growth: The Economics of Sustainable Development.* Beacon, 1996, p. 11.
65. *Limits to Growth: The 30-Year Update,* p. 54.
66. The 2001 Sustainable Development Commission cited in Jonathon Porritt. *Capitalism as if the World Matters.* Bath Press, 2005, p. 291.
67. Jonathan Freedland. "We would be fools to banish business from the great climate battle." *Guardian Unlimited,* December 5, 2007. [cited April 8, 2008]. guardian.co.uk/commentisfree/story/0,,2222074,00.html.
68. Porritt, p. 232.
69. As H.G. Wells put it in his *Outline of History* (1920): "Human history becomes more and more a race between education and catastrophe."
70 *BBC News,* July 18, 2008. "Gore challenges US to ditch oil." [cited July 18, 2008]. news.bbc.co.uk/2/hi/americas/7513002.stm

Chapter 8

1. For example, a 2003 UN report stated that crime was increasing in almost every country measured. Crime is highest in places where there is the combination of a high percentage of young males and easy access to drugs and guns. Economic inequality and urbanization also increase crime rates. However, there appears to be no correlation between poverty and crime or involvement in religious activities and crime: Moisés Naim. "The Hidden Pandemic." *Foreign Policy* (July/August 2007). [cited April 8, 2008]. foreignpolicy.com/story/cms.php?story_id=3879.
2. C. Wright Mills. *The Sociological Imagination.* Oxford, 1959, p. 6.
3. These concepts are from Alastair M. Taylor. *Time-Space-Technics.* 1999. [cited April 8, 2008]. bestfutures.org/content/view/16/34.
4. Different theories argue whether historical change is driven primarily by the introduction of new means of production or the introduction of new ideas. The position advanced here is that it is never possible to completely separate economic and social factors, since a new societal system cannot function without a complete set of congruent material and societal technics. These interacting subsystems evolve together, rather than in a linear sequence.
5. "Dalits in conversion ceremony." *BBC News,* October 14, 2006. [cited April 8, 2008]. news.bbc.co.uk/2/hi/south_asia/6050408.stm.
6. American policies in the Middle East have a similar contracting dynamic since the public position of every administration has been to both promote democratic ideals and defend American economic interests. As a result the US officially supports democratic movements in the name of freedom while simultaneously arming dictatorships in the name of economic stability and national security. It is not surprising that people in the region are both confused and cynical about American values and objectives.
7. This is only possible in resource-rich countries like Saudi Arabia. Because oil revenues allow the Saudis to import finished goods and skilled labor from abroad, the

country can provide its citizens with consumer lifestyles without having to modernize their social structures.

8. Michael Stevens. "The Unanticipated Consequences of Globalization: Contextualizing Terrorism." in Chris Stout, ed. *The Psychology of Terrorism: Theoretical Understandings and Perspectives*. Praeger, 2002.
9. Under conditions of intense stress, fear and alienation, group identity can become more important than individual identity. People then tend to see the world in black and white terms (we are threatened by *them*) and are often willing to fight and sacrifice their lives for their nation, religion or group. These feelings can be mobilized by political or religious leaders through invoking memories of the group's past traumas and glories — the suffering they historically endured at the hands of the others, and the times when they were victorious: Vamik D. Volkan. *The Third Reich in the Unconscious*. Routledge, 2002, pp. 39-47.
10. The challenge for ruling elites in developing countries is to develop an open market economy and a highly educated and mobile workforce without losing economic and political power. The industrialized countries have solved this problem through creating democratic capitalist political institutions. While these institutions reduce social tensions through checking corruption and spreading political power among competing interests, they also enable wealthy elites to retain inordinate influence through their control of the economy and the media. In the West political democracy coexists with economic dictatorship: people can vote every few years for political parties, but they have no say over who will be their boss at work.
11. *Cell Phones and Globalization*. The Globalist, 2007. [cited May 1, 2008]. theglobalist.com/globalicons/syndication/sample.htm.
12. "Chinese outsurf US web users." *Couriermail.com.au*, March 14, 2008. [cited April 8, 2008]. news.com.au/couriermail/story/0,23739,23376130-954,00.html.
13. 715 million people from around the world tuned into the final game between Italy and France: *2006 FIFA World Cup in numbers*. FIFA.com website, 2007. [cited April 8, 2008]. fifa.com/aboutfifa/marketingtv/factsfigures/numbers.html.
14. "Q&A: What are the most spoken languages on Earth?" *Mongabay.com*, July 25, 2005. [cited May 1, 2008]. news.mongabay.com/2005/0724-unesco.html.
15. Approximately 3% of the world's population are migrants: Economic and Social Research Council. "Global Migration." 2007. [cited April 8, 2008]. esrcsocietytoday.ac.uk/ESRCInfoCentre/facts/international/migration.aspx?ComponentId=15051&SourcePageId=14912.
16. Ronald Inglehart. *Modernization and Postmodernization*. Princeton University Press, 1997.
17. Courtesy of Ronald Inglehart. The Inglehart-Welzel Cultural Map of the World (2003) is available from The World Values Survey Organization at worldvaluessurvey.org.
18. Ronald Inglehart. *Commentary on The Inglehart-Welzel Cultural Map of the World*. [cited April 9, 2008]. worldvaluessurvey.org/.

19. The middle class youth of the 1960s in the West were the first large group of people who didn't have to worry about where their next meal was coming from — the first generation with post-modern values. This is why the 1960s was a period of major cultural change.
20. Data: Jeremy Rifkin. *The European Dream: How Europe's Vision of the Future Is Quietly Eclipsing the American Dream.* Tarcher, 2004.
21. For example, 140,000 people in Britain in 2006 voluntarily paid to offset the carbon dioxide produced by the flights they purchased: "Tracking carbon through your phone." *BBC News,* August 22, 2007. [cited April 9, 2008]. news.bbc.co.uk/2/hi/technology/6957235.stm.
22. Most people in the world believe that the US, followed by China, bears the most responsibility for global warming: Pew Global Attutides Project. "Global Unease With Major World Powers: Rising Environmental Concern in 47-Nation Survey." June 27, 2007. [cited April 9, 2008]. pewglobal.org/reports/display.php?ReportID=256.
23. "Most ready for 'green sacrifices'." *BBC News,* November 9, 2007. [cited April 9, 2008]. news.bbc.co.uk/2/hi/7075759.stm. At the same time many people are cynical about whether governments are levying new *green taxes* to solve problems or simply to raise more revenues: Colin Brown. "The green tax revolt: Britons 'will not foot bill to same planet'." *The Independent,* May 2, 2008. [cited May 29, 2008]. independent.co.uk/environment/climate-change/green-tax-revolt-britons-will-not-foot-bill-to-save-planet-819703.html.
24. An example of a social movement with a long history is the animal rights movement in the West. Studies show that the more people accept violence and cruelty to animals, the more likely they are to accept violence and cruelty to humans: "Warning over child animal cruelty" *BBC News,* November 22, 2001. [cited April 9, 2008]. news.bbc.co.uk/2/hi/health/1670025.stm.
25. Paul Hawken. From the Introduction to *Blessed Unrest: How the Largest Movement in the World Came into Being and Why No One Saw It Coming.* Viking, 2007. [cited May 2, 2008]. gristmill.grist.org/story/2007/5/15/16245/1159.
26. Find WiserEarth at wiserearth.org/ and The Natural Capital Institute at naturalcapital.org/whoweare.htm.
27. For example, the Northern Ireland Women's Coalition played a pivotal role in ending decades of warfare between Protestants and Catholics: Anne Carr. "A Northern Ireland lesson." *Open Democracy,* June 5, 2007. [cited April 9, 2008]. opendemocracy.net/democracy-fifty/ireland_lesson_4673.jsp.
28. Frances Harrison. "Women graduates challenge Iran." *BBC News,* September 19, 2006. [cited April 9, 2008]. news.bbc.co.uk/2/hi/middle_east/5359672.stm; Alex Kingsbury. "Many Colleges Reject Women at Higher Rates Than for Men." *US News and World Report,* June 17, 2007. [cited April 9, 2008]. usnews.com/usnews/edu/articles/070617/25gender.htm.
29. "Cluster bomb ban treaty approved." *BBC News,* May 28, 2008. [cited May 29, 2008]. news.bbc.co.uk/2/hi/europe/7423714.stm.

30. Of course the public will only support wars if they believe that they are necessary. Most Americans believed that defeating the Axis powers was vital to world security; after a few years in Vietnam and Iraq, only a minority still believed that these wars were necessary.
31. Andrew J. Bacevich. "Bushed Army." *The American Conservative*, June 4, 2007. [cited April 9, 2008]. amconmag.com/2007/2007_06_04/article1.html.
32. By October, 2007, 105 countries were members of the court. Another 41 countries had signed the agreement but not yet ratified it. The main opposition to the court comes from the United States, China and India, who believe that the court infringes on national sovereignty. See icc-cpi.int/about.html.
33. John Elkington. "Rising to the challenge?" *The Guardian*, March 28, 2007. [cited April 9, 2008]. environment.guardian.co.uk/climatechange/story/0,,2044046,00.html.
34. World music is another example of an emerging integrative art form.
35. In 1690 John Locke attacked the concept of absolute monarchy and the divine right of kings in *An Essay Concerning Human Understanding*. Adam Smith attacked mercantilism and supported free trade in his 1776 book *An Inquiry into the Nature and Causes of the Wealth of Nations*.
36. We have all grown up talking about weather systems, political systems, electrical systems, digestive systems, computer systems and so on. But our industrial worldview means that we perceive every system as existing in isolation from every other system. As a result we don't have universities but multiversities; in the absence of a unifying theoretical foundation economists, psychologists, historians, physicists, biochemists and other scientists have little in common.
37. Angus Taylor. *The Nature of History: Dialectical Materialism and General Systems Theory*. 2007. [cited April 9, 2008]. bestfutures.org/content/view/17/35/.
38. Ken Wilber compares some of the literature on the evolution of culture and consciousness in *Integral Psychology: Consciousness, Spirit, Psychology, Therapy*. Shambhala, 2000.
39. The structure of a holistic system is discussed in Chapter 11.
40. It is quite significant that Sir Tim Berners-Lee, the inventor of the Internet, insisted that the web should be made free. This allowed it to spread rapidly: Darren Waters. "Web in infancy, says Berners-Lee." *BBC News*, April 30, 2008. [cited May 2, 2008]. news.bbc.co.uk/2/hi/technology/7371660.stm.
41. See articles by Michel Bauwens at others at p2pfoundation.net/Main_Page.

Chapter 9

1. Barbara Tuchman. *The Guns of August*. Presidio, 1962.
2. Paul Roberts. *The End of Oil*. Mariner, 2005, pp. 307-308.
3. An old saying regarding the accuracy of data is: "Garbage in, garbage out."
4. "The stresses and multipliers are a lethal mixture that sharply boosts the risk of collapse of political, social and economic order in individual countries and globally —an outcome I call synchronous failure. This would be destructive — not creative —

catastrophe.": Thomas Homer-Dixon. *The Upside of Down: Catastrophe, Creativity and the Renewal of Civilization*. Island Press, 2006, p. 16.

5. Alejandro Litovsky. "Energy poverty and political vision." *openDemocracy*, September 4, 2007. [cited April 10, 2008]. opendemocracy.net/node/34494.
6. Geoffrey Lean. "Oil and gas may run short by 2015, say industry experts." *The Independent*, July 22, 2007. [cited April 10, 2008]. environment.independent.co.uk/ climate_change/article2790960.ece.
7. "Opec warns oil could reach $200." *BBC News*, April 28, 2008. [cited May 2, 2008]. news.bbc.co.uk/2/hi/business/7370441.stm. In June 2008, Alexey Miller, the head of Gazprom, the world's largest energy company, predicted that oil prices would hit $250 per barrel "in the foreseeable future." Danny Fortson. "An ominous warning that the rapid rise in oil prices has only just begun". *Independent*, June 11, 2008. [cited July 17, 2008]. independent.co.uk/news/uk/home-news/ an-ominous-warning-that-the-rapid-rise-in-oil-prices-has-only-just-begun-844217.html.
8. Elisabeth Rosenthal. "Europe Turns Back to Coal, Raising Climate Fears." *The New York Times* April 23, 2008. [cited May 2, 2008]. nytimes.com/2008/04/23/ world/europe/23coal.html?scp=7&sq=April+23%2C+2008&st=nyt.
9. Paul Rogers. "Iraq: the dissonance effect." *openDemocracy*, August 30, 2007. [cited April 10, 2008]. opendemocracy.net/node/34470.
10. Ira Chernus. "The Democrats' Iraqi Dilemma." *Mother Jones*, July 24, 2007. [cited April 10, 2008]. motherjones.com/commentary/tomdispatch/2007/07/iraqi_dilemma.html. However, American views may be changing. Paul Roberts argued that "the greatest casualty of the Iraq war may be the very idea of energy security .... it is now clear that even the most powerful military entity in world history cannot stabilize a country at will or 'make' it produce oil ...": *The End of Oil*, p. 337.
11. This paragraph summarizes data provided in Chapters 2 and 3.
12. Photo credit UN Photo.
13. "Global cereal supply and demand brief ." *FAO Crop Prospects and Food Situation #1* (February, 2008). [cited May 2, 2008]. fao.org/docrep/010/ah881e/ah881e04.htm.
14. Bracewell. *World Food Shortages: Pushing Interest Rates Up*. June 20, 2007. [cited April 10, 2008]. bracewell.livejournal.com/297282.html.
15. Angela Balakrishnan. "UN body urges agriculture reforms to stave off food crisis." *The Guardian*, April 15, 2008. [cited May 2, 2008]. http://www.guardian.co.uk/ environment/2008/apr/15/food.unitednations1.
16. Rosslyn Beeby. "UN issues warning of critical food shortages 'The livelihoods of billions of people will be severely challenged." *Canberra Times*, December 26, 2007. [cited April 10, 2008]. canberra.yourguide.com.au/news/local/news-features/ un-issues-warning-of-critical-food-shortages-the-livelihoods-of-billions-of-people-will-be-se/1151741.html.
17. Jorn Madslien. "The end of cheap clothes is near." *BBC News*, April 23, 2008. [cited May 2, 2008]. news.bbc.co.uk/2/hi/business/7362343.stm.

18. Joachim von Braun. *Poverty, Climate Change, Rising Food Prices and the Small Farmers.* International Food Policy Research Institute, April 22, 2008, slide 41. [cited May 2, 2008]. ifad.org/gbdocs/repl/8/ii/e/presentations/IFAD_21-04-08.pps#40.
19. David Loyn. "Task force faces major challenge." *BBC News*, April 28, 2008. [cited May 21, 2008]. news.bbc.co.uk/2/hi/americas/7374006.stm.
20. Bryan Walsh. "Global warming's next victim: wheat." *Time*, August 27, 2007. [cited April 10, 2008]. time.com/time/world/article/0,8599,1656570,00.html.
21. *Lowest food supplies in 50 or 100 years: global food crisis emerging.* National Farmers Union press release, May 11, 2007. [cited April 10, 2008]. nfu.ca/press_2007.html.
22. "UN sets up food crisis task force." *BBC News*, April 29, 2008. [cited May 5, 2008]. news.bbc.co.uk/2/hi/europe/7372393.stm.
23. Mark Kinver. "Eco-farming 'helps world's poor'." *BBC News*, February 15, 2008. [cited April 10, 2008]. news.bbc.co.uk/2/hi/science/nature/4716224.stm.
24. "The recipe for food rights." *Al Jazeera*, April 11, 2008. [cited May 21, 2008]. english.aljazeera.net/NR/exeres/D1347B14-5364-45C5-8C46-950C6541801D.htm?FRAMELESS=true&NRNODEGUID=%7bD1347B14-5364-45C5-8C46-950C6541801D%7d.
25. Mark Bittman. "Rethinking the Meat-Guzzler." *New York Times*, January 27, 2008. [cited May 21, 2008]. nytimes.com/2008/01/27/weekinreview/27bittman.html?_r=1&scp=2&sq=rethinking+the+meat-guzzler&st=nyt&oref=slogin.
26. Lester Brown. "Water tables falling and rivers running dry." July 24, 2007. [cited April 10, 2008]. earth-policy.org/Books/Seg/PB2ch03_ss2.htm.
27. Peter S. Goodman. "The Dollar: Shrinkable but (So Far) Unsinkable." *New York Times*, May 11, 2008. [cited May 21, 2008]. nytimes.com/2008/05/11/weekinreview/11goodman.html?scp=1&sq=the+US+dollar%3A+shrinkable+but+%28so+far%29+unsinkable&st=nyt.
28. Alan Tonelson. "U.S. 2007 trade deficit drops on falling imports." *AmericanEconomicAlert*, February 20, 2008. [cited April 10, 2008]. americaneconomicalert.org/view_art.asp?Prod_ID=2945.
29. See note 27.
30. Stephen Roach "Double Bubble Trouble." *New York Times*, March 5, 2008. [cited May 5, 2008]. nytimes.com/2008/03/05/opinion/05roach.html?scp=1&sq=Double+Bubble+Trouble&st=nyt.
31. Eric Janszen. "The next bubble: priming the markets for tomorrow's big crash." *Harper's*, February 2008. [cited April 10, 2008]. harpers.org/archive/2008/02/0081908.
32. Jeffrey Garten. "Beware the Weak Dollar." *Newsweek*, September 10, 2007. [cited April 10, 2008]. newsweek.com/id/40830.
33. Lau Nai-keung. "It's time to take seriously a US-Led global recession." *China Daily*, October 6, 2005. [cited April 10, 2008]. chinadaily.com.cn/english/doc/2005-10/06/content_482807.htm.
34. Jamie Robertson. "Markets brave recession fears." *BBC News*, April 1, 2008. [cited May 5, 2008]. news.bbc.co.uk/2/hi/business/7324923.stm.

35. The mix of poverty and inequality means that extreme social tensions already exist in many countries. For example, entrenched poverty in India means that one in six Indians lives in an area where there is armed insurgency: Andrew Buncombe. "India at 60: special report." *The Independent*, August 10, 2007. [cited April 10, 2008]. news.independent.co.uk/world/asia/article2851536.ece.
36. Keith Bradsher. "High Rice Cost Creating Fears of Asia Unrest." *New York Times*, March 29, 2008. [cited April 10, 2008]. nytimes.com/2008/03/29/business/worldbusiness/29rice.html?_r=1&scp=1&sq=keith+bradsher+rice+shortages&st=nyt&oref=slogin.
37. Peter Biles. "Tensions erupt in city of promise." *BBC News*, May 21, 2008. [cited May 29, 2008]. news.bbc.co.uk/2/hi/africa/7413009.stm.
38. James Reynolds. "China in Africa: Developing ties." *BBC News*, July 3, 2007. [cited April 10, 2008]. news.bbc.co.uk/2/hi/asia-pacific/6264476.stm.
39. Countries are currently jockeying to gain military, economic and political advantages — for example India is building strategic relationships with the US on some issues and with Iran and China on others. Subir Bhaumik. "Five-nation naval exercise begins." *BBC News*, September 4, 2007. [cited April 10, 2008]. news.bbc.co.uk/2/hi/south_asia/6977376.stm; "Energy dominates Shanghai summit." *BBC News*, August 17, 2007. [cited April 10, 2008]. news.bbc.co.uk/2/hi/asia-pacific/6949021.stm.
40. Zbigniew Brzezinski. *The Grand Chessboard: American Primacy and Its Geostrategic Imperatives*. Basic Books, 1997.
41. Trita Parsi. "Long division." *The American Conservative*, September 10, 2007. [cited April 10, 2008]. amconmag.com/2007/2007_09_10/article3.html. Dr. Parsi describes the struggle between the US and Iran for hegemony in the Middle East.
42. It will be much cheaper to prevent wars than fight them. By 2008 the Iraq war was estimated to have cost the United States between $1 trillion and $4 trillion: David Herszenhorn. "Estimates of Iraq War Cost Were Not Close to Ballpark." *New York Times*, March 19, 2008. [cited April 10, 2008]. nytimes.com/2008/03/19/washington/19cost.html?partner=rssnyt&emc=rss. In June 2008, Senator John Kerry proposed that the US take the lead in creating a nuclear weapons free world. John Kerry. "America looks to a nuclear-free world". *Financial Times*, June 24, 2008. [cited June 30, 2008]. ft.com/cms/s/0/d24eb694-424f-11dd-a5e8-000779fd2ac.html?nclick_check=1.
43. Robin McKie. "Climate wars threaten billions." *The Observer*, November 4, 2007. [cited May 29, 2008]. guardian.co.uk/environment/2007/nov/04/climatechange.scienceofclimatechange.
44. Richard Norton-Taylor. "Al-Qaida has revived." *The Guardian*, September 13, 2007. [cited April 10, 2008]. guardian.co.uk/alqaida/story/0,,2167923,00.html.
45. For example, Al Gore and the Alliance for Climate Protection has launched one of the most ambitious public advocacy campaigns in US history to mobilize Americans to push for aggressive reductions in greenhouse gas emissions: Juliet Eilperin. "Gore Launches Ambitious Advocacy Campaign on Climate." *Washington Post*, March 31,

2008. [cited May 21, 2008]. washingtonpost.com/wp-dyn/content/article/2008/03/30/AR2008033001880.html.

46. For example, the World Conservation Union has recently moved western lowland gorillas from the Endangered Species Category to the Critically Endangered category, where they join another of our closest animal relatives, the orangutans: Richard Black, "Gorillas head race to extinction." *BBC News*, September 12, 2007. [cited April 10, 2008]. news.bbc.co.uk/2/hi/science/nature/6990095.stm.
47. Courtesy David Wasdell. The original with valuable background information may be viewed at meridian.org.uk/Resources/Global%20Dynamics/TippingPoint/index.htm.
48. Steve Connor. "Global warming 'past the point of no return'." *The Independent*, September 15, 2005. [cited April 10, 2008]. news.independent.co.uk/sci_tech/article312997.ece.
49. James Lovelock. *The Revenge of Gaia*. Allen Lane, 2006, pp. 51-52.
50. David Wasdell. *Beyond the Tipping Point: Positive Feedback and the Acceleration of Climate Change. 2006.* [cited April 10, 2008]. meridian.org.uk/Resources/Global%20Dynamics/TippingPoint/index1.htm.
51. The global climate will eventually re-equilibrate at a higher temperature, as it has in the past. However, most of the species that are alive today will not be adapted to this climate and it will take millions of years for new flora and fauna to evolve. If current trends continue, concentrations of greenhouse gases will be much higher than during the Paleocene-Eocene Thermal Maximum, a mass extinction event which occurred 55 million years ago.

Chapter 10

1. Eamon O'Hara. "Focus on carbon 'missing the point'." *BBC News*, July 30, 2007. [cited May 6, 2008]. news.bbc.co.uk/2/hi/science/nature/6922065.stm.ok
2. US National Intelligence Council. *The Contradictions of Globalization*. 2004, p. 34. [cited April 18, 2008]. http://www.dni.gov/nic/NIC_globaltrend2020_s1.html#tech.
3. "The context here is straightforward: the Millennium Development Goals are not going to be reached, the gap between the richest fifth of the global population and the rest is widening, yet improved education and communications are leading to a wider self-perception of marginalisation. This results in bitterness and frustration and aids the evolution of radical and extreme social movements.": Paul Rogers. "The SWISH Report (3)." *openDemocracy*, May 19, 2005. [cited April 18, 2008]. opendemocracy.net/conflict/swish_2523.jsp.
4. Steven R. Weisman. "Finance Ministers Emphasize Food Crisis Over Credit Crisis." *New York Times*, April 14, 2008. [cited April 18, 2008]. nytimes.com/2008/04/14/business/14finance.html?scp=1&sq=finance+ministers+emphasize+food+crisis&st=nyt.
5. A dangerous trend is the rapid growth of Internet censorship. At least 25 countries now censor the net in order to control political dissent and social values: "Global net

censorship 'growing'." *BBC News*, May 18, 2007. [cited April 18, 2008]. news.bbc.co.uk/2/hi/technology/6665945.stm.

6. "World Bank fears Pakistan crisis" *BBC News*, March 28, 2008. [cited April 18, 2008]. news.bbc.co.uk/2/hi/business/7316517.stm.
7. American foreign policy reflects this contradiction. While the US advocates the development of more politically and economically open societies, it has often helped to overthrow democratically elected governments that opposed American interests. For example, in 2006 the United States actively undermined the Hamas government in Palestine — which was, whatever its policies, the only freely elected Arab government in the Middle East — while providing massive military support to dictatorships in Egypt and Saudi Arabia. It is not surprising that most Muslims believe that American troops invaded Iraq to control the oil fields, not to establish a free and independent country.
8. For example, see Robert Burns' report on President Bush's visit to Iraq: "AP Analysis: Bush calls Anbar a success." *Washington Post*, September 4, 2007. (online – link expired).
9. The misuse of the term *sustainable security* is similar to the misuse of the term *sustainable development*. Sustainable development means developing sustainability; it does not mean maintaining unsustainable material growth. Similarly, sustainable security means creating a sustainable society; it does not mean maintaining environmentally unsustainable material growth. Sustainable security removes the causes of violent conflicts through meeting essential biophysical and human needs.
10. Chris Abbott. "Beyond terrorism: towards sustainable security." *openDemocracy*, April 4, 2007. [cited April 18, 2008]. opendemocracy.net/globalization-institutions_government/beyond_terrorism_4532.jsp.
11. Sandra Polaski. *U.S. living standards in an Era of Globalization*. Carnegie Endowment for International Peace, July 2007. [cited April 18, 2008]. carnegieendowment.org/publications/index.cfm?fa=print&id=19442.
12. Kevin Phillips. *Wealth and Democracy: A Political History of the American Rich*. Broadway, 2003, p. 422.
13. Frank O'Donnell of Clean Air Watch commented that "The danger is that the three sacred cows of the US economy, coal, cars and corn, will hijack the new mood of environmental awareness.": Leonard Doyle. "Democrats in $7bn plan to turn US green." *The Independent*, August 22, 2007. [cited April 18, 2008]. news.independent.co.uk/world/americas/article2883854.ece.
14. "Editorial: China's Economic Puzzle." *New York Times*, October 19, 2007. [cited May 6, 2008]. nytimes.com/2007/10/19/opinion/19fri2.html?scp=1&sq=China%27s+Economic+Puzzle&st=nyt.
15. Liu Mingkang, the head of China's Banking Regulatory Commission, warned that the Chinese economy is fundamentally unbalanced and too dependent on exports — personal consumption makes up only 50% of Chinese output. He stated that the social security net has been weakened and the government now spends too little on health, education, housing and pensions: Steve Schifferes. "The promise and perils of

globalisation." *BBC News,* June 11, 2007. [cited April 18, 2008]. news.bbc.co.uk/2/hi/business/6733755.stm.

16. "'Severe' jobless problem in China." *BBC News,* March 9, 2008. [cited April 18, 2008]. news.bbc.co.uk/2/hi/asia-pacific/7286024.stm.
17. Joseph Kahn and Jim Yardley. "As China Roars, Pollution Reaches Deadly Extremes." *New York Times,* August 26, 2007. [cited April 18, 2008]. nytimes.com/2007/08/26/world/asia/26china.html?_r=1&oref=slogin.
18. Li Datong. "China's media change: talking with Angela Merkel." *openDemocracy,* September 6, 2007. [cited April 18, 2008]. opendemocracy.net/node/34509.
19. Robert Mackey. "Answers from Orville Schell" in the series "Choking on growth." *New York Times,* August 29, 2007. [cited April 18, 2008]. china.blogs.nytimes.com/.
20. Robert Mackey. "Answers from Dr. Elizabeth Economy" in the series "Choking on growth." *New York Times,* August 27, 2007. [cited April 18, 2008] china.blogs.nytimes.com/.
21. "Mozambique riots push fuel down." *BBC News,* February 13, 2008. [cited April 18, 2008]. news.bbc.co.uk/2/hi/africa/7242323.stm.
22. Roger Harrabin. "China 'now top carbon polluter'." *BBC News,* April 14, 2008. [cited April 18, 2008]. news.bbc.co.uk/2/hi/asia-pacific/7347638.stm.
23. David Adam and John Vidal. "Rich states failing to lead on emissions, says UN climate chief." *The Guardian,* April 14, 2008. [cited April 18, 2008]. guardian.co.uk/environment/2008/apr/14/climatechange.carbonemissions.
24. See note 22.
25. Adam Morton. "Stern gets sterner on emissions." *The Age,* May 2, 2008. [cited May 6, 2008]. theage.com.au/news/environment/stern-gets-sterner-on-emissions/2008/05/01/1209235059204.html.
26. Ibid.
27. Andrew Dobson. "A climate of crisis: towards the eco-state." *openDemocracy,* September 19, 2007. [cited May 29, 2008]. opendemocracy.net/article/globalisation/politics_climate_change/state.
28. For example, in May, 2008 Marina Silva, the Brazilian Environment Minister, resigned stating that her efforts to protect the Amazon rainforest were being blocked within the government by powerful agribusiness lobbies: Daniel Howden. "I give up, says Brazilian minister who fought to save the rainforest." *The Independent,* May 15, 2008. [cited May 22, 2008] independent.co.uk/environment/climate-change/i-give-up-says-brazilian-minister-who-fought-to-save-the-rainforest-828310.html.
29. Cahal Milmo. "Biofuel: the burning question." *The Independent,* April 15, 2008. [cited April 21, 2008]. independent.co.uk/environment/climate-change/biofuel-the-burning-question-808959.html.
30. "Security fears as food prices soar." *Al Jazeera,* April 12, 2008. [cited April 21, 2008]. english.aljazeera.net/NR/exeres/BC286730-045A-48BD-834F-451F5E71EDEC.htm?FRAMELESS=true&NRNODEGUID=%7bBC286730-045A-48BD-834F-451F5E71EDEC%7d.

31. The current Russian agenda is re-establishing itself as a great power. Its 20th century values and goals are at odds with values of its European neighbors and with the needs of the 21st century: Ivan Krastev. "Russia vs Europe: the sovereignty wars." *openDemocracy*, September 5, 2007. [cited April 21, 2008]. opendemocracy.net/node/34504.
32. Richard Black. "Tackling the fossil fuel juggernaut." *BBC News*, November 17, 2007. [cited April 21, 2008]. news.bbc.co.uk/2/hi/science/nature/7100039.stm.
33. Alix Kroeger. "EU warns of climate change threat." *BBC News*, March 10, 2008. [cited April 21, 2008]. news.bbc.co.uk/2/hi/europe/7287168.stm.
34. Hans-Jürgen Schlamp et al. "Merkel caught between Industry and the Climate." *Spiegel Online*, January 22, 2008. [cited April 21, 2008]. spiegel.de/international/europe/0,1518,druck-530162,00.html.
35. Andrew Sparrow and Patrick Wintour. "Nuclear is UK's new North Sea oil — minister." *The Guardian*, March 26, 2008. [cited April 21, 2008]. guardian.co.uk/environment/2008/mar/26/nuclearpower.energy.
36. Graeme Taylor. "Evaluating nuclear energy in the context of global sustainability." *Social Alternatives* Vol. 26 #2 (Second Quarter, 2007), pp. 12-17.
37. See note 34.
38. "EU agrees climate plan deadline." *BBC News*, March 14, 2008. [cited April 21, 2008]. news.bbc.co.uk/2/hi/europe/7296564.stm.
39. Richard Black. "New crops needed to avoid famines." *BBC News*, December 3, 2006. [cited April 21, 2008]. news.bbc.co.uk/2/hi/science/nature/6200114.stm.
40. "Sarkozy details green French plan." *BBC News*, October 25, 2007. [cited April 21,2008]. news.bbc.co.uk/2/hi/europe/7062577.stm.
41. Daniel Howden. "Take over our rainforest." *The Independent*, November 24, 2007. [cited April 21, 2008]. environment.independent.co.uk/climate_change/article3191500.ece.
42. Juliet Eilperin. "New focus on global warming." *Washington Post*, April 6, 2008. [cited April 21, 2008]. washingtonpost.com/wp-dyn/content/article/2008/04/05/AR2008040501136.html.
43. UNEP. "Drying up of Lake Chad." Our Changing Climate poster series, 2005. [cited May 6, 2008]. unep.org/Themes/climatechange/multimedia/posters.asp?poster=Chad.
44. This means that our most urgent scientific priorities are to discover the tipping points at which major ecosystems collapse and runaway climate change occurs. However, the need for these studies cannot be an excuse for not taking immediate action to prevent further ecological destruction.
45. UNEP. *Meltdown in the mountains.* Press release, March 16, 2008. [cited April 21, 2008]. unep.org/Documents.Multilingual/Default.asp?DocumentID=530&ArticleID=5760&l=e.
46. Elisabeth Rosenthal and James Kanter. "Alarming UN report on climate change too rosy, many say." *International Herald Tribune*, November 18, 2007. [cited April 21, 2008]. iht.com/articles/2007/11/18/europe/climate.php.

47. James Hansen et al. *Target Atmospheric CO2: Where Should Humanity Aim?* April 7, 2008. [cited April, 2008]. columbia.edu/~jeh1/2008/TargetCO2_20080407.pdf.
48. Paul Rogers. "A century on the edge: 1945-2045." *openDemocracy*, December 29, 2007. [cited April 21, 2008]. opendemocracy.net/article/conflicts/global_security/century_change.
49. The painting is in the Royal Albert Museum in the UK.
50. David Spratt and Philip Sutton. *Climate code red: the case for a sustainability emergency.* 2008. [cited April 21, 2008]. climatecodered.net.
51. Greg Cran. "How It all Ends." 2007. [cited May 22, 2008]. http://www.youtube.com/watch?v=mF_anaVcCXg&feature=related.
52. Institute for Intercultural Studies. *Frequently Asked Questions About Mead/Bateson.* [cited April 21, 2008]. interculturalstudies.org/faq.html.

Chapter 11

1. "Evo Morales." *BetterWorldHeroes.com.* [cited April 22, 2008]. betterworld.net/heroes/pages-m/morales-quotes.htm.
2. Riane Eisler. *The Real Wealth of Nations: Creating a Caring Economics.* Berrett-Koehler, 2007.
3. Quoted in *Deep Democracy Explained.* Deep Democracy Institute, June 2, 2007. [cited May 7, 2008]. deepdemocracyinstitute.org/blog/section/05%20what%20is%20ddi/.
4. Duncan Taylor and Graeme Taylor. "The requirements of a sustainable planetary civilization." *Social Alternatives*, Vol. 26#3 (Third Quarter, 2007), pp. 10-16.
5. FutureGenerations. *SEED-SCALE Process of Change Overview.* [cited April 22, 2008]. future.org/pages/04_process_of_change/01_processofchange.html.
6. Jeff Vail. "The Design Imperative." *Rhizome*, April 8, 2007. [cited April 22, 2008]. jeffvail.net/2007/04/design-imperative.html.
7. Jeff Vail. "Envisioning a Hamlet Economy: Topology of Sustainability and Fulfilled Ontogeny." *Rhizome*, April 12, 2006. [cited April 22, 2008]. jeffvail.net/2006/04/envisioning-hamlet-economy-topology-of.html.
8. For example, see Michael Levenston. "How Far Can Urban Agriculture Go? Bogota, Columbia." *City Farmer News*, April 10, 2008. [cited May 7, 2008]. cityfarmer.info/how-far-can-urban-agriculture-go%C2%A0-bogota-columbia/.
9. Bill Mollison. *Permaculture: A Designer's Manual.* Tagari, 1988.
10. "Vernacular Architecture." *greenhomebuilding.com.* [cited April 22, 2008]. greenhomebuilding.com/vernacular.htm.
11. Slow Food International website: slowfood.com.
12. "Great Victory! The Sustainable Communities Bill Becomes Law. We've Won!" *Local Works*, July 31, 2007. [cited April 22, 2008]. localworks.org.
13. Cited in "Interesting Quotes: On topics from: Peace to Religion". *Religious Tolerance.* [cited May 29, 2008]. religioustolerance.org/quotes5.htm.
14. Jack Santa Barbara. "Non-Violence." *Resurgence Magazine* #245 (November-December 2007). [cited May 22, 2008]. truthforce.info/?q=node/view/2419.

15. The Commission on Global Governance. *Our Global Neighborhood*. Oxford, 1995. A summary by Henry Lamb is available online at sovereignty.net/p/gov/gganalysis.htm.
16. Mikhail Gorbachev. "A New World Political Architecture." *Kosmos Journal*, Vol. VI #2 (Spring/Summer 2007). [cited April 22, 2008]. kosmosjournal.org/kjo/backissue/s2007/newworldpol.shtml.
17. Ivan Krastev and Mark Leonard. "The world's choice: super, soft or herbivorous power?" *openDemocracy*, October 26, 2007. [cited April 22, 2008]. opendemocracy.net/article/globalisation/visions_reflections/global_poll_IBSA.
18. George Schöpflin. "The European Union's troubled birthday." *openDemocracy*, March 23, 2007. [cited April 22, 2008]. opendemocracy.net/democracy-europe_constitution/EU_Birthday_4463.jsp.
19. Ibid.
20. John Galtung. "Violence, Peace and Peace Research." *Journal of Peace Research*, Vol. 6#3 (1969), pp. 167-191.
21. Karl Polanyi. *The Great Transformation*. Beacon, 1957.
22. Erik Erikson. *Identity: Youth and Crisis*. Norton, 1968.
23. Martin Luther King, Jr. *Remaining Awake Through a Great Revolution*. March 31, 1968. The Martin Luther King, Jr., Research and Education Institute [cited May 7, 2008]. stanford.edu/group/King/publications/sermons/680331.000_Remaining_Awake.html.
24. Bruce Alexander. *The Roots of Addiction in Free Market Society*. Canadian Centre for Policy Alternatives, 2001, p. 1.
25. Vamik Volkan et al. *The Third Reich in the Unconscious*. Routledge, 2002.
26. Peter Levine. *Waking the Tiger: Healing Trauma*. North Atlantic, 1997, p. 225.
27. Robert Putnam. *Bowling Alone: The Collapse and Revival of American Community*. Simon and Schuster, 2000.
28. Morris Berman. *Dark Ages America: the Final Phase of Empire*. Norton, 2006, pp. 45-46.
29. See Chapter 6, Figures 3 and 6.

Chapter 12

1. The Dalai Lama. "Love, Compassion, and Tolerance" in Richard Carlson and Benjamin Shield, eds. *Handbook for the Spirit*. New World Library, 2008, p. 3.
2. Pope Benedict. "Pope leads eco-friendly festival." *BBC News*, September 2, 2007. [cited April 28, 2008]. news.bbc.co.uk/2/hi/europe/6974475.stm.
3. Like other spiritual leaders, Pope Benedict connects environmental destruction with greed, selfishness and "unbridled capitalism": "The emergencies of famine and the environment demonstrate with growing clarity that the logic of profit, if predominant, increases the disproportion between the rich and the poor and leads to a ruinous exploitation of the planet.": "Climate change a moral obligation: Benedict." *Catholic News*, September 24, 2007. [cited April 28, 2008]. cathnews.com/news/709/126.php.
4. Richard Black. "Church to step up climate fight." *BBC News*, March 10, 2008. [cited April 28, 2008]. news.bbc.co.uk/2/hi/science/nature/7287484.stm.
5. Diagram by Neil Davidson and Graeme Taylor 2007.

6. This is a very general description of a complex and dynamic process. Individuals and groups can simultaneously support some innovative ideas while opposing others.
7. Useful ideas for designing organizations to encourage constructive engagement are found in Cathy Costantino and Christina Merchant. *Designing Conflict Management Systems : A Guide to Creating Productive and Healthy Organizations.* Jossey-Bass, 1995.
8. Donella H. Meadows. "Places to Intervene in a System: Strategic Levers for Managing Change in Human Systems." *Whole Earth Review*, Winter 1997. [cited April 28, 2008]. wholeearth.com/issue/91/article/27/places.to.intervene.in.a.system.
9. The Earth Charter Initiative: www.earthcharter.org.
10. Mirian Vilela. "The Earth Charter endeavour: building more just and sustainable societies through a new level of consciousness." *Social Alternatives*, Vol. 26 #3 (2007), p. 35.
11. Ann Pettifor. "Green movement forgets its politics." *BBC News*, May 6, 2008. [cited May 8, 2008]. news.bbc.co.uk/nolpda/ukfs_news/hi/newsid_7385000/7385615.stm.
12. "One of the critically important implications to the triple bottom line decision-making process where society wants to pursue a range of goals simultaneously is that it must be based on the principle of 'no-major-trade-offs'. Logically, if society is committed to sustaining something, it cannot trade-off the continued existence of that thing or attribute in order to meet other goals.": Karlson Hargroves and Michael Smith, eds. *The Natural Advantage of Nations: Business Opportunities, Innovation and Governance in the 21st Century.* Earthscan, 2005, Chapter 3 (Part 3). [cited April 28, 2008]. naturaledgeproject.net/NAON1Chapter3.3.aspx.
13. Morton Deutsch. *The Resolution of Conflict.* Vail-Ballou, 1973. The basic principles of interest-based negotiating are described in Roger Fisher and William Ury. *Getting to Yes: Negotiating Agreement Without Giving In.* Penguin, 1991.
14. For an excellent introduction to conflict resolution see Kenneth Cloke. *Mediating Dangerously: The Frontiers of Conflict Resolution.* Jossey-Bass, 2001.
15. "The Choice Between Love and Fear". *A Passion for Peace.com.* [cited May 29, 2008]. apassionforpeace.com/2005/11/the_choice_betw.html.
16. Peter Levine. *Waking the Tiger: Healing Trauma.* North Atlantic, 1997, pp. 193-194.
17. Marshall Rosenberg. *Nonviolent Communication: A Language of Life.* PuddleDancer, 1999.
18. "Quoted in *L.A. Times*, 30 July 1989. According to the Gandhi Institute for Nonviolence, this has not been traced in Gandhi's writings but 'the Gandhi family states that M. K. Gandhi was known to say this verse many times in his lifetime and believes it to be original with him.'" Fred R. Shapiro, ed. *Yale Book of Quotations.* Yale, 2006, p. 299.
19. While the effectiveness of violence in solving problems is usually overestimated, the effectiveness of non-violent approaches to change is usually underestimated. Nevertheless, non-violence does not always work — for example, non-violence is often ineffectual when dealing with brutal dictators like Hitler. An excellent analysis of the issue is provided by Kurt Schock in *Unarmed Insurrections: People Power Movements in Nondemocracies.* University of Minnesota Press, 2005.

20. Susan Scott. *Fierce Conversations: Achieving Success at Work & in Life, One Conversation at a Time.* Berkley, 2002, p. 7.
21. Diagram by Neil Davidson and Graeme Taylor 2008.
22. "Executive Summary." *OECD Environmental Outlook to 2030.* OECD, 2008, p. 3. [cited April 28, 2008]. oecd.org/document/20/0,3343,en_2649_37465_39676628_1_1_1_37465,00.html.
23. David Spratt and Philip Sutton. *Climate code red.*, p. 70. [cited April 29, 2008]. climatecodered.net.
24 David Herszenhorn. "Estimates of Iraq War Cost Were Not Close to Ballpark." *New York Times,* March 19, 2008. [cited April 29, 2008]. nytimes.com/2008/03/19/washington/19cost.html?_r=1&sq=estimates%20of%20iraq%20war%20cost%20were%20not%20close%20to%20ballpark&st=nyt&adxnnl=1&oref=slogin&scp=1&adxnnlx=1209381543-465jNEo1yEzVz0htWCOPtw.
25. Illustration courtesy of US National Archives. "Rosie the Riveter: Women's Roles in World War II." ARC ID 535413, Title: We Can Do It!, ca. 1942 - ca. 1943. [cited May 8, 2008]. archives.gov/research/arc/education/national-history-day-2007.html.

Index

Page numbers in **bold** indicate figures.

A
Abbott, Chris, 194
abuse, child, 74
accounting, 123, **123, 124,** 124–125. *see also* economy
acid rain, 205
action *versus* inaction, 210–211
addiction and fear, 66–69
addictions, 221–222
Adeel, Zafar, 51
adjusting existing system scenario, 204–205
advertising/marketing: and artificial scarcity, 69–70; consumption promotion, 150; expenditures, **60**, 124; happiness, 68; and lifestyle/wealth, 67; media, 74–75, 77; and selfishness, 74–75
affordable energy, 31–33
Afghanistan, 81, 82, 110, 193, 202
Africa: child mortality, 64; consumption, 24; desertification, 49; economic challenges, 198; food production, 51; global crises, 193; life expectancy, 72–73; Mozambique, 18–19; Nigeria, 49; resources, 185; society collapse, 110; soil conditions, 51; Somalia, 202
Africa Rice Center, 56
Age of Reason, 96–97
Agrarian Age, 130, 155–156, **164**. *see also* agrarian civilization; agrarian societies
agrarian civilization. *see also* other civlizations: and cultural destruction, 83; ecological footprint of, **89**; energy, renewable, 97–98; evolution, **6**; expansion limits, 105–106; global footprint expansion, **88**; population growth, 15–17, **16**
agrarian societies. *see also* other societies: Asia, 94–95; cultural destruction, 83; preservation, 159; social structure, 65; social values, 162–163; societal growth, 99
agreement, global, 229–232
Agricultural Age, 94–95
agriculture. *see also* crops; farming; fertilizers; grain; irrigation; soil: food production and, 46; food shortages and, 44–45, 180; high yield crops, 46; land, 20–21, 105; overharvesting, 52; social system and, 157; species extinction and, 52; a systems approach, 143; water supplies and, 42; wheat harvest, 48; wheat imports, 180
air pollution, 49, **54**. *see also* carbon dioxide (CO_2); emissions; greenhouse gases; pollution
al-Baghdadi, Omar, 81
alcohol expenditures, **60**
Alexander, Bruce, 221–222
algae as a bio-fuel, 136
'Alice in Wonderland' approach, 125
Alliance for Zero Extinction, 55
allies and opponents, **238**
al-Qaeda, 81
ambulance story, 29–30
American Dream, 165–166
American Revolution, 97, 159
American Transcendentalists, 171
Amnesty International, 169
Amritanandamayi, Mata, 226

Amu Darya, 42
ancient civilizations, 103–114. *see also* other civilizations (ie. industrial civilization)
Angkor, 104
Anielski, Mark, 125
animals, **45**, 45–46, 129
annual global natural capital, **92**
annual global natural income, **88, 89, 91**
anti-corruption, 169
Anti-Slavery International, 211
anti-slavery movement, 211
Anti-Slavery Society, 211
apathy, global crises, 112–113
aquaculture, 50. *see also* fishing
aquifers. *see* water
arable land, loss of, 49
Aral Sea, 42
architecture, 143
Arctic ice caps, **35**, 209
Arctic Sea, 188
Argentina products, 141
Aristotle, 172
armed conflict, 81
arms race, 79, 186
art, 170
artificial organisms, 139
Asia, desertification of, 49
atomic phenomena, 131
Australia, 69, 141, 145
Australian Climate Institute, 199
automobile industry, 141, 147, 204
awareness, 1, 161–162, 203, 214–215, 229–230. *see also* behavior

B

bacteria example, 113–114
Bakan, Joel, 100
Bali Action Plan, 40–41
ballistic missiles, 79
Bangladesh, 18
bankruptcy, ecological, 64, 148–149
Barber, Benjamin, 83
Barlow, John Perry, 142
Bartlett, Albert, 113–114
basic human needs, **60**
Beddington, John, 46
bees and flowering plants, 128
behavior. *see also* awareness: allies and opponents, **238**; apathy, global crises, 112–113; commitment and courage, 237–238; constructive values and, 236; courage and commitment, 237–238; fear, 66–69, 79–80, 128; greed, 64–65; individual and societal, **220**; love, 225–226; love and faith, 241–242; moral imperative, 226–227; personal interests, 237–238; psychosocial integration, 221; self-awareness, 214–215; socially structured, 65–66; trust, 235–237; violent, 77
Benedict XVI, Pope, 226
Benyus, Janine, 138
Berman, Morris, 223
biodiversity. *see* biological systems
bio-energy, 134
biofuels, 33, 48, 135–136, 202. *see also* fuel
biologically productive land and sea, 23
biological systems. *see also* ecosystems; physical systems; systems: biodiversity, 55, 56, 186–188; coevolution, 2, 2–3; essential needs, 118; evolution, **6**; global economy, 55; parts of, **126**
biomimicry, 137–141
biophysical limits. *see also* biophysical systems: ecological footprints and, 23–27, **25, 26**; economic activity within, 149; food and water shortages, 182–183; living within, 85; social system collapse and, 106; systems, 13–15; of water supply, 42–45
biophysical systems. *see also* biological systems; biophysical limits; ecosystems; physical systems; systems: dependencies, **118**; equilibrium and species extinction, 55; interrelated with other systems, 217; needs, 120–121, 127; viability of, 117–118
biotechnology, 137–141
Blessed Unrest (Hawken), 115, 168
Boulding, Kenneth, 14
Bowling Alone study (Putman), 222
Bradford University, 179
Braungart, Michael, 146
Brazil, 38, 198
British International Institute for Strategic Studies (ISS), 187
Brown, Lester, 47
Brundtland Report, 118
Brzezinski, Zbigniew, 186
Buddhism, 158. *see also* religion
Buffet, Warren, 65
bulbs, compact fluorescent, 137
bulbs, incandescent, 136–137
Bulletin of the Atomic Scientist (BAS), 112
Bush, George W., 169, 195–196
Bush Administration. *see* United States
businesses/corporations. *see also* institutions: conserver economy and, 151–152; distribution of power, 215; global responsibility, 100; neglect of communities, 124; responsibilities, 115; as usual scenario, 203–204

C

Cambodia society collapse, 110
Campaign for Nuclear Disarmament, 168
Campbell, Colin, 30, 33
Canada, 24, 39, **57**, 57–58, 68
capital, natural global, 21–23, **22**
capital expansion, 99
capitalism: industrial society, 63; inequality and, 70; market forces, 147; rise of, 96; standards of living, 63; success of, 91–92; sustainable economy and, 151; wealth distribution and, 66
Capitalism as if the World Matters (Porritt), 152
Capra, Fritjof, 131, 145
carbon capture, **140**, 140–141
carbon dioxide (CO_2). *see also* air pollution; emissions; global warming; greenhouse gases: and biofuel production, 202; climate change, 35–36; emissions reduction, 38, 135, 140, 210; levels, 53; species extinction, 53
carbon sequestration, industrial, **140**
carbon tax, 40
Cardiff Business School, 147
carrying capacity, 21, 125, 128
cars, hydrogen-electric, 141
Carson, Rachel, 14
Carter, John Cain, 147
cascading crises, 206–207
Catholicism, 96
censorship, information, 197
Center for Nonviolent Communication, 236
centralized/decentralized societal system, 215
centralized production, 138
CERES Principles, 149
change. *see also* paradigms; tipping points: allies and opponents, **238**; constructive, 211, 229–230, 232–233, 237–238, **240**, 240–241; incremental, 175–176; non-linear, 175–176; synergistic, 7, 203
Charles, Prince of Wales, 112
chemicals, toxic, 53, 112
Chevron, 32–33
child(ren): childhood, 221–222; death rates, 73–74; immunization, **70**; labor, 112, 167; mortality reduction of African, 64; rearing, 164; as slaves and workers, 74
China: African resources, 185; air pollution, **54**; ancient civilizations of, 104; automobiles, 204; carbon dioxide emission reduction, 38; consumption, 24; corn imports, 180; crop yields and climate change, 49–50; economic challenges, 198; economy, 185; energy, solar/wind, 134; gender discrimination, 76; global warming and, 40; greenhouse gases, 40, 199; inequality, wealth, 72; leadership, 196–197; social harmony, 196; soil erosion, 49; water supplies, 43; wealth inequality, 72; weapon development, 79; wheat harvest, 47–48
China Daily newspaper, 185
Chinese Empire, decline of, 103
chlorofluorocarbons, 205
Christensen, Diana, 137
Christianity, 96, 99, 158. *see also* religion
city design, 143
civic engagement, 222
civil conflict and poverty, 82
civilizations, failure/collapse of, 103–104. *see also* other civilizations (ie. industrial civilizations)
civil movements, 232
Clean Air Act, 205
climate change. *see also* global warming; temperature increases: biggest threat, 209–210; deforestation and accelerating, 53; desertification, 51; Doomsday Clock, 112, 114; environmental movement and, 169–170; food production and, 49; food security, 49–50; increasing, 35–41; projected impacts, 36; responses to, 186–188; social system collapse and, 105; species extinction and, 52; water supply and, 42; weather events and, **36**
Climate Code Red: The Case for Emergency Action (Spratt, Sutton), 37, 210–211
Climatic Research Unit, 38
closed-loop manufacturing, **140**
coal, 32–33, 33
coalitions, 232–233, **233**
cod fishery, Grand Banks, **57**, 57–58
coevolution, **2**, 2–3
collapse: causes of, 104–106; delayed, 191; earlier civilizations, 103–104; as a main trend, 189–191; possible scenarios, 191–192; 'seeing' the danger, 112–114; trend towards, 4
Collapse (Diamond), 104, 105
The Collapse of Complex Societies (Tainter), 105–106
Columbia, 81, 82
combined heat and power generation (CHP), 137
Commercial Revolution, 95–97
commitment and courage, 237–238
communication, international, 171
communications technology, 132–133
communism, 80
communities, 84, 124, 215, 221–222. *see also* relationships
compact fluorescent bulbs, 137
competition, 95–101, 127–128, 151, 186
compound interest, 113–114
computers, 142, 161, **161**. *see also* Internet; technologies

computer scientists, 131
Condorcet, 97
conflict, violence and, 77–78
conflict resolution, 234–235
conservation, 136–137, 141, 148
conserver economy, 151–152
conserver lifestyle, 136
constructive: change, 211, 229–230, 232–233, 237–238, **240**, 240–241; interventions, 200–203; trends, 191; values, 236
consumer society. *see also* consumption: concept of, 85; ecological footprint, **92**; focuses of, 69; global support, 202; global system sustainability and, 64–65; industrial worldviews and, 122–123; selfishness, 74–75; values and needs, 127–128
consumption. *see also* consumer society: agrarian and consumer economies, 99; as an addiction, 68; ecological footprints, 23–27; expansion of, 204; expenditures, **60**; food and food production, 48; goods and services, **17**; grain and meat, **45**; growth of, 17–20, **19**; The Industrial Revolution and, 88–89; lifestyle/wealth and, 67; natural gas, **31**; non-renewable resources, 149; oil discovery and, **32**; passive, 122; promotion and marketing, 150; reducing meat, 51; reduction for conservation, 136–137; renewable resources, 149; sustainability of, **19**, 94; tobacco, **60**; water, 42; wealth distribution and, 66
contraction and convergence (C&C) principle, 199
Convention on International Trade in Endangered Species (CITES), 56
cooking fuels, 51
cooperative economics, 150–151
coral reefs, species extinction and, 53
corn imports, 180
corn production, 48
corporations/businesses, 100, 115, 124, 151–152, 203–204
costs, 59–60, 124, **124**, 152, 210–211. *see also* expenditures
courage and commitment, 237–238
Cradle to Cradle: Remaking the Way We Make Things (McDonough, Braungart), 146
crime expenditures, 124
crises: cascading, 177, 191, 206; growing, 56–59; land degradation, 105; resource, 105
crops. *see also* agriculture; grain: climate change, 49–50; creation of new, 139; genetically modified (GM) soya and cotton yields, 47; high yield, 46; Norwegian rare, 56; overharvesting, 52; wheat harvest, 48; wheat imports, 180; yields and rising temperatures, 49–50
Cubism, 170
cultural. *see also* culture: change, process of, **227**; coevolution, **2**, 2–3; destruction, **81**, **83**, 83–84; diversity, 81; relevance, 215; traditions, 221–222; transformation, 174
culture. *see also* cultural: individual and societal behaviors, **220**; industrial social development of, 95–97; pattern, 223–224, **224**; 'rurban' communities, 216; of violence, 77–78
currency, economic crises and, 184–185
cyberspace. *see* Internet
Czisch, Gregor, 135

D
Dafur conflict, 105
Dalai Lama, 226
Dallimore, Audrey, 38
Daly, Herman E., 13, 149
Darwin, Charles, 131
Datong, Li, 196–197
death rates, child, 59
decentralized distributed production, 141–142
deep democracy, 173–174
deficit, United States, 183
deforestation, 52–53, 55–56, 112
degradation, ecosystem, 55
de Jong, Carmen, 43
delayed collapse, 191
democracy, 173–174, 195, 215. *see also* governance; politics
depression, global, 179, 183
desalinating, 44
desertification, 49, 51
design, whole systems, **144**
destructive: activities, costs of, **124**, 124–126; interventions, 200–203; trends, 191; values, 192
Deutsche Bank, 180
developed countries: access to electricity, 30–31; consumer society of, 159; culture of, 69; ecological footprints, personal, 24–25; education and jobs, 75; emotional scarcity, 69; environmental problems, 198–199; happiness and health of, 69; population growth in, 16–17; relationships/health in, 70
developing countries: access to electricity, 30–31; basic needs, 70; child abuse, 74; ecological footprints, personal, 24–25; education and jobs, 75; environmental problems, 198–199; food shortages in, 45; globalization, 74; global warming and, 40; industrialization benefits for, 159; inequality, wealth, 74; material scarcity, 69; physical violence, 74–75; population growth in,

16–17; resource shortages, 198; water supply and, 41–42
development, sustainable, 117–120
developmental needs, human, 122
Diamond, Jared, 104–105
diminishing returns, law of, 106
discrimination and hierarchy, 75–76
disruptive innovations, 130–134, 132–133
distributed production, 138, 141–142
Dobson, Andrew, 201
domestication, plant and animal, 129
dominant force, industrial system and, 159
Doomsday Clock, 112, 114
dot com boom/bubble, 133, 184
downcycling, 146
drip irrigation, 51
droughts, 42, 49, 180
drug addiction, 82
drug development expenditures, 124
drug trade, 81–82
Duke Energy, 206
Dyer, Gwynne, 50
dynamics, non-linear, 175–176

E

Earth Charter, 149
Earth Policy Institute, 47, 55, 59
Earthshare. *see also* ecological footprint: assignment, 199–200; average per person, 24; per capita, **25**; sustainable, **27**; United States and fair, **26**
The Earth Charter, 7, 228–229
East Anglia, University of, 38
ecodesign, 145–146
ecological. *see also* ecological footprint: bankruptcy, 64, 148–149; knowledge, 198; worldview, 215
ecological footprint. *see also* Earthshare; ecological: 1650 C.E., **88**; 1750 C.E., **89**; 1850 C.E., **90**; 1950 C.E., **91**; 2000 C.E., **92**; 2050 C.E., **93**; agrarian civilization, **89**; assignment, 199–200; consumer culture, **92**; expansion of, **88**, **89**, **90**, **91**, **92**; industrial civilizations, **91**; per capita, **25**; sustainable, 24, **27**
ecologists, 131
'ecology of organizations,' 145
economic. *see also* economic systems; economy; global economy: activity, 128, 149, 152; crises, 183–185, 184–185; efficiency, 148; expansion and sustainability, 64–65; gap between rich and poor, 70–71; growth, 13–15, **17**, 140; inequality, 194; justice, 199; law of diminishing returns, 106; power, 76; priorities, 84; renewable energy, 140; 'rurban' communities, 216
The Economics of Happiness (Anielski), 125
economic systems. *see also* accounting; economic; economy; global economy: agrarian and consumer, 99; cooperative, 150–151; distortions, 124–125; The Industrial Revolution, 94–95; market, 183–184; survival and, **123**; technology innovations and, 130
economy. *see also* economic; economic systems; global economy: biophysical equilibrium, 55; capital expansion, 99; conserver, 151–152; consumer to military, 239; European, 95; growth of, 18; industrial social cultures, 95; industrial system worldview, 99–100; redesign costs, 152; species extinction, 55; worldview, 206
Economy, Elizabeth, 197
economy, global. *see* global economy
ecosystems. *see also* biological systems; biophysical systems; physical systems; systems: climate change projected impacts, **36**; collapse of, 110; degradation, 55; failure of, 23, 186–188; functional integrity, **120**; global warming, 36–37, 52; health, 214; natural capital, 23; needs of the planet, 127; protection, 55; species extinction, 52; sustainability of, 117–118, 119
education, 74–75, 75–76, 167. *see also* information; knowledge
Egypt, 104, 202
Ehrlich, Paul, 13
Einstein, Albert, 1, 16, 18, 131, 172
electricity, access to, 30–31. *see also* energy
electromagnetic force fields, 131
electronic devices, 148
Emerson, Ralph Waldo, 171
Emirate of Abu Dhabi's Masdar Initiative, 133
emissions, 39, 59, 149, 199. *see also* air pollution; carbon dioxide (CO_2); greenhouse gases
emotional scarcity, 69
emotional/spiritual needs, 84
empire system evolution, **6**
employment, access to, 75–76
energy. *see also* fossil fuels; natural gas; oil; renewable energy: affordable, 31–33, **178**; coal, 32–33, 33; conservation opposition, 39; electricity, access to, 30–31; forecasts, 177; fossil fuels, 32–33; home, 141; hydro power, 134; incandescent bulbs, 136–137; increasing renewable, 133; industrial technologies, 97–98; needs prediction, 176–177; non-renewable, 97–98; oil, 32–33; oil production, 109; requirements, **139**; shortages, 30–34, 178–179; slaves, **30**; societal systems and, **139**; solar, 134, 135, 138–139, 141, **190**; sources, 32–33; war and, 179

Engels, Friedreich, 172
engine, internal-combustion, 130
England industrial social cultures, 96
English language, 162
Enlightenment, 98
Enlightenment thinkers, 97
Envirofit International, 137
environment. *see also* environmental: acid rain, 205; avoidance of issues, 197; constructive-destructive interventions, 200–203; governments/business, 124; human survival and, 117–118; protection, 59; social neglect of, 124
environmental. *see also* environment: collapse and technological advances, **85**; conservation, 60–61; degradation, 23; destruction, 229–230; extra spending, **60**; food production, 181; global agreement, 229–230; The Industrial Revolution impact, 87–88; lightening the load, 133; moral imperative, 226–227; movement, 169–170, 171; needs, **60**; preservation, 226–227; problems and system collapse, 104–105; reforms, blocked, 112–113; relevance, 215; sustainability, **17**, **190**
environmentalism, 147
equality, women's, 59, 76, 168
Equiano, Olaudah, **210**
Erhlich, Paul, 16, 46
Erikson, Erik, 221
Esty, Daniel, 137
ethanol, 48, 135
ethics, 64–65, 119, 139, 237–238, **238**
Europe. *see also* European Union (EU): changing values, 165; consumption, 24; economy, 95; industrial social cultures, 95; policy differences, **166**; religious persecutions, 96
European Commission, 135
European Dream, 165–166
European Union (EU), 166, 205, 218–219, 232. *see also* Europe
evolution: biological and social systems, 233–234; discovery of, 131; industrial system, 94–95; life on Earth, 2, 2–3; societal, 1–3, 157–159, 241; societal systems, 6; species extinction and, 52; supporting, 238–241; transformation of society, 152–153
expectations, rising, 161–162
expenditures. *see also* costs: advertising/marketing, 124; alcohol, **60**; crime, 124; discretionary and essential, **60**; drug development, 124; luxury, **60**, **70**; pesticides, 124; on war, 78
exploitation and selfishness, 74–75
export, grains, 48
ExternE, 135
extinction, 38, 51–56, 52–53, 83–84. *see also* species extinction
ExxonMobil, 32, 39
Exxon Valdez oil spill, 125

F

Facebook, 173
faith and love, 241–242
family, 124, 151
Faraday, Michael, 131
farming. *see also* agriculture; crops; fertilizers; grain; irrigation; soil: fish production, 50; food shortages and, 180; land, 20–21, 105; loss of land, 49; organic, 124, 137, 182; overharvesting, 52; rising costs of, 49; subsidies, 182; sustainable, 146; wheat harvest, 48; wheat imports, 180
fear, 66–69, 79–80, 128
feedback, **220**
Feeding the World (Smil), 50
Ferrini, Paul, 235
fertilizers, 47, 49, 124, 182. *see also* agriculture; farming
First World War. *see* World War I
fishing: destructive and species extinction, 53; farmed, 50; Grand Banks cod fishery, **57**, 57–58; industry catch reduction, 58; marine reserves, 55; polyculture, 50; scientists warnings, 58; species decline, **49**, 49; subsidies, 58–59, 182
fixed reality, 131
Flasbarth, Jochen, 56
floods, 49, 180
flowering plants and bees, 128
food. *see also* food production; food shortages: junk, **60**; prices, 46, 209; processing, 51; riots, 202; security and climate change, 49–50; security and rice production, 48; sources, **36**; storage, 51
food production. *see also* food; food shortages: aquaculture, 50; consumption and, 48; environmental degradation, 181; floods, 49; pesticides and, 47; plant photosynthesis, 47; pollution, 49; prices, 209; waste elimination, 51
food shortages. *see also* food; food production: floods, **180**; growing, 44–51; hunger, 180; responses to, 179–183; time frame, 209
footprint, ecological. *see* ecological footprint
forecast, future, 175–177
fossil fuels, 32–33, 59, **60**, 135. *see also* energy; fuel; oil
France, 97, 205
Franklin, Ursula, 217
free markets, 221
French Revolution, 97

fresh water, 43
freshwater species decline, 52
fuel, 48, 135, 136. *see also* biofuels; fossil fuels
Fuller, Buckminster, 5
future forecast, 175–177

G

Gabriel, Sigmar, 51, 55
Galtung, John, 221
Gandhi, Mahatma, 72, 236
Ganges River, 42
Garten Jeffrey, 184
gas, natural. *see* natural gas
Gates, Bill, 65, 147
GDI accounting, **124**
GDP accounting, **124**
General Electric, 100
General Motors, 39
genetically modified (GM) soya and cotton yields, 47
genetic diversity, 83
genocide, Rwanda, **78**
Genuine Progress Indicators (GDI), 125
George Mason University, 108
geo-thermal energy, 134
German Environment Ministry, 205
Germany, 69, 80, 104, 205
Ghazi, Abdul Rashid, 82
Gillet, Stephan, 138
glaciers, 209
global. *see also* globalization: agreement, 229–232; natural capital, **88**, **89**, **91**, **92**, **93**; personal wealth distribution, **73**; 'race to the bottom,' 115; resources availability, 20–23, **21**; system strengths/weaknesses, 192–200; system structure, 65–66; values, 162–167, **164**; village, 162; wealth and poverty, 84
global economy. *see also* economy: biological diversity, 55; capital expansion and, 99; climate change projected impacts, **36**; collapse of, 110; crises, 107; development of, 161; ecological footprint, **92**; expansion limits, 64; expansion of, 133; growth of, 18; holistic paradigm, 213–214; Industrial Age and the, 160; inequality and violent culture reduction, 82; resources, 185; social system resilience, 106–107; sustainable, 213–214; US Gross Domestic Product (GDP), 125; vitals signs, 29–30
globalization, 3–4, 74, 161, 177. *see also* global
global warming. *see also* carbon dioxide (CO_2); climate change; temperature increases: Arctic ice caps, **35**; carbon sequestration, industrial, **140**; catastrophic, 39; coal consumption and, 33; food shortages, 180; industrial carbon sequestration, **140**; species extinction/ ecosystem disruption, 52; time to turn around, 112; tipping points, **140**, 187, **187**, 187–188; tragedy of the commons and, 100
Glorious Revolution of 1688, 96
Gobi Desert, 49
God and worldview, 131
Goddard Institute for Space Studies, 39
Goldstone, Jack, 108
goods, supply of, 185
goods and services, 18–19
Google, 135
Gore, Al, 152
governance, 171, 173–174, 216–219, 218–219. *see also* democracy; government(s); politics
government(s). *see also* governance; institutions; politics: collaboration, 201; developed countries, 198–199; developing countries, 198–199; distribution of power, 215; economic control, 152; environmental issues, 60–61, 124, 198–199; fishing industry, 57–58, 58–59; food and water shortages, 182; future scenarios, 193; global warming and, 39; new technologies and, 133; social issues, 124; during World War II, 152
grain, 45, **45**, 47–48, 112, 180. *see also* agriculture; crops; farming
Grand Banks cod fishery, **57**, 57–58
The Grand Chessboard (Brzezinski), 186
Great Britain, 66, 68, 77, 94–95, 103, 141, 147. *see also* United Kingdom
Great Chain of Being (Aristotle), 172
The Great Depression (1930s), 110
Greece worldview, 95
greed, 64–65
green building, 137, 143, 216. *see also* sustainability; sustainable
Green Gross Domestic Product (GDP), 196
greenhouse gases, 35–36. *see also* air pollution; carbon dioxide (CO_2); emissions; carbon tax on, 59; China/United States, 40, 199; deforestation and, 53; emissions, 199, 205; European Union (EU), 205; Kyoto Protocol, 40; methane, 39; Organization for Economic Co-operation and Development (OECD), 239; reduction, 137, 140
Greenpeace, 168
Green Revolution, 46
Green to Gold (Esty, Winston), 137
Gross Domestic Product (GDP), 18, 60, 125, 196
Gross National Product (GNP), 125
groundwater supplies, 43

growth, 13–15, 150
Guyana rainforest preservation, 205

H
habitats and species extinction, 52
handcrafts, 95
Hanover Principles, 149
Hansen, James, 39, 209–210
happiness, 66–67, **67**, 122
Hawken, Paul, 115, 168
health. *see also* health, human: consumption of goods and, 68; education and, 59; income levels and, 66–67; individual and societal, 219–223; lifestyles, 124; natural, 214; species extinction, 53; of systems, 119; toxic chemicals impacts, 53; wealth, 66
health, human. *see also* health: healthcare, 70, 92, 167; life expectancy, 73; longevity, 66; mental, 221; natural, 214; obesity, 145; species extinction and, 53; whole-systems design and, 145
herder-cultivator societies, **6**, 15–17, **16**
hierarchy, human needs, **121**
hierarchy and discrimination, 75–76
Hinduism, 158. *see also* religion
historical ages: Agrarian Age, 130, 155–156, **164**; Agricultural Age, 94–95; Industrial Age, **89**, 94–95, 109, 130, **139**, 160, 162, **164**, 213–214, **224**; Information Age, **6**, **139**, 173, 214–215, **224**
history of change, 155–156
history of social movements, 167–169
Hitler, Adolph, 80
holarchical communities, **6**
holarchical structure, 216
holistic systems. *see also* systems: culture, 170, 203; culture pattern, 223–224, **224**; ecologists, 14; global economy, 213–214; human needs, 127; shift to, 5; worldview, 171–172
home energy, 141
Homer-Dixon, Thomas, 113
Hong Kong, **178**
housework, 124–125
housing design, 143
human. *see also* health, human: brain evolution, 121–122; interrelated systems, 217; needs, 121–122, 122, 125–126, **126**; rights and social justice movements, 169; societal systems dependencies, **118**; survival and the environment, 117–118
Human Development Index (HDI), 82
Human Development Report (United Nations), 39, 59, 73, 82
hunger, 59, 64, **70**, 143, 179–183. *see also* poverty
hunter-gatherer societies, 3, **6**, 15–17, **16**, 30
hydrogen-electric cars, 141
hydrogen gas, 136
hydro power, 134

I
Iceland, 136
identity, sense of, 162
ignorance to global crises, 112–113
immigration, 162
immunization, **70**
importing/exporting efficiency, 141
Impressionist painters, 170
incandescent bulbs, 136–137
include and transcend, 233–235, **234**
income. *see also* wealth: distribution and health, 66–67; ecological footprint and personal, **25**; happiness and, 122; natural global, 21–23; natural global and capital, **22**; personal and happiness, 66–67, **67**; United States and Bangladesh, 18
incremental technological changes, 130
India. *see also* Pakistan: African resources, 185; automobiles, 204; carbon dioxide emission reduction, 38; child abuse, 74; consumption, 24; economic challenges, 198; gender discrimination, 76; global crises, 193; grain consumption, 46; grain production, 48; inequality, wealth, 72; water management, 182; water supplies, 43; weapon development, 79; wheat imports, 180
individual: behaviors, **220**; health, 219–223; needs, 122, 126
Indonesia, 56, 107, 198
Indus River, **42**
Industrial Age. *see also* industrial civilization; industrial economies; industrialization; Industrial Revolution; industrial system: changing global values, 164, **164**; culture pattern, **224**; ecological footprint os, **89**; energy requirements, **139**; global economy and the, 160; human identity of, 162; oil and mechanization, 109; technology, 130; worldview, 94–95, 213–214
industrial carbon sequestration, **140**
industrial civilization. *see also* agrarian civilization; Industrial Age; industrial economies; industrialization; industrial system; other civilizations: capitalism, 63; collapse of, **93**; cultural destruction, **83**, 83–84; culture and conflict, 80–81; ecological footprint, **91**, **93**; energy, non-rewnewable, 97–98; expansion limits, 105–106; population growth, 15–17, **16**; production expansion, 204; social cultures development of, 95–97; social structure of, 65; societal growth,

99; sustainability and, 69; values, 221; worldview, 113
industrial economies, 30, **92**, 98
industrialization. *see also* Industrial Age; industrial civilization; industrial economies; industrial system: agriculture, 48; collapse or transformation and, 189–190; consumer society worldviews, 122–123; ecological footprint, **90**; energy slaves, 30; Islam resistance to, 159–160; land and water use, 48; nation states evolution, **6**; rate of change, 155–156; societal change and, 163; standards of living, 63; technologies development, 97–98; technology innovations, 130; worldview and technology innovations, 130
Industrial Revolution, 87–88, 161. *see also* Industrial Age
industrial system. *see also* Industrial Age; industrial civilization; industrial economies; industrialization; social systems; societal systems; systems: computers and the Internet, 142; dominant force and the, 159; evolution of, 94–95, 158; expansion, 192; growth limits, 98–99; hierarchy and, 75–76; historical expansion of, 87–94; The Industrial Revolution, 87–88; oil production and, 109; production, 95; standards of living improvement, 84; strengths/weaknesses, 192–200; values, 100–101; violence, 82; violent cultures and structures, 82; whole-systems design and the, 145; worldviews, 95, 99–100
inequality, 71–74, 75, 82, 194, 195
information, 173, 197. *see also* education; knowledge
Information Age, **224**; economy, **6**; efficiency, **139**; self-awareness, 214–215; worldview, 173, 214
information and communications technology, 133
information technologies, **161**
infrastructure and population growth, 17
Ingelhart, Ronald, 163–164
The Ingenuity Gap (Homer-Dixon), 113
Inglehart-Welzel Cultural Map of the World, **164**, 165
innovations, 130, **132**, 144, 158. *see also* technologies
institutions. *see also* businesses/corporations; government(s): culture patterns, **224**; international, 215, 219; needs, **126**; self-organization and, 214; social systems, 127, **220**, 220–222; transformation of society, 236
interactive triune brain, **121**
interest, compound interest, 113–114
interests, common, 235
The Intergovernmental Panel on Climate Change (IPCC): climate change, 209; emissions reductions, 199; food production, 49–50; global temperature increase, 35–36; sea levels, 39; water shortages, 43; water supplies and, 42
internal-combustion engine, 130
International Assessment of Agricultural Science and Technology for Development, 47
International Criminal Court, 169
International Energy Agency (IEA), 34, 39, 178
International Monetary Fund, 192
international organizations, 170–171
International Telecommunications Union, 170–171
The International Rice Research Institute, 48
International Water Management Institute, 182
Internet, 142, 161, **161**, 173–174. *see also* computers
interventions, contructive/destructive, 200–203
invasive species and extinction, 52
Iran, 43
Iraq: collapse of, 202; cost of, 239; peace movement and, 169; society collapse, 110; US energy supplies and, 186; US Gross Domestic Product (GDP), 125; US occupation of, 81–82; women's rights, 168
irrigation, 42, 51, 157. *see also* agriculture; farming
Islam, 83, 99, 158, 159–160. *see also* religion
Italy, 104

J
Jackson, Erwin, 199
Jagdeo, Bharrat, 205
Japan, 24, 76, 104
Jevons paradox, 148
Jiabao, Wen, 197
Jihad vs. McWorld, 83
Jintao, Hu, 196
Johnson, Allan, 77
Johnson, Lyndon B., 169
junk food, **60**

K
Kennedy, John F., 7
Kennedy, Robert F., 125
Khelil, Chakid, 178
Khmer Empire collapse, 104
King, Martin Luther Jr., 222
kinship system, **6**
knowledge, 161, 173, 215. *see also* education; information
Korea, 107
Kumar, Satish, 85
Kyoto agreements, 188
Kyoto Protocol, 39
Kyoto Strategy, **187**

L

Lake Chad, 42
Lal, Rattan, 51
Lama, Dalai, 225, 226
Lamarck, Jean Baptiste, 131
land, 20–21, 105
land and sea, biologically productive, **21**, 23
languages, 83–84, 162
La Niña, 35–36
Laszio, Ervin, 172
Latin America, 107
Law of Minimum (Liebig's Law), 108–109
lean design. *see* whole-systems design
Lebow, Victor, 98
Levine, Peter, 222, 236
Lewis, Michael, 180
Liebig's Law (Law of Minimum), 108–109
life expectancy, 73
light-emitting diodes (LEDs), 137
lighting efficiency, 136–137
light weapons, 80
Limits to Growth study, 14, 28
literacy, **70**
living social systems, **126**
living standards, 156
living systems, 5–6
Locke, John, 171, 172
logging, illegal, 55–56
love, 225–226
love and faith, 241–242
Lovelock, James, 188
Luther, Martin, 96
luxury expenditures, **60**, **70**
Lynas, Mark, 37

M

manufacturing, **140**, 141, 153. *see also* production
marine reserves, 55
market economies, 183–184
market forces, 146–148
marketing. *see* advertising/marketing
Marx, Karl, 98, 172
mass extinction, 39, 54
material: needs, human, 122; scarcity, 69; technics, 157, **157**; technologies, 156–157
materialistic age, 84
mathematicians worldview, 131
Maxwell, James Clerk, 131
Mayan nobles, **81**
McDonough, William, 146
Mead, Margaret, 211
Meadows, Donella, 228
meaning, need for, 122
meat consumption, **45**, 46
mechanical force, 131
mechanization, 98, 109
media, 74–75, 77
mental violence, 74
mercantilism, 97
mercury levels, human, **54**
Meridian Programme, 188
Merkel, Angela, 205
metals, availability of, 20–21
methane, 38
Mexico, 24, 43
Middle East, 82, 179, 194
migration, species extinction and, 53
Milanovic, Branko, 72
military, **60**, 80–81, 82, 186. *see also* war; weapons
Millennium Development Goals (MDGs), 59
Millennium Ecosystem Assessment, 42, 43, 44, 55
Mindell, Arnold, 214
mobile phone use, 162
modernization, 159, 190
modern values, 163
monarchical absolutism, 97
money, investment of, 99
Montfort Boy's Town, 146
Montreal Protocol, 205
moon landing, 7
Morales, Evo, 213
moral imperative, 226–227
moral issues, 64
moratoriums, fishing, 58
movements, social, 167–170
movies, 170
Mozambique, 18–19
Muir, John, 171
Müller, Michael, 205
murder rates, 77
MySpace, 173

N

Nai-keung, Lau, 185
Nanosolar, 138
nanotechnology, 137–141, 153
National Climate Assessment Report (China), 40
National Farmers Union of Canada, 181
National Intelligence Council (NIC), 177, 192
National Petroleum Council, 178
nation-state system, **6**, 96, 100, 217
NATO, 79
natural capital, 21, **22**, **88**, **89**, **91**, **92**, **93**
Natural Capital Institute, 168

Natural Capitalism, 146–148
Natural Capitalism Principles, 149
Natural Edge Project, 133
natural gas, **31**, 32–33, 209. *see also* energy; fossil fuels
natural income, annual global, **88**, **89**, **91**, **93**
Natural Resource Canada, 38
natural resources, 20–23
Natural Step Principles, 149
needs: advertising and, 68; basic, 64–65; biophysical, 120–121; conflict and, 79; human, 119, 121–122, 125–126, **126**, 127, 158; to innovate, 133; institutional, **126**; material and values, 122–123; sustainability and, 118
Neolithic Chinese culture, 38
Netherlands, 96
networking, 173–174
New Deal, 195, 203
Newton, Isaac, 171, 172
Newtonian models, 131
Next Industrial Revolution Principles, 149
Nigeria, 49
Nile River, 42
nitrous oxide, 35–36
non-governmental organizations (NGOs), 168
non-linear dynamics, 175–176
non-renewable energy. *see* energy
non-renewable resources, 20–23, 33, 149
nonviolence, 236–237
North Sea cod stocks, 58
Northwestern University, 38
Norwegian, 56
nuclear power stations, 210–211
nuclear power subsidies, 135
nuclear weapons, 79, 112

O
Oak Ridge National Laboratory, 137
obesity, 145
ocean sediments, 38
Ogallala aquifer, 43
O'Hara, Eamon, 189
Ohio State University, 51
oil. *see also* energy; fossil fuels: as an energy source, 32–33; discovery and consumption, **32**; peak, 33, 34, 176, 178; price, 176–177; production, 34, 109, **109**, 112; shortages, 209; supplies, 32–33
OPEC (Organization of the Petroluem Exporting Countries), 32–33, 177
open systems, 5, 119
organic farming, 124, 137, 182
organic systems coevolution, **2**, 2–3
Organization for Economic Co-operation and Development (OECD), 239
Organization of the Petroleum Exporting Countries (OPEC), 32–33, 177, 178
Orkut, 173
orthodox economists, 14
overharvesting, 52
overshoot, 27–28
Oxfam, 169
Oxford Research Group, 194

P
Pachauri, Rajendra, 199, 209
Pakistan, 43, 82, 193. *see also* India
Paleotechnical Era, 138
paradigms. *see also* change: artificial organisms, 139; competition to cooperation, 150–152; holistic, 213–214; internal-combustion engine, 130; scientific, 129–131; theoretical, 171
paramedic story, 29–30
parameters, 107–110, 119, 125
parenting, 124–125
Peabody Energy, 39
peace: conflict resolution, 234–235; disarmament, 79; keeping, 79; movement, 168–169; nonviolence, 236–237; tragedy of the commons and, 100; violence and conflict, 77–78
peak oil, 33, 34, 176, 178. *see also* oil
peer-to-peer projects, 173
pensions, 92
People's Daily newspaper, 197
perfect storm, 107–110
permafrost, 38, 188
personal wealth distribution, **73**
Peru, 202
pesticides, 47, 124
Petitfor, Ann, 230
pharmaceutical companies, 147
Philippines, 202
Phillips, Kevin, 195
photosynthesis, 47
physical systems, **2**, 2–3, 5–6, **118**, 217. *see also* other systems
physical violence, 74–75
physics, modern, 131
Pickens, T. Boone, 134–135, 135
planetary system evolution, **6**
plants, domestication, 129
plants and bees, 128
poaching, 55–56
polar bears and pollution, **54**
policy differences, **166**

political and economic inequality, 194
politics. *see also* democracy; governance: fishing industry collapse and, 58; global governance, 171; The Industrial Revolution, 94–95; international organizations, 170–171; system collapse, 104; system priorities, 84; wealth and, 195
pollution. *see also* air pollution: emission controls, 149; food production and, 49; globalization and, 74; mercury levels, human, **54**; polar bears, **54**; pregnancy and, **54**; reduction, 147; species extinction, 52–53, **54**; spreading, **54**
polycentric planetary civilizations, 216
polyculture, fish, 50
Pope Benedict XVI, 226
population, 87–88, 108–109, 110, 162. *see also* population growth
population growth, 14, 15–17, **16**, 27–28. *see also* population
Porritt, Jonathan, 152
post-industrialization, 189–190
Postmodern Age, 164, **164**
poverty. *see also* hunger: civil conflict and, 82; economic gap and rich, 70–71; future of, 64; of illegal crop farmers, 82; income and happiness, 122; increases with global wealth, 84; life expectancy, 73; mental violence, 74; reduction, 59; social justice movements and, 169; tragedy of the commons and, 100
predictions, future, 175–177
pregnancy, **54**, 73
pre-industrial societies, 190
pre-industrial technologies, 198
pre-modern and modern values, 163
prices, rising of goods, 13–14
printing press, 98
problems, collective, 100
production. *see also* manufacturing: centralized, 138; decentralized, 215; distributed, 138, 141–142; goods and services, 133; industrial, development of, 97–98; oil, 109, **109**; waste reduction in, 146
productivity, 88–89, 133
products, durability of, 150
prostitution, 74
Protestant Reformation, 96
proxy wars, 80–81
psychosocial integration, 221
public programs, 92
Putman, Robert, 222

Q

qualitative development, 128
qualitative growth, 150
Qualman, Darren, 181
quantitative growth, 150
quantum theory, 131

R

'race to the bottom,' 115, 218
rainforest deforestation, 52–53
rapid collapse, 191
rate of change, 155–156
rationalist perspective, 96–97
rationalist worldview, 97
reactive trend, 159–160
reality, fixed, 131
recession, 184
reciprocity, relationship, 128
reform, environmental/social, 113–114
reindeer, 27–28
RE C initiative, 135
relationships. *see also* communities: based on reciprocity, 128; family, 124, 151; happiness and, 69; health of in developed countries, 70; parenting, 124–125
relativity theory, 131
religion, **88**, 131, 158, 226. *see also* Buddhism; Christianity; Hinduism; Islam; religious
religious persecutions, 96
religious worldviews, **88**, 95
renewable energy. *see also* energy: agrarian societies, 97–98; capturing, 134–136; economic growth and, 140; electricity, **140**; The Industrial Revolution, 87–88; instead of war, 239–240; technologies, 134–136
renewable resources, 20–23, 33, 133, 149. *see also* resources
resilience, loss of, 106–107
resilience, systems, 121
resource. *see also* renewable resources; resources: conservation, 136–137; demand reduction, 136; quality and availability, crises of, 105; sharing, 136, 147; shortages, 13–14, 104–105, 109, 185, 198
resources. *see also* renewable resources; resource: availability of, 20–23, 105; common, 100; conflict over, 185–186; global economy and, 185; increasing renewable energy, 133; non-renewable consumption of, 149; oil production, 109; population growth and, 17; responsibility for, 100; supplies of critical, 119; waste, 124
Resurgence magazine, 85
Revolutionists, 97
Rhys, Garel, 147
rice harvests declining, 48

Riftkin, Jeremy, 165–166
Rio Declaration on Environment and Development, 149
Rogers, James, 206
Rogers, Paul, 179, 210
Roman Catholic Church, 96
Roman Empire collapse, 105–106
Roosevelt, Franklin D., 65, 169, 195, 203
Rosenberg, Marshall, 236
Rosie the Riveter, **240**
The Rule of 70, **19**
'rurban' communities, 215–216
Russia, 24, 79, 107
Rwanda genocide, 78
Ryskin, Gregory, 38

S
Samba, Sall, 58
Sanders, Richard, 148
Sarkozy, Nicholas, 205
Saudi Arabia, 39
scarcity, 42, 69–70, 127, 147–148, 177
scenarios, future, 203–204, 204–205, 206, 208–210
Schell, Orville, 197
science worldviews, 96–97, 131
scientific paradigms, 129–131
scientists warnings, 58
Scott, Lee, 169
seafood species, 49, **49**
sea levels rising, 38
Seattle, Chief, 51
sea water, biologically productive, 20–21, **21**
Second World War. *see* World War II
secular-rational values, 164, **164**
secular society and rationalist perspective, 96–97
security, sustainable, 194
seed preservation, 56
SEED-SCALE Process for Community Change, 215
self-awareness, 214–215
self-expression values, 164, **164**
selfishness and exploitation, 74–75
self-organization, 214–216
September 11 terrorist attacks, 59
Setser, Brad, 183
sexual violence against women, 77
Shah, Tushaar, 182
Sheel Foundation, 137
Sheeran, Josette, 46, 143
Shell Oil, 32–33
Shiva, Vandana, 182
shortages: energy, 30–34, 178–179; food, 44–51, 179–183, 180, **180**, 209; resource, 13–14, 104–105, 109, 185, 198; water, 20–21, **41**, 41–45, 179–183
Sierra Club, 171
signs, vital, 29–30
Silent Spring (Carson), 14
Simon, Julian, 13
Singer, Jerome, 77
Singh, Charan, 84
Six Degrees (Lynas), 37
slavery, **210**, 211
slaves, energy, **30**
Smil, Vaclav, 50
Smith, Adam, 98, 147, 171
social. *see also* social systems: change, 4–5, 227–228; existence needs, 122; industrial cultures, 95–97; institutions, 4–5, **220**; justice movements, 169; movements, 167–170; reforms, blocked, 112–113; structures, 95, 214–216; sustainability, 69; technics, 157, **157**; technologies, 156–157; transformation, 156–161; unrest, **190**, 193; values, 162–163
socially structured behaviors, 65–66
social systems. *see also* other systems: agriculture and irrigation, 157; collapse and political system collapse, 104; essential needs, 118; institutions, 127; living, **126**; parameters, 107–110; resilience, loss of, 106–107; stress, 107–109; structure of, 65–66; sustainability of, 119; technology innovations and, 130
societal. *see also* societal systems: behaviors, **220**; dynamics of change, 206–208; evolution, 157–159, 241; health, 219–223; modeling change, **207**
societal systems. *see also* other systems: centralized/decentralized, 215; coevolution, 2, 2–3; computers and the Internet, 161; energy requirements, **139**; evolutionary change, 131–132; evolution of, **6**; holarchial structure, 216; and industrialization, 163; Stone Age change, 129–130; sustainability requirements, 126–128; transformation of, 152–153
soil, 49, 51, 105. *see also* agriculture; farming
solar energy, 134, **190**
solar panels, 141, **190**
solar power, 138–139
solar power generating plants, 135
Somalia, 202
Soros, George, 65
South America, 49
South Asia, 49, 64, 76
Soviet Union, 79, 103, 110
space-based laser weapons, 79

spaceship Earth, 61
Spain, 141
species. *see also* species extinction: diversity and, 120–121; habitat parameters, 125; loss of, 83; protection, 55; relationships based on reciprocity, 128; seafood, **49**; stability of keystone, 119
species extinction. *see also* extinction; species: accelerating rates of, 51–56; biodiversity, 52–56; chemicals, toxic, 53; climate change, 209–210; invasive, 52; politicians and, 187–188; pollution, 53; rainforest deforestation, 52–53
spending. *see* expenditures
spiritual needs, human, 122
sporting events, 162
Spratt, David, 37, 210–211
springboard effect, **207**, 208
St. Matthew Island, 27–28
standards of living, 84, 161, 167, 185
state, independence of, 97
state separation, church and, 96
steam engines, 98
Steiner, Achim, 40–41
Stern, Sir Nicholas, 34, 37, 199
The Stern Review, 38
Stirling Energy Systems, 135
Stone Age, **6**, **16**, 104, 129
stoves, 137
Strahan, David, 34
strategic alliances, 186
Strauss-Kahn, Dominique, 192
stress, 66–67, 107–109, 120–121, 222
structural change, 233–234
structural inequality, 71–74
structural violence, 71–74, 221
subsidies, farming, 182
subsidies, fishing, 58–59, 182
sufficiency, real needs and, 85
supply of goods, 185
survival and economics, **123**
survival values, 164, **164**
sustainability. *see also* green building; sustainable: achievement of global, 127–128; consumer culture, 85; of current consumption, 94; of current dominant world system, 64; ecological footprint, **88**; ecosystems, 119; ethics and, 64–65; global economy expansion, 133; of human economy, 94; of industrial civilization, 69; industrial system growth limits, 98–99; The Industrial Revolution, 87–88; limits of environmental, 17; qualitative development, 128; social systems, 119; societal system requirements, 126–128; worldview and, 149
sustainable. *see also* green building; sustainability: civilization culture pattern, 223–224, **224**; development, defined, 117–120; development planning, 148–150; ecological footprint, 24, **27**, **89**, **91**; economy and capitalism, 151; farming, 146; global economy, 213–214; security, 186, 194, 235; systems, environmentally, **190**
Sustainable Communities Bill, 216–217
Sutton, Philip, 37, 210–211
sweatshops, 74
systems. *see also* biological systems; biophysical systems; ecosystems; industrial system; physical systems; social systems; societal systems: abundance of, **118**; adjusting to existing, 204–205; approach to forecasting, 177; biological and social evolution, 233–234; change/transformation, 228; decentralized, 215–216; dysfunctional, 127; environmentally sustainable, **190**; failure/collapse, 103–114; failure consequences, 110–112; global responsibility for, 100–101; health, 119; holistic (see holistic systems); individual and societal behaviors, **220**; interrelated, 217; nation-state, 6, 96, 100, 217; non-linear dynamics, 175–176; open, 5, 119; resilience, 121; self-awareness, 214–215; springboard effect, **207**, 208; state changes, 176; strengths/weaknesses, 192–200; transformation, **111**; village, **6**; whole-systems design, 142–145, **144**

T

Tainter, Joseph, 105–106
tax, carbon, 39
taxes, value subtracted, 152
Taylor, Alastair M., 5
technics, material and social, 157, **157**, 160
technologies. *see also* computers; innovations; technology: advances, **85**; environmental collapse and, **85**; information, **161**; material, 129–153, 156–157; new, **85**, 106, 132–134, 203; pre-industrial, 198; saving civilization, 84; social, 156–157; systems-based, 215
technology, 49, 58, 130, 192–193. *see also* computers; technologies
Tehran, **178**
temperature increases. *see also* climate change; global warming: allowing, 209–210; carbon emission reduction, 140; crop failures, 52; impacts of, 36–37; and species extinction, 52; tipping point, 188; yields, crop, 49–50
terrestrial species decline, 52
terrorism, 81–82, 139, 160, 194
Thailand, 43, 107

The Stern Review, 38
30 Years War, 96
Thoreau, Henry David, 171
tidal energy, 134
tipping points. *see also* change: climate change, **36**, 140; global warming, 37–38, 140, **187**, 187–188; innovations and, 130; transformation and, 3–4
tobacco consumption, **60**
Total (oil company), 32–33
trade, world, 162
trade deficit, 183
traditional groups, 190
traditional values, 164, **164**
tragedy of the commons, 100
transformation: cultural, 174; evolutionary, 129–131; as a future scenario, 192; social, 156–161; system, **111**; tipping points and, 3–4; trend towards, 4, 189–191; trust, 235–237
transformational change scenario, 206
transformative technological changes, 130
Trans-Mediterrannean Renewable Energy Cooperation (TREC), 135
Transparency International, 169
trauma, 222, 236
travel, 162
Treaty of Westphalia, 96
trends, 4, 175–177, 191
tribal societies, 83
tropical rainforests, 52–53
trust, 235–237
Ts'lal-la-kum, 51
turbines, wind, **85**

U

United Kingdom, 18, 66, 73, 79, 149–150. *see also* Great Britain
United Nations (UN): desertification, 49; The Earth Charter, 228–229; Environment Program, 209; Food and Agriculture Organization, 45; food shortages, 180; Global Compact, 149; global responsibility, 100; Human Development Report, 39, 59, 73, 82; Millennium Ecosystem Assessment, 42, 43; peacekeeping budget, 79; species extinction, 52; University's International Network on Water, Environment and Health, 51; water scarcity, 177; World Food Program, 143
United States: Afghanistan war, 81; American Dream, 165–166; Bush, George W., 169, 195–196; Bush Administration, 79, 184; civic engagement, 222; Clean Air Act, 205; conflict/war, 81; corn production for biofuels, 48; deficit, 183; disarmament, 79; drug addiction, 82; ecological footprint, personal, **26**; economic growth limits, 18; energy supplies, 186; grain consumption, **45**; greenhouse gas emissions, 199; Gross Domestic Product (GDP), 18, 125; happiness and income levels, **67**; health and happiness, 68; health/longevity and wages, 66; imports, 194; inequality within, 195; Iraq war, 81, 186, 239–240; Kyoto Protocol, 40; media violence, 77; mercantilism, 97; Middle East control of, 179; moon landing, 7; National Intelligence Council (NIC), 176, 177, 192; occupation of Iraq, 81; personal incomes, 68; policy differences, **166**; political and economic inequality, 194; recession, 184; spending on war in Pakistan, 82; as superpower, 183; violent conflict/murder rate, 77; 'war on drugs,' 81, 82; 'war on terror,' 81–82, 82, 160, 194; water scarcity, 177; water supplies, 43; women's rights, 168
Universal Culture Pattern, **126**, 223
Universal Postal Union, 170–171
University of Colorado, 113–114
University of East Anglia, 38
University of Kassel, 135
University's International Network on Water, Environment and Health, United Nations, 51
unpaid work, **124**, 124–125
unsustainable: consumption growth rates, **19**; defined, 117–118; ecological footprint, **92**, **93**
urbanization, 48, 83
Ustinov, Peter, 100

V

Vail, Jeff, 101, 215
values: changing, 156, 162–167, **164**; coalitions and shared, **233**; computers and the Internet, **161**; constructive, 236; destructive, 192; family, 151; include and transcend, 233–235, **234**; industrial civilization, 221; industrial system, 100–101; leaders/corporations, 113; materialistic, changing, 84–85; materials needs, 122–123; pre-modern and modern, 163; speed of change, 4–5; sustainability and, 119; taxes on subtracted, 152; of unpaid work, **124**, 124–125
Vietnam, 24, 81, 169
village system, **6**
Viner, David, 38
violence, 71–75, 77–78, 82, 221
virtuous cycle, **231**
vision, common, 229–230, 231
vision, world, 59–60
vital signs, 29–30
Voltaire, 97

W

wages, 92–93, 167
Wal-Mart, 169
war. *see also* military; weapons: changing values and, 165; domination and, 78–82; energy and, 179; post-traumatic stress, 222; protests, 168; 'war on drugs,' 81, 82; 'war on terror,' 81–82, 82, 160, 194
Warsaw Pact, 79
Wasdell, David, 188
waste: conservation of materials, 141; ecodesign and reduction of, 145; elimination in food production, 51; and lowering consumption, 136; reduction, 124, 136; resources, 124
water: aquifers, 43, 48; availability of, 20–21; biologically productive, 20–21, **21**; clean drinking, **70**; climate change projected impacts, **36**; consumption, 42; desalinating, 44; drip irrigation, 51; economics and survival, **123**; food production and, 49; fresh, 43; fresh water quality, 44; global supply, **41**; grain production and, 48; groundwater and, 43; scarcity, 177; shortages, 179–183; shortages, growing, 41–45; tables, 49; use in agriculture, 51
watershed, **187**
Watson, Bob, 47
Watson, Harlan, 40
Watson, Robert, 143
watt com boom, 133
wave energy, 134
wealth. *see also* income: distribution, 195; distribution and capitalist system, 66; economic gap and poor, 70–71; health/longevity and, 66; materialistic age and, 84; personal distribution, **73**; redistribution, 59; transferred from poor to rich, 75
weapons, 79, 81, 112, 186. *see also* military; war
weather events, **36**
weather patterns, 35–36
Weber, Max, 98
weight loss, 69
well-being, human, **121**
Western capitalist democracies, 92
Western military interventions, 82
Western Roman Empire collapse, 103–104
wheat harvest, 48
wheat imports, 180
whole-systems design, 142–145, **144**
Wikipedia, 173
Wilber, Ken, 172
Wilde, Oscar, 122
wind energy, 134
wind turbines, **85**
Winston, Andrew, 137
Wiser Earth, 168
Witoelar, Rachmat, 55–56
women(s): access to education and jobs, 75–76; being abused, 74; equality, 59, 76, 168; health care, **70**, 73; pollution and, **54**; pregnancy, **54**, 73; rights movement, 168; sexual violence against, 77; weight loss, 69
work, unpaid, **124**
workforces in industrial economies, 98
World Bank, 72, 160, 180, 193
World Commission on Water, 44
World Cup 2006, 162
world peace. *see* peace
world population growth, **16**
worldviews: consumer society, 122–123; development of science and, 96–97; ecological, 131, 215; economy and nature, 206; emerging, 131–132; holistic, 171–172; include and transcend, **234**; Industrial Age, 94–95, 213–214; industrial and technology innovations, 130; industrial civilizations, 122–123; industrial system, 95, 99–100; Information Age, 173; integral, 170–173; leaders/corporations, 113; mathematicians, 131; religious, 95; secular, 95; sustainability and, 119, 149
World War I, 80, 165
World War II: communism, 80; government economic control, 152; military production, 239; peace movement and, 169; political system collapse, 104; public support, 203; resource shortages, 186; Rosie the Riveter, **240**
Worm, Boris, 58

Y

Yale University, 184
Yamani, Sheik Ahmed Zaki, 133
Yangtze River, 43
Yellow River, 42
yields, crop, 49–50
Yugoslavia, 110

Z

Zero Emissions Research and Initiatives (ZERI), 146

About the Author and Illustrator

The author, Graeme M. Taylor

Following a career as an emergency paramedic in British Columbia, Graeme completed a Master's degree in Conflict Analysis and Management. Since 2003 he has been the coordinator of BEST Futures (www.bestfutures.org), a project researching sustainable solutions to global problems. A writer and speaker, he currently lives in Brisbane, Australia with his wife Ferie.

The illustrator, Fereshteh M. Sadeghi

Fereshteh (Ferie) was born in Persia, where she graduated and worked as an Interior Architect. In 1989 she moved to New Zealand and then to Australia, starting a new career as a Graphic Designer. Her passion is watercolour painting. Her wish is to use her art and skills to help people and solve planetary problems.